RIVER FLOWS AND
CHANNEL FORMS

River Flows and Channel Forms

SELECTED EXTRACTS FROM

THE RIVERS HANDBOOK

EDITED BY

GEOFFREY PETTS

Environmental Research and Management
The University of Birmingham
Edgbaston, Birmingham

AND

PETER CALOW

Department of Animal and Plant Sciences
University of Sheffield, Sheffield

b

**Blackwell
Science**

© 1996 by
Blackwell Science Ltd
Editorial Offices:
Osney Mead, Oxford OX2 0EL
25 John Street, London WC1N 2BL
23 Ainslie Place, Edinburgh EH3 6AJ
238 Main Street, Cambridge
 Massachusetts 02142, USA
54 University Street, Carlton
 Victoria 3053, Australia

Other Editorial Offices:
Arnette Blackwell SA
 224, Boulevard Saint Germain
 75007 Paris,
 France

Blackwell Wissenschafts-Verlag GmbH
 Kurfürstendamm 57
 10707 Berlin, Germany

 Zehetnergasse 6
 A-1140 Wien
 Austria

First published 1996

Set by Setrite Typesetters, Hong Kong
Printed and bound in Great Britain
at the Alden Press Limited,
Oxford and Northampton

The Blackwell Science logo is a
trade mark of Blackwell Science Ltd,
registered at the United Kingdom
Trade Marks Registry

DISTRIBUTORS

Marston Book Services Ltd
PO Box 269
Abingdon
Oxon OX14 4YN
(*Orders*: Tel: 01235 465500
 Fax: 01235 465555)

USA
Blackwell Science, Inc.
238 Main Street
Cambridge, MA 02142
(*Orders*: Tel: 800 215-1000
 617 876-7000
 Fax: 617 492-5263)

Canada
Copp Clark, Ltd
2775 Matheson Blvd East
Mississauga, Ontario
Canada, L4W 4P7
(*Orders*: Tel: 800 263-4374
 905 238-6074)

Australia
Blackwell Science Pty Ltd
54 University Street
Carlton, Victoria 3053
(*Orders*: Tel: 03 9347 0300
 Fax: 03 9349 3016)

A catalogue record for this title
is available from the British Library
and the Library of Congress

ISBN 0-86542-920-0

Library of Congress
Cataloging-in-Publication Data

Rivers handbook. Selections.
 River flows and channel forms:
 selected extracts from The rivers handbook /
 edited by Geoffrey Petts and Peter Calow.
  p. cm.
 Includes bibliographical references and index.
 ISBN 0-86542-920-0
 1. Streamflow.
 2. River channels.
 I. Petts. Geoffrey E.
II. Calow. Peter. III. Title.
G81207.R592 1996
551.4'83—dc20   56-11446
                        CIP

Contents

Plate 8.1 falls between pp. 170 and 171

List of Contributors

N. W. ARNELL *Department of Geography, University of Southampton, Highfield, Southampton, Hampshire SO17 1BJ, UK*

A. BROOKES *National Rivers Authority Thames Region, Kings Meadow House, Kings Meadow Road, Reading, Berkshire RG1 8DQ, UK*

T. P. BURT *School of Geography, University of Oxford, Mansfield Road, Oxford, OX1 3TB, UK*

P. A. CARLING *Hydrodynamics and Sedimentology Laboratory, Lancaster University, Department of Geography, Lancaster LA1 4YB, UK*

M. CHURCH *Department of Geography, University of British Columbia, Vancouver, British Columbia V6T 1Z2, Canada*

N. P. FAWTHROP *National Rivers Authority Anglian Region, Kingfisher House, Goldhay Way, Orton Goldhay, Peterborough PE2 5ZR, UK*

A. GUSTARD *Institute of Hydrology, Wallingford, Oxon. OX10 8BB, UK*

J. LEWIN *Institute of Earth Studies, University College of Wales Aberystwyth SY23 3DB, UK*

J. D. NEWBOLD *Stroud Water Research Center, Academy of Natural Sciences of Philadelphia, 512 Spencer Road, Avondale, PA 19311, USA*

D. E. WALLING *Department of Geography, University of Exeter, Exeter EX4 4RJ, UK*

B. W. WEBB *Department of Geography, University of Exeter, Exeter EX4 4RJ, UK*

Preface

If the extent to which books are 'borrowed' from libraries is a measure of their success, then personal experience with the extent to which *The Rivers Handbook* volumes have gone missing from our own shelves has to be encouraging! Another factor is, of course, price; and we have been very much aware that the *Handbook* in its original form was not affordable for most students. So following some encouragement we have decided to produce less expensive versions. In doing this we could have simply published softback editions of the originals. But, again following encouragement, we opted instead to take a selection of chapters from the *Handbooks* and reorganize these into three groupings intended to be especially helpful for supporting course work (for undergraduate and postgraduate programmes) in river ecology. The result has been the development of three books: *River Biota; River Flows and Channel Forms* and *River Restoration* of which this is one.

Each book opens with a completely new chapter, presenting general principles and indicating how the rest of the book is structured around these. All the other chapters are taken, with some updating, from the parent *Handbooks*. Our hope, therefore, is that the repackaging brings benefits in terms of both availability and convenience for a broader readership and we would welcome feedback on this. As ever, we are grateful to authors of individual chapters for their co-operation in compiling the reorganized versions and especially to the Publishers for their encouragement and support in this venture.

Geoffrey E. Petts
Peter Calow

1: Fluvial Hydrosystems: the Physical Basis

G. E. PETTS AND P. CALOW

1.1 INTRODUCTION

Many of the problems for river management today relate to landuse changes, water resource developments and industrial expansion which have altered the pattern of runoff, the quality of river water, and the load of sediment delivered to river channels. Channel engineering — for flood control, land drainage and navigation — has fixed the river into an artificial course. These catchment-wide changes have had severe impacts not only on river ecosystems but also on biodiversity at the landscape scale (e.g. Petts *et al* 1989; Naiman & Decamps 1990; Petts & Amoros 1996).

A goal of sustainable development requires integrated land and water planning based upon the river catchment as the fundamental unit, incorporating a new environmental ethic that gives due regard to ecological needs, and seeks to promote long-term objectives. Modern river management faces three challenges: (i) to meet the growing demands for water resources, especially for irrigation agriculture; (ii) to control water quality and to achieve effective waste management; and (iii) to sustain water and land resources, including biodiversity and nature conservation. To respond to these challenges, scientists must have a firm understanding of the physical and chemical processes that drive river ecosystems. This volume seeks to provide this. A companion volume focuses on the biota of river ecosystems and their functional links — food webs — (Petts & Calow 1996a). A third volume applies this information to restoring degraded river systems (Petts & Calow, 1996b). The chapters that comprise these edited volumes were first published with additional chapters, including a range of case studies, as *The Rivers Handbook* (see Calow & Petts 1994) in two volumes. The first volume explores the scientific principles and the second puts these principles into practice.

1.2 RIVER ECOSYSTEMS

From a biological point of view, flowing water has a number of advantages over still water: it is constantly mixed by turbulence providing nutrients, exchange of respiratory gases, and removal of wastes. Some species of aquatic macrophyte, for example, are highly adapted to the improved metabolic conditions of flowing water. Flowing water is fundamental for the downstream and lateral (e.g. into floodplain lakes and backwaters) movement of plants and animals. However, the character of the flow changes from the headwaters to the river mouth, especially in temperature and depth (i.e. light penetration), and this leads to a characteristic zonation in the biota of rivers. Thus, the 'River Continuum Concept' (Vannote *et al* 1980) describes progressive changes of stream conditions (Fig. 1.1). The fauna of relatively cool, litter-dominated headwater streams shaded by trees along the river banks, contrasts with that of the relatively wide and shallow mid-sectors where light and nutrients favour algal production on the channel bed, and the lower river where food chains are based on high levels of fine particulate organic matter from upstream and from floodplain inputs. In forested headwater streams, coarse woody debris plays an important role in sustaining the diversity of habitats, in regulating flood flows and in controlling sediment movement (Maser & Sedell 1994).

Primary production by biofilms and macrophytes

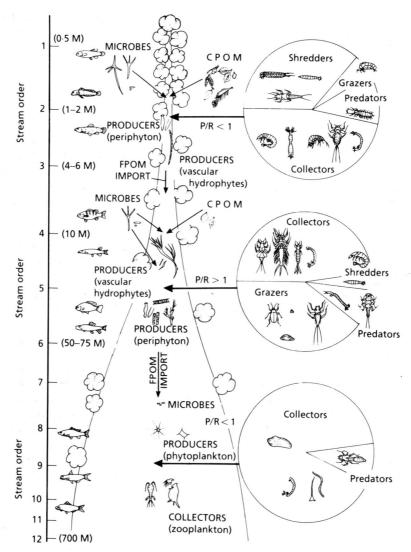

Fig. 1.1 The River Continuum Concept (RCC; e.g. Vannote *et al* 1980) a generalized model of the shifts in the relative abundances of invertebrate functional groups along a river tributary system from headwaters to mouth. The river system is shown as a single stem of increasing order and width. The headwaters (orders 1–3) are depicted as dominated by riparian shading and litter inputs resulting in a heterotrophic $P/R < 1$. The invertebrates are dominated by shredders which utilize riparian litter as their food resource once it has been appropriately conditioned by aquatic micro-organisms (especially aquatic hyphomycete fungi) and collectors that feed on fine particulate organic matter (FPOM). The mid-reaches (orders 4–6) are less dependent upon direct riparian litter input and with increased width and reduced canopy shading they are autotrophic with a $P/R > 1$. The shredders are reduced and the scrapers are relatively more important as attached microalgae become more abundant. The larger rivers are dominated by FPOM (and therefore collectors), and the increased transport load of this material together with the increased depth results in reduced light penetration, and the system is again characterized by a $P/R < 1$ (after Cummins 1975).

can be important in most running waters. However, running water food webs are dominated by energy (organic matter) inputs from catchment sources: hillslopes, the river margin and floodplain. In headwater streams dissolved organic matter can be supplied by surface runoff, interflow through soils and groundwater. In floodplain rivers, the recession of the annual flood delivers high levels of dissolved organic carbon and detritus (wood, leaves, seeds etc.) to the main channel. This lateral connectivity, so important for sustaining the integrity of large floodplain rivers, is not included in the river continuum concept.

A regular, annual flood is of particular advantage to aquatic systems along large floodplain rivers (Junk *et al* 1989). Many species of aquatic fauna are adapted to this annual flood pulse which connects the main river channel to floodplain backwaters and food resources (Fig. 1.2). Aquatic organisms colonize the floodplain at rising and high water levels because of the breeding and feeding opportunities that arise. Large floodplain rivers derive most of their animal biomass from within the floodplain. Without the flood pulse, production within the river ecosystem is drastically reduced, and community composition and energy pathways are radically changed.

Despite the resource advantages of flowing water, rivers and streams are hazardous environments for biota. Statzner and Higler (1986) argued that the structure and function of most aquatic communities is determined, to a large extent, by the stability or predictability of hydrological patterns and instream hydraulic conditions. The diversity of biota has been shown to be highest where there is a high diversity of hydraulic conditions.

Disturbance plays a critical role in organizing communities and ecosystems (Reice *et al* 1990). Disturbances in rivers include: erosion and abrasion, siltation and burial, desiccation, and extremes of water quality. Only a few spring-fed streams fit a disturbance-free model of river ecosystems dominated by biotic interactions. The mosaic of vegetation patches along a river corridor, for example, is related to the rejuvenation of successions associated with channel erosion and deposition. In braided rivers, frequent disturbance inhibits long plant successions and pioneer populations dominate. In contrast, less active sectors are characterized by hardwood forests.

1.3 THE PHYSICAL PROCESSES

In addition to the strong longitudinal and lateral dimensions described above, rivers are characterized by an important vertical dimension — the interactions between the surface-water and groundwater systems that are especially important in sectors with major alluvial aquifers (Stanford & Ward 1988). These exchanges exert an important influence on the spatial distribution of fauna and

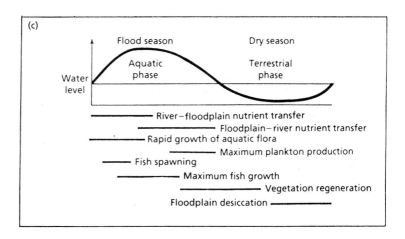

Fig. 1.2 Roles of the floodpulse in sustaining the ecological integrity of floodplain rivers (based on Ward 1989).

flora, communities reflecting water source (e.g. upwelling or downwelling). Thus, we prefer to consider rivers as 'fluvial hydrosystems' (Petts & Amoros 1996): three-dimensional corridors that extend laterally to the edge of the floodplain and vertically to the limit of surface-water—groundwater interaction.

Within this context, fluvial hydrosystems are in large part determined by a range of hydrological, water-quality, hydraulic and morphological variables. Rivers are characterized by changing flows, variable water-quality conditions, unsteady hydraulics, and changing channel forms. These dynamic conditions can be detected over periods as short as a minute and as long as 10 000 years or more. Regular diurnal and seasonal patterns contrast with storm-related and annual (weather-pattern related) changes that appear as random variations about some medium-term (e.g. 100 year) average condition that may fit a long-term trend related to climate or sea-level change. Water quality and fine sediment dynamics fingerprint headwater catchment characteristics (Foster *et al* 1995) and determine the environment within which biota live. However, it is the nature of river channel changes (Gurnell & Petts 1995) that determines the availability of suitable habitat for biota.

Today, the dynamics of river systems throughout much of the world reflect human intervention, either within the catchment or directly within the river corridor. Most rivers in the developed world are intensively regulated, many are hydraulically optimal but ecologically poor, often eutrophic and organically enriched, and their once ecologically diverse and productive corridors have been eliminated. In the developing countries, continuing population growth implies an increasing need for water security in agriculture as well as increasing water needs for domestic and industrial supply. In developing an approach to river management that embraces conservation objectives, including the conservation of genetic diversity and ecological processes, and the restoration of ecosystems degraded by the process of industrialization, decisions and approaches must be founded in a sound understanding of hydrology and fluvial geomorphology.

1.4 THE STRUCTURE OF THE BOOK

This book presents the fundamental physical basis for understanding river systems. The first part of the book focuses on hydrological processes within headwater catchments (Chapter 2) — the primary source areas — and moves on to consider flow variations within larger rivers and at the regional scale (Chapter 3), before considering the modelling of these processes within a single river system (Chapter 4). The second part examines the fundamental controls on water-quality variations: the physical characteristics (Chapter 5), chemical dynamics (Chapter 6), and the specific case of nutrient spiraling within rivers (Chapter 7).

Flows and sediments interact to produce the range of channel forms that characterize different rivers and different sectors of rivers. The third section of the book begins with a discussion of sediment transport processes (Chapter 8). This leads into an examination of the different types of channel (Chapter 9) and their associated floodplains (Chapter 10).

Finally, the book concludes with two chapters that review the ecological impacts of past changes in flow regime and channel form (Chapter 11) and examine the potential impacts of future climate changes on the fundamental hydrological processes within given basins (Chapter 12).

REFERENCES

Calow P, Petts GE. (1994) *The Rivers Handbook*, Vols. 1 & 2. Blackwell Scientific Publications, Oxford. [1.1]

Cummins KW. (1975) The ecology of running waters; theory and practice. In: Baker DB, Jackson WB, Prater BL (eds) *Proceedings of the Sandusky River Basin Symposium, International Joint Commission, Great Lakes Pollution*, pp 277—93. Environmental Protection Agency, Washington DC. [1.2]

Foster IDL, Gurnell AM, Webb B. (eds) (1995) *Sediment and Water Quality in River Channels*. John Wiley and Sons, Chichester. [1.3]

Gurnell AM, Petts GE. (eds) (1995) *Changing River Channels*. John Wiley and Sons, Chichester. [1.3]

Junk WJ, Bayley PB, Sparks RE. (1989) The flood-pulse concept in river-floodplain systems. In: Dodge DP (ed.) *Proceedings of the International Large River Symposium (LARS)*, pp 110—27. Canadian Special Publication of Fisheries and Aquatic Sciences 106. [1.2]

Maser C, Sedell JR. (1994) From the forest to the sea. In: *The Ecology of Wood in Streams, Rivers, Estuaries and Oceans.* St Lucie Press, Florida. [1.2]

Naiman RJ, Decamps H. (eds) (1990) *The Ecology and Management of Aquatic–Terrestrial Ecotones.* UNESCO, Paris; Parthenon, Carnforth, UK. [1.1]

Petts GE, Amoros C. (eds) (1996). *Fluvial Hydrosystems.* Chapman and Hall, London. [1.1, 1.3]

Petts GE, Calow P. (1966a) *River Biota.* Blackwell Science, Oxford. [1.1]

Petts GE, Calow P. (1996b) *River Restoration.* Blackwell Science, Oxford. [1.1]

Petts GE, Moller H, Roux AL. (eds) (1989) *Historical Change of Large Alluvial Rivers: Western Europe.* John Wiley and Sons, Chichester, UK. [1.1]

Reice SR, Wissmar RC, Naiman RJ. (1990) Disturbance regimes, resilience and recovery of animal communities and habitats in lotic ecosystems. *Environmental Management* **14**: 647–60. [1.2]

Stanford JA, Ward JV. (1988) The hyporheic habitat of river ecosystems. *Nature* **335**: 64–6. [1.3]

Statzner B, Higler B. (1986) Stream hydraulics as a major determinant of benthic invertebrate zonation patterns. *Freshwater Biology* **16**: 127–39. [1.2]

Vannote RL, Minshall GW, Cummins KW, Sedell JR, Cushing CE. (1980) The river continuum concept. *Canadian Journal of Fisheries and Aquatic Sciences* **37**: 130–7. [1.2]

Ward JV. (1989) The four-dimensional nature of lotic ecosystems. *Journal of the North American Benthological Society* **8**: 2–8 [1.2]

2: The Hydrology of Headwater Catchments

T.P.BURT

2.1 BACKGROUND

The need for process studies

Until recently, hydrologists took the view that the headwaters of a drainage basin were nothing more than source areas for runoff. Since their concern was largely with forecasting water resources or floods at downstream locations, they felt able to ignore the physical characteristics of the headwaters and the exact processes responsible for generating the runoff. This philosophy was reflected in the lumped models traditionally employed to forecast flood runoff and is perhaps best exemplified by the unit hydrograph model of Sherman (1932), which is still much used today. At about the same time that Sherman introduced his model, Horton (1933) proposed his classical theory of hillslope hydrology in which he assumed that the sole source of flood runoff was excess water which was unable to infiltrate the soil. Water that could infiltrate would eventually recharge groundwater and so provide a source for baseflow. Horton's theory of runoff production provided a physical basis for Sherman's model. Together, their ideas dominated hydrology for several decades.

It became apparent, however, that Horton's model was inappropriate in many areas because the infiltration capacity of many soils was too high to produce infiltration-excess overland flow. In particular, researchers such as Hursh and Hewlett at the Coweeta Hydrologic Laboratory in North Carolina, demonstrated that subsurface stormflow was dominant in forested basins with permeable soils and that overland flow, if it did occur at all, was limited in extent and generated by quite different processes than those described by Horton (Swank & Crossley 1988).

Since the 1960s, many field studies have been conducted in headwater catchments, with the purpose of describing the full range of runoff processes that may occur, and identifying and explaining their location in time and space. Such studies have made possible the development of distributed runoff models (Beven 1985); these have the capability of forecasting the spatial pattern of hydrological conditions within the catchment as well as simple outflows and storage volumes. Field studies have also identified clear links between hydrological pathways and streamwater quality; these are described in detail in Chapters 6 and 8.

How large is a headwater catchment?

Any simple definition of a 'headwater catchment' is likely to be inadequate. The approach taken here is to assume that the pattern of outflow discharge from headwater catchments is strongly related to runoff production at the hillslope scale. In 'large' river basins, flow patterns will be dominantly controlled by channel and floodplain storage and by the routing of runoff through the channel network. In addition, mixing of runoff from widely differing source areas (e.g. in terms of precipitation inputs or land cover) will obscure the process–response relations that can be identified in smaller basins. Most field research on runoff production has been conducted in small catchments, often less than 10 km^2 in area; detailed study of small runoff plots on hillslopes has provided the foundation for such work. There is surprisingly little work in physical hydrology that

6

has attempted to relate runoff production in small catchments to the flood response of larger basins. Most of the examples discussed here refer to basins of less than 100 km² in area, but this must be seen as an arbitrary division.

Headwater catchments can be found in any geographical location; they are not necessarily small upland valleys despite their frequent portrayal as such. Equally, the headwaters of a large river could be found in flat agricultural land or even within an urban area. Whatever its characteristics, the importance of the headwater region is that it determines both the quantity and quality of water received downstream and as such may impose important constraints on resource development and hazard control on the main river.

Terminology

In studying headwater catchments, it is important to distinguish between unconcentrated flows on hillslopes and concentrated flows in channels. Atkinson (1978) identified three methods of measuring moisture movement on hillslopes: methods involving interception of the flow, in which all or part of the flow is intercepted into a measuring device in order to determine its discharge; methods involving the addition of tracers in order to determine the velocity of flow; and indirect methods whereby measurements of moisture content and hydraulic potential over the slope profile or experimental plot are used to calculate moisture flux in the soil matrix. These methods are fully reviewed in Atkinson (1978) and Goudie (1990).

A variety of terms have been used to describe the movement of water in channels: streamflow, runoff and discharge being the most common. The rate of flow (i.e. discharge) is the product of a cross-sectional area of flowing water and its velocity; it is usually expressed in terms of volume per unit time (e.g. cubic metres per second, or 'cumecs'). Often, the discharge is divided by the area of the catchment so that the runoff can be expressed as depth per unit time (e.g. mm per hour); this allows runoff rates to be compared easily with other variables such as rainfall intensity or infiltration rate which are also expressed

in depth units. Gauged flows are usually calculated by converting the record of stage, or water level, using a stage–discharge relation, often referred to as the rating curve. Stage is measured and recorded through time by instruments placed in a stilling well; traditionally floats have been used but pressure transducers are becoming increasingly common. Stage is recorded continuously by pen and chart or digitally by a datalogger. The stage–discharge relation is obtained either by installing a gauging structure, usually a weir or flume, with known hydraulic characteristics, or by measuring the stream velocity and cross-sectional area of flow at a site where the river channel is stable. Some methods of gauging streamflow are particularly well suited for use on small streams in headwater catchments (e.g. dilution gauging, thin-plate v-notch weirs). Methods for stream gauging are fully reviewed in Gregory and Walling (1973), Rodda *et al* (1976), Herschy (1978) and British Standards Institution BS 3680 (1964).

A graph of discharge plotted against time is called a hydrograph. It is clear from the above discussion of runoff generation and from Fig. 2.1

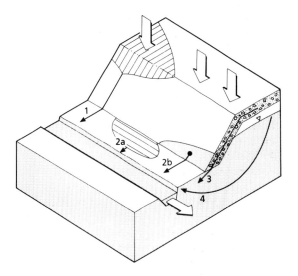

Fig. 2.1 Hydrological pathways. 1, Infiltration-excess overland flow (IOF). 2, Saturation-excess overland flow (SOF); 2a, direct runoff from saturated soil; 2b, return flow. 3, Subsurface stormflow (SSF). 4, Groundwater flow.

that many different pathways exist by which hill-slope runoff can reach a stream. Some precipitation (or snowmelt) takes a rapid route to the stream channel and is often described as quickflow or stormflow; quickflow is usually associated with high discharge. Subsurface flow moves at much lower velocities, often by longer flow paths, and, although it may contribute to stormflow as noted above, its main effect is to maintain streamflow during dry periods through the sustained release of water stored within soil and bedrock; this is termed low flow or baseflow. As a result, a hydrograph typically consists of episodes of high discharge separated by longer periods of low, gradually declining, flow. For some purposes it is necessary to separate arbitrarily stormflow and baseflow; given that stormflow can be produced by subsurface as well as by surface runoff, this line of separation has no physical basis, but may be helpful nevertheless (e.g. see the discussion relating to Fig. 2.8). The value selected by Hewlett and Hibbert (1967) of 0.05 cubic feet per second per square mile per hour (0.000546 cumecs per square kilometre per hour, or 0.0472 mm per day; Ward & Robinson 1990) has often been adopted, although this may not be appropriate for all basins. Figure 2.2 shows the terms normally used to describe a storm hydrograph and its (arbitrary) separation from the baseflow component.

When studying quickflow hydrographs in small basins, it is usually necessary to plot instantaneous discharge values; on larger rivers hourly mean or daily mean flows may suffice. However, even when studying small basins, it is often more convenient to analyse certain aspects of streamflow (e.g. seasonal variations in flow) using discharge totals or mean discharge rather than instantaneous values.

2.2 STORM RUNOFF MECHANISMS

The nature of the soil and bedrock determine the pathways by which hillslope runoff will reach a stream channel. The paths taken by water (see Fig. 2.1) determine many of the characteristics of the landscape, the uses to which land can be put and the strategies required for wise land-use management (Dunne 1978). Much work has been done in the temperate and warmer latitudes where rainfall is the primary hydrological input; hillslope hydrology during the snowmelt season has received less attention and the hillslopes of frigid regions are rarely considered (Church & Woo 1990).

The dominance of runoff theories based on the occurrence of infiltration-excess overland flow (IOF) (Horton 1933, 1945) meant that research into subsurface flow mechanisms was neglected. However, despite modifications such as the partial area concept proposed by Betson (1964), it

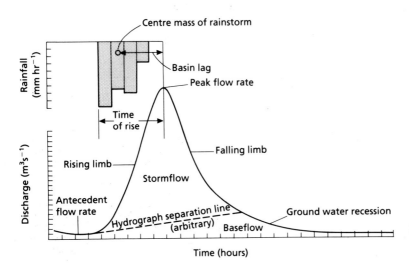

Fig. 2.2 Terms commonly used to describe a storm hydrograph (after Hewlett 1982).

became apparent that Horton's model was inappropriate in many locations. Where permeable soils overlie impermeable bedrock, subsurface stormflow (SSF) within the soil can account for most of the flood runoff leaving the catchment. When the soil profile becomes completely saturated, saturation-excess overland flow (SOF) may also occur (Dunne 1978). Both SOF and SSF may occur at rainfall intensities well below those required to generate IOF. SOF and SSF will be produced from source areas that are limited, although variable, in size and different in location to the source areas for IOF. The notion of localized sources of storm runoff which may vary in area both seasonally and during precipitation events provides the basis for the variable source area model first outlined by Hewlett (1961). Since then, this concept has come to dominate hillslope hydrology; subsurface flow is regarded as *the* major mechanism controlling the generation of storm runoff, both because of its influence on the generation of 'return flow' (a component of SOF; Dunne & Black 1970) and as an important process in its own right (Anderson & Burt 1978; Burt 1986). Figure 2.3 classifies storm runoff mechanisms on the basis of the two main models; these mechanisms are reviewed briefly below.

Infiltration-excess overland flow (IOF)

Central to the theory of hillslope hydrology put forward by Horton (1933) was his view concerning the role of infiltration processes at the soil surface. Horton considered that infiltration divides rainfall into two parts, which thereafter pursue different courses. One part goes via overland flow to the stream channel as surface runoff; the other goes initially into the soil and thence through the groundwater flow again to the stream or else is returned to the air by evaporative processes.

Horton established that infiltration capacity, the maximum rate at which a soil can absorb falling rain (or meltwater), decreases asymptotically over time as saturation of the surface soil causes a reduction in hydraulic gradients near the surface. Changes in the surface of the soil (e.g. swelling of clay particles, in-washing of fine particles into pores, compaction by rainbeat) may also reduce infiltration capacity through the course of a storm. Figure 2.4 shows the way in which rainfall intensity and infiltration capacity may interact during a storm to produce overland flow. At the beginning of the storm, infiltration capacity exceeds rainfall intensity and there is no surface ponding. However, when later the rainfall

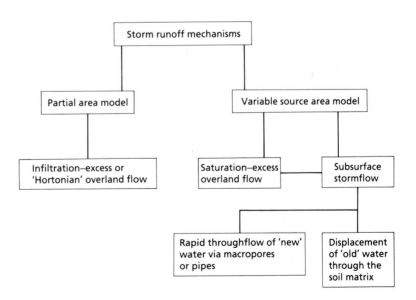

Fig. 2.3 Storm runoff mechanisms.

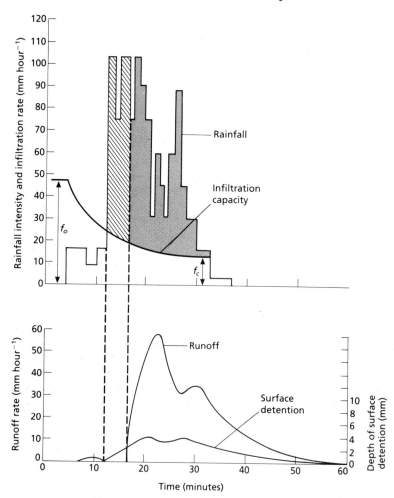

Fig. 2.4 Relationship of infiltration-excess overland flow (IOF) and surface detention to rainfall intensity and infiltration capacity for a storm of varying intensity. The lightly shaded portion represents depression storage which must be full before runoff, darkly shaded, is generated (after Horton 1940; Dunne 1978).

intensity exceeds infiltration capacity, the excess begins to fill up surface depressions. When these are full, the excess rainfall overflows downslope and surface runoff begins.

Infiltration capacity is not necessarily constant even within a small catchment, and Betson (1964) used the idea that IOF might be produced from only part of the basin area to improve his predictions of the volumes of storm runoff. In identifying relatively impermeable surfaces within a basin, Betson's partial area model remains the best guide to the location of source areas for IOF. Recent evidence suggests that IOF is not necessarily so rare as proponents of SSF have argued in the past (Heathwaite *et al* 1990). Where IOF is the

dominant producer of storm runoff, overland flow may be generated across large areas of hillside.

Subsurface stormflow (SSF)

As Fig. 2.3 shows, SSF may be generated by two mechanisms: by non-Darcian flow through large voids such as macropores or pipes, and by Darcian flow through the micropores of the soil matrix. Hillslope hydrologists have, until recently, emphasized micropore flow, although there has been a continued interest in pipeflow (Gilman & Newson 1980; Jones 1981) and in the influence of underdrainage on the flood response of small catchments (Robinson & Beven 1983). Field evidence relating to macropores (e.g. Whipkey 1965) was somewhat ignored until studies such as those of Mosley (1979), Beven and Germann (1982) and Kneale (1986) described the hydrological effects of rapid infiltration down macropores. Kneale and White (1984) studied infiltration into 9-cm cores of dry cracked clay-loam soil. Bypassing flow occurred down the cracks once the rainfall intensity exceeded the infiltration capacity of the soil peds (2.2 mm per hour). For the Oxford (UK) region where the soils were collected, their results imply that 10–20% of summer rainfall would bypass the root zone. Coles and Trudgill (1985) and Germann (1986) have identified important thresholds governing macropore flow. In addition to the infiltration threshold of the peds already described, they show that antecedent soil moisture is also an important control: if the soil is too dry, no macropore flow will happen whatever the rainfall rate; if the soil is at field capacity, then all rainfall must bypass the soil peds. Such results have clear implications for the production of storm runoff; Robinson and Beven (1983) showed that flow through cracks in a clay soil produced higher peak flows compared with an uncracked soil, in summer when the soil was dry. One unresolved question remains the connectivity of macropores in the downslope direction.

Lateral subsurface flow through the soil matrix will occur in any soil in which the hydraulic conductivity declines with depth (Zaslavsky & Sinai 1981). Where a layer of low permeability is found at depth in the soil profile, the flow direction in the more permeable soil above is parallel to the slope (Burt 1986). On the other hand, if both soil and bedrock are permeable at depth, percolation remains vertical and no lateral flow occurs within the soil layers; under these circumstances, infiltrating water serves only to recharge groundwater storage and to provide baseflow.

In itself, the occurrence of subsurface flow is not enough to produce storm runoff and it used to be assumed that micropore flow was too slow to provide this. However, rapid subsurface flow can occur in several ways: if the hydraulic conductivity of the soil is high, infiltration can lead to rapid recharge of the saturated zone at the base of the soil profile; macropore flow can have the same effect. Soils close to saturation (the 'capillary fringe'; Abdul & Gillham 1984) may require only a small amount of infiltration to produce a significant rise in the water-table. In all cases, if the saturated hydraulic conductivity of the soil is high, then large amounts of SSF will be produced as a result of a rapid rise in the water-table. In addition to this immediate effect, SSF may also occur in the form of a delayed hydrograph peaking several days after the rainfall input. Anderson and Burt (1978) and Burt and Butcher (1985a, 1985b) showed that convergent flow of soil moisture into hillslope hollows causes extensive development of soil saturation at such locations (the 'saturated wedge'). Figure 2.5 shows changes in soil water potential during the period of a double-peaked storm hydrograph. The mapped hillslope measured 70 × 50 m; soil moisture conditions are shown at a depth of 600 mm for a hillslope hollow and its adjacent spurs. The maps show that rainfall causes a widespread but shallow zone of saturation to develop at the base of the slope. The generation of the delayed peak is associated with convergence of soil water into the lower part of the hollow from further upslope and from the adjacent slopes. This causes the deepest saturation during this period, although the saturated wedge is now spatially less extensive than during the rainfall. SSF contributes to the first peak in stream discharge; the second peak is entirely SSF. Such runoff is strongly seasonal, being largely confined to winter when soil moisture deficits have been recharged. That SSF can dominate the stormflow response of some catchments is a point discussed further below.

Saturation-excess overland flow (SOF)

It is impossible to divorce the generation of SSF from the production of SOF. The variable source area model (Hewlett 1961; Troendle 1985) is based on the assumption that water moves downslope through the soil. The source areas for SSF and SOF are therefore essentially identical (Burt & Arkell 1986). Three locations within a catchment may be identified where maximum soil moisture levels will be reached (Kirkby & Chorley 1967; Burt 1986): at the foot of any slope, particularly those that are concave in profile; in areas of thin soil where soil moisture storage is reduced; and,

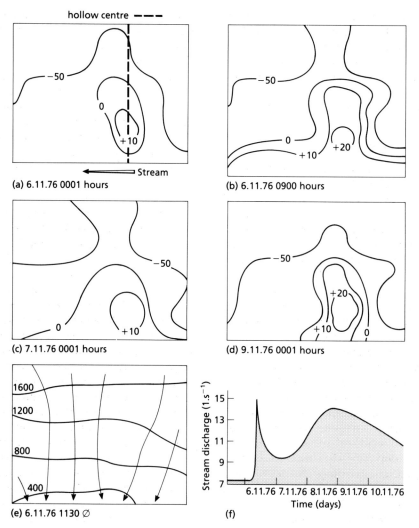

Fig. 2.5 Soil-water potential changes on an instrumented hillslope at Bicknoller Combe during the period of a double-peaked hydrograph: maps (a)–(d) show soil-water potential changes at a depth of 600 mm; map (e) shows the flow net; map (f) shows the stream hydrograph (after Burt 1978).

most importantly, in hillslope hollows where, as described above, convergence of flow lines favours the accumulation of soil water. The extent of the saturated area depends upon soil wetness and so source areas vary seasonally and during storms (Dunne 1978). Several authors have attempted to predict the distribution of soil moisture in relation to topography; in all cases upslope drainage area is the crucial control (Kirkby 1978; O'Loughlin 1981, 1986; Burt & Butcher 1985a; Thorne *et al*

1987). Where surface saturation occurs to any great extent (e.g. where wide valley bottoms exist), SOF will dominate the stormflow response with higher peak discharges and lower lag times than are characteristic for SSF (Dunne 1978). If soils are permeable, the source areas for SOF are likely to remain a relatively small percentage of basin area and the ratio of storm runoff to rainfall may be low (commonly below 5%). Where soils are impermeable (e.g. peat), surface saturation will be extensive, with much more rainfall translated into runoff (Burt & Gardiner 1984). SOF will be a mixture of return flow ('old' soil water) and direct runoff (water unable to infiltrate into saturated soil) so that its solute and sediment load may differ widely from that produced by IOF (which is essentially direct runoff alone).

Snowmelt and storm runoff

Snowmelt runoff occurs when energy is added to a snowpack that is at 0°C. The principal source of energy for snowmelt is direct solar radiation, but long-wave radiation and fluxes of sensible and latent heat may also be important on occasions. A strong diurnal rhythm dominates snowmelt. The form of the diurnal melt flood is controlled by the pattern of melt at the surface, by the passage of water through the snowpack, and by the saturated flow at the base of the snow to an open channel (Church 1988). In the UK, snowmelt occurs relatively infrequently and, additionally, its effect is often overlooked because the flood is often associated only with rainfall at lower elevations; even so, melt-related events have generated the most extreme floods recorded on many basins. Over much of Canada, snow cover persists sufficiently long to alter the pattern of runoff from that which would otherwise be determined by the distribution of precipitation. Here, engineering designs and water management procedures must be modified to take account of snow accumulation and snowmelt (Church 1988).

Runoff from a snowpack is the last occurrence in a series of events beginning with snowfall itself. The density of new snow varies from 0.05 to 0.20 depending on air temperature during its precipitation. The density of the snowpack gradually increases due to settling and compaction, melting and recrystallization, and rainfall, and may be 0.30−0.50 by springtime. An important element in snowpack evolution is 'ripening' by which the temperature of the snowpack rises to reach 0°C and its liquid content is maximized. A deep layer of snow may be well below freezing initially, but when milder weather sets in some melting will take place at the surface. This meltwater percolates into the pack where much of it may refreeze, liberating latent heat of fusion in so

doing; at this stage little of the melt may form runoff. In addition, heat is added to the snowpack from the overlying air. The pack warms continually and eventually reaches 0°C. With continued melting, the liquid content of the snowpack rises to its field capacity and is now said to be 'ripe'; further melt will generate significant runoff (Dunne & Leopold 1978). Most snowpacks contain strata of varying texture and permeability; during the ripening stage, these serve to diffuse the meltwave and reduce peak flow somewhat. In shallow snow, or when the snowpack is ripe, runoff delay due to passage through the snow becomes less important, and stream hydrographs become more peaked (Church 1988). Church considers potential meltwater production rates (in the absence of rainfall) and shows that the size of a meltwater flood is strictly limited by the energy available. Near the solstices, up to 40 mm per day may melt if all available radiant energy is utilized; in late winter between 10 and 20 mm per day is more likely. Such rates may be important in a large basin such as the Fraser River which is so large that only snowmelt can generate runoff over most of it at any one time; rainfall events and other synoptic weather variations serve only to modify details of the seasonal melt hydrograph. On the other hand, in small basins, although melt may continue over several months, individual peak flows are usually rain-on-snow events (Church 1988). In addition, thermally induced snowmelt may produce a distinctive diurnal rhythm of runoff in small drainage areas in direct response to energy inputs for melt; Church (1988) provides examples.

Once the snowpack is ripe, rapid snowmelt is most frequently brought about by a sudden influx of warm, moist, unstable air, and rain often accompanies the melt. With an air mass of high humidity, much of the energy for melt may be provided by the release of latent energy when there is condensation on to the snow surface. None the less, even though the heat energy supplied by warm, moist air and the rainfall itself can be important in a large, warm rainstorm, the volume of water directly contributed by rainfall will often far outweigh the amount of melting (Dunne & Leopold 1978). Critical situations can arise when rainfall combines with snowmelt, especially if the soil below is frozen, producing a flood which may be extreme even in a small basin; on larger rivers it may even be the major design criterion (NERC 1975, pp. 502−4). This is particularly the case in temperate, coastal mountains where autumn and early winter storms are the most vigorous of the year and frequently bring heavy rain on to shallow snow (Church 1988). For example, the UK Flood Studies Report (NERC 1975) showed that 14–20% of all floods on the South Tyne were snowmelt related over the period 1959–69; for 53 of 329 gauging stations, the record peak discharge was associated with snowmelt.

Once runoff from the snowpack begins, the pathways by which meltwater may reach a stream are identical to those already discussed. Where the topsoil has remained frozen during the winter, infiltration capacity may be effectively zero and large quantities of IOF will be generated (Dunne & Black 1971; Dunne *et al* 1976). However, in many cases the ground below a snowpack is unfrozen or covered only in porous 'needle ice', in which case infiltration capacity is unaffected and meltwater is able to enter the soil (Stephenson & Freeze 1974).

Peak runoff from glacierized basins is displaced into middle or late summer when seasonal snow is much reduced, glacial drainage paths are well integrated and, most importantly, glacier ice with low albedo is exposed to melt (Church 1988). Episodic runoff events are associated either with meteorological events identical to those described for snowmelt, or with the catastrophic release of water from within, or dammed up by the glacier.

Controls of storm runoff

Climate, soil type and bedrock lithology control which type of runoff mechanism will dominate in a particular catchment, with vegetation cover and topography as an important secondary control at the hillslope scale (Dunne 1978; Whipkey & Kirkby 1978). Kirkby (1978) recognized the crucial role of hydraulic conductivity in relation to rainfall intensity and identified two runoff domains: where rainfall intensities commonly exceed infiltration capacity, IOF will be the dominant storm runoff mechanism; where the infiltration capacity of the soil is higher than the rainfall rate, SSF and SOF will be important, the balance between the two being determined by the permeability of soil and bedrock, and by catchment topography (Dunne 1978; Anderson & Burt 1990). Church and Woo (1990) have noted that spatial variation in rainfall intensity is much less than the range of soil permeabilities, so that soil may be a more important control of runoff generation than climate. IOF is often associated with arid environments, encouraged by a combination of intense rainfall and bare ground (Horton 1945; Langbein & Schumm 1958). Conversely, in humid temperate environments, the combination of low rainfall intensity, complete vegetation cover and high infiltration capacity favours SSF and SOF. This is, however, an oversimplification; runoff may be absent in arid areas if rainfall intensity is low or if soils are permeable (Yair & Lavee 1985). In humid areas, many soil types have a naturally low infiltration capacity (e.g. peat; Burt & Gardiner 1984), and in other cases the soil surface may have become less permeable due to compaction by poor management (Heathwaite *et al* 1990), so that IOF may be more common than is often thought. Thus, although it has been represented as a zonal phenomenon, the propensity for IOF to take place may be mainly a function of soil properties (Church & Woo 1990).

Figure 2.6 summarizes these points. The occurrence of IOF depends on the relative magnitude of rainfall intensity and the hydraulic conductivity of the upper soil layer. The occurrence of SSF depends on the balance between rainfall intensity and the hydraulic conductivity of the lower layer. SOF is most likely to occur in soils of medium to low hydraulic conductivity where drainage is slow and storage limited, in thin soils, or in deep permeable soils where flow convergence leads to soil saturation. Vegetation cover may significantly influence production of IOF, soils in forests tending to have much higher infiltration capacities than those of farmland. Land use and techniques of land management may strongly influence the hydrological response of a catchment; some effects of a change in land use are described at the end of this section.

The information given in Fig. 2.6 defines the domains for stormflow, but gives no indication of

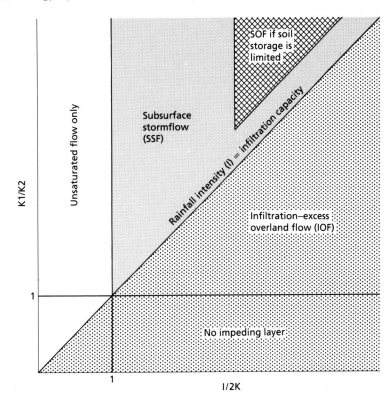

Fig. 2.6 Regimes of flow for a two-layer soil in relation to the hydraulic conductivity of the upper (K1) and lower (K2) layers and rainfall intensity (I) (after Whipkey & Kirkby 1978; Burt 1986).

the magnitude and timing of storm runoff production; Fig. 2.7 provides this information. In general, the largest and quickest responses are generated by surface runoff. Where source areas for surface runoff are very limited in extent or absent altogether, stormflow is generated by SSF alone. Since SSF can effectively provide two peaks in stream discharge (see Fig. 2.5), the lag times to peak on Fig. 2.7 show a wide range for a given basin area.

2.3 BASEFLOW GENERATION

The two main sources of baseflow are ice and snow melt, and drainage of soil and ground water. Stream response to melt depends on meteorological inputs to the snowpack (energy and matter), the state of the snowpack, the delivery of melt to the ground surface beneath it, and the routing of meltwater over or through the soil to the stream. In small basins, flood peaks involving snowmelt are most likely to be generated by rain-on-snow events. Since continued melt in mountain en- vironments is essentially a thermal process, the resulting pattern of streamflow may bear little relation to the temporal pattern of precipitation input, although it may be diurnally quite regular. Given the low rates of melt that are likely to occur (up to 40 mm per day; Church 1988) in comparison with the infiltration capacity of many soils, it is likely that snowmelt will augment baseflow since it is likely to recharge soil and ground water, rather than generate surface or near-surface runoff.

Water that percolates to groundwater moves at much lower velocities by longer paths and reaches the stream slowly over long periods of time, sustaining streamflow during rainless periods (see Fig. 2.1; Dunne & Leopold 1978). Any layer of rock or unconsolidated material that can yield significant quantities of water is known as an aquifer; a stratum through which water cannot move except at negligible rates is known as an aquiclude or aquitard. Although its porosity determines how much water a rock can store, its specific yield, the amount of water released by

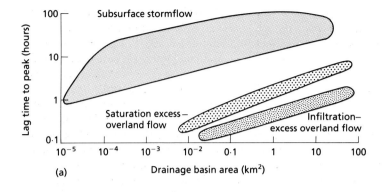

(a)

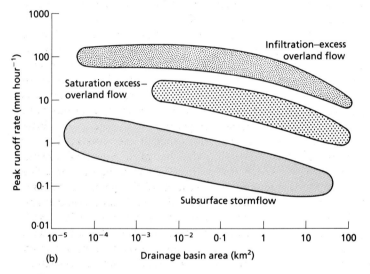

(b)

Fig. 2.7 Responses of catchments to hillslope flow processes: (a) lag times; and (b) peak runoff rates (after Dunne 1978; Kirkby 1985; Anderson & Burt 1990).

drainage under gravity, is of more relevance in supplying water to local rivers. This, together with its permeability and thickness, determines the importance of any aquifer. As described below, the nature of the porosity may also be influential; a porous aquifer may have a very different pattern of drainage compared with a well-fractured or pervious rock. In headwater catchments, the main mode of recharge is infiltration of water from the ground surface; influent seepage from river beds and lakes is likely to assume greater importance downstream. The main process of groundwater discharge is by spring flow and effluent seepage into rivers and lakes.

Depending on the nature of the aquifer, peak baseflow discharge may lag significantly behind precipitation inputs. Impermeable rocks are as-sociated with strongly peaked hydrographs because there is little subsurface storage and rapid surface or near-surface runoff. Basins composed of highly permeable formations such as limestones or basalts tend to have flatter, broader, and more delayed responses (Ford & Williams 1989). There is a continuum of rainfall-runoff responses, with cavernous (karst) limestones providing the most rapid, peaked groundwater hydrographs, while deep, porous aquifers, such as the chalk limestone of southern England, yield the most subdued hydrographs. Except for cavernous limestones, it is unusual for aquifers to provide flood hydrographs. If there are no source areas for overland flow, floods may be generated by groundwater. Otherwise high discharges are generated by surface and near-surface processes, and groundwater provides the prolonged recession flow associated with the falling limb of the hydrograph (see Fig. 2.2). The flood hydrographs of vadose (above the water-table) cave systems tend to be peaked and similar to surface streams, but should the cave streams flow into a flooded phreatic (below the water-table) zone before their waters emerge at a spring, then the influence of the inflow hydrograph on the composite outflow hydrograph of a spring is similar to that of a tributary flowing into a lake — the outflow is a muted, delayed reflection of the inflow (Ford & Williams 1989). Karstic limestones, being massive, low-porosity limestones with mainly conduit permeability, have low baseflow because little water is held in storage. Deep, porous aquifers such as chalk limestone have low flows, only slightly lower than the mean flow because the response time for discharging water in storage is comparable with the time between wet and dry seasons. In such aquifers, there will be a long lag before a response occurs, a slow rate of rise to peak groundwater discharge, and a low rate of recession. Flow regimes which reflect the influence of groundwater discharge are exemplified below in section 2.4.

Hewlett and Hibbert (1963) showed that, in the absence of an aquifer, drainage from a soil profile can maintain streamflow over long periods. Drainage of an isolated block of soil during the first few days represents outflow from a thinning and flattening saturated wedge. Thereafter, discharge is produced by drainage from the unsaturated zone which is transmitted by the thin saturated zone at the outlet (R.D. Moore, personal communication). Anderson and Burt (1977) argued that in some situations unsaturated flow remains insignificant so that the outflow is provided primarily by drainage of the saturated wedge. As noted in section 2.4 below, in catchments that predominantly produce subsurface flow through the soil layers, the runoff regime is more seasonally variable than that of an aquifer.

Nevertheless, many catchments lack significant storage of soil or ground water. Even where rainfall is relatively uniform throughout the year, large rates of evaporation in summer mean that many headwater catchments have a characteristic drought, even in humid climates (see also comments on river regimes and flow duration curves

in section 2.4 below). Conditions are most critical in late summer, or late in the dry season in the tropics and subtropics. Agricultural underdrainage and reclamation of wetlands have exacerbated this problem in highly developed areas such as the UK. As a result, many headwater streams are seasonally ephemeral. This is well known even on reliable aquifers such as the chalk: prolonged summer discharge lowers the water-table to such an extent that the source of the river migrates several kilometres downstream. Autumn recharge eventually raises the water-table and the intermittent ('bourne') stream reappears, although perhaps not until several months after the main period of rainfall. Where the abstraction of groundwater occurs, such natural effects may be aggravated. A numerical model of the Lambourn, a typical bourne stream on the chalk in southern England, showed that if abstracted water was piped directly out of the basin, a pronounced downstream migration of the perennial head would occur. However, if the abstracted water was piped into the river at the perennial streamhead, although less efficient in supply terms, this preserved the location of the perennial streamhead, a highly desirable result on ecological and amenity grounds (Oakes & Pontin 1976). Land-use changes in headwater catchments may have similar effects. Afforestation will significantly decrease baseflow during the summer to such an extent that headwater streams may dry up. This effect is further elaborated in the last paragraphs of section 2.4 below.

2.4 PATTERNS OF STREAMFLOW IN HEADWATER CATCHMENTS

In the following paragraphs, a selection of commonly used methods of analysing streamflow are presented; in general, as the timescale of analysis increases, the use of aggregate discharge figures becomes more convenient. Some of the data used below were compiled by the Surface Water Archive (IH 1989), an extremely useful source of hydrological information for UK catchments, both large and small.

Daily mean flow

This is calculated by averaging the instantaneous flows occurring throughout the water-day (in the UK from 09.00 to 09.00 hours Greenwich Mean Time) obtained from the continuous record. For comparatively small basins that respond rapidly to rainfall or melt events, hydrographs of daily mean flow provide a useful summary of flow variations during a given year. In particular, quickflow and baseflow contributions are easily separated by eye on such plots. As discussed below, the relative proportions of quickflow and baseflow are an important diagnostic feature of headwater catchments.

Figure 2.8 shows the annual hydrograph for 1978 for two lowland basins (Foston Beck and Blackfoss Beck) that are about 30 km apart in east Yorkshire, England; total precipitation for the year was about 700 mm in both cases. In addition, flow figures are shown for a third basin (Snaizeholme Beck), a small upland basin in the same region where total rainfall was 1600 mm over the same period. For the two lowland basins, the large contrast in their annual hydrographs must relate to differences in catchment characteristics rather than to differences in climate. Blackfoss Beck has very subdued topography; an extensive network of ditches and tile drains provides the necessary drainage of the clay soil; the clay bedrock below is totally impermeable. Infiltration capacities are reasonably high, because of the presence of macropores in the clay, but may fall to very low levels if compacted by agriculture. Foston Beck lies on chalk limestone: high infiltration capacities and permeable bedrock mean that little storm runoff is generated despite the hillier terrain; groundwater provides almost all of the runoff. The flashy response of Blackfoss Beck is typical of a basin which produces little baseflow but much quickflow; the high peak discharge and symmetry of the quickflow hydrographs suggest that much surface or near-surface runoff is generated during storm events. Once the quickflow has finished, there is almost no baseflow to maintain dry weather flow. By contrast, the annual hydrograph for Foston Beck is typical of a catchment dominated by baseflow production with very small quickflow hydrographs and a very protrac-ted rise and fall of groundwater discharge during the year. Thus, streamflow at Blackfoss Beck is intimately correlated with rainfall inputs whilst at Foston Beck the pattern of streamflow bears little relation to the pattern of rainfall, peak baseflow occurring some 2 months after the main period of rainfall. The hydrograph for Snaizeholme Beck is also typical of a catchment which produces much surface runoff and little baseflow; soils here are peaty and permanently saturated for most of the year. The higher number of quickflow peaks reflects the higher number of precipitation events in the Pennine uplands, compared with the much drier Vale of York which lies to the east in its rain shadow.

The storm hydrograph

As might be expected from section 2.2, the shape and dimensions of a storm hydrograph in a headwater catchment reflect closely the runoff mechanisms operating. Figure 2.2 shows the elements of the storm hydrograph. Surface runoff tends to produce the highest peak runoff rate and the shortest lag time to peak (see Fig. 2.7); in cases where little surface runoff is produced, SSF will dominate the flood hydrograph. In dry periods the recession limb of the storm hydrograph may be very steep because of the lack of subsurface flow. However, as successive rainstorms recharge the soil moisture and groundwater, the recession becomes more gradual and baseflow increases. In winter when recharge is complete, SSF may be important enough to generate delayed peaks in storm discharge (see Fig. 2.5) and groundwater discharge can maintain baseflow at high levels for several months (Fig. 2.8).

Figure 2.9 shows a variety of hydrographs for small, headwater catchments. The flashy response of Shiny Brook, another Pennine stream, is typical for a small basin that produces little but surface runoff; the symmetry of the hydrograph shows the almost complete lack of subsurface flow, a contribution that normally causes the falling limb to be more gentle than the rising limb. Peat-covered catchments produce very little baseflow indeed, a point well illustrated in this example (*cf.* Snaizeholme Beck in Fig. 2.8). The runoff response of the Shiny Brook catchment

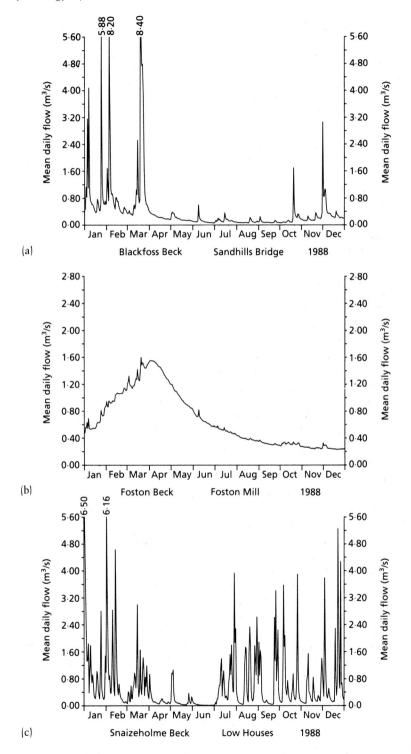

Fig. 2.8 Annual hydrographs for three basins. Data supplied by Yorkshire Region, National Rivers Authority, UK.

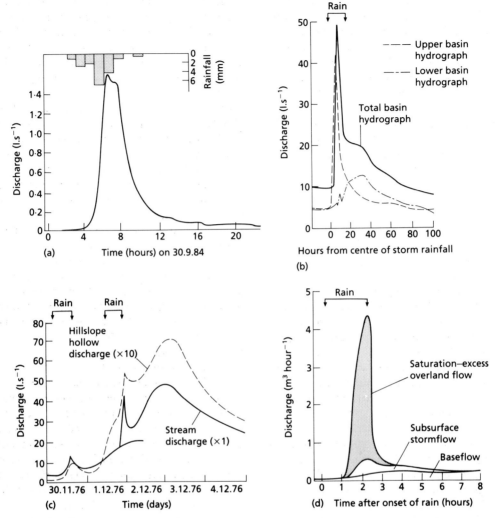

Fig. 2.9 Storm hydrographs for four catchments: (a) Shiny Brook; (b) East Twin Brook; (c) Bicknoller Combe; and (d) Sleepers River. See text for details and source references.

has been described in some detail by Burt and Gardiner (1984) and by Burt *et al* (1990). The hydrograph for the East Twin Brook (Weyman 1973) shows that this small basin may be divided into two distinct sections: in the upper basin, which is infilled by peat, the response is very like Shiny Brook in that much surface runoff is generated; in the lower basin, where permeable loamy soils overlie impermeable sandstone, the subsurface runoff is much delayed compared with the surface flow. Together, the total basin hydrograph reflects these two contributions, the gentle re-

cession denoting the contribution of subsurface drainage. At Bicknoller Combe (Anderson & Burt 1978), little surface runoff is produced and the runoff response is dominated by the delayed peak in stream discharge (see also Burt & Butcher 1985a, 1985b and Fig. 2.5). The subsurface discharge from the hillslope hollow mirrors that of the stream and shows that even the 'first' peak in stream discharge for each event must contain subsurface as well as surface runoff. In the Sleepers River basin (Dunne & Black 1970), SSF is restricted by complete saturation of the soil profile, so that SOF dominates the runoff response; once again, where surface runoff is most important, a sharp peak in discharge tends to result.

Clearly, these four examples can only begin to demonstrate the range of hydrographs which are possible in headwater catchments. In all four cases, interpretation was made possible by knowledge of the hillslope hydrology. Where only the stream hydrograph is available, little can be said with confidence about the *exact* runoff mechanisms operating since many different processes might generate a given response.

Floods

As noted in the introduction to this section, forecasting the flood hydrograph, either in its entirety or specific aspects such as the peak discharge, is a long-established practice in hydrology. Hydrological forecasting has been fully reviewed by Anderson and Burt (1985). Standard methods, such as the unit hydrograph or extreme frequency analysis, are described in most hydrology texts, including Shaw (1988). For the UK, the Flood Studies Report (NERC 1975) has provided a comprehensive guide to flood prediction in ungauged basins throughout the country, although the Report did not use much information from very small basins (<10 km^2). Often, the unit hydrograph or even a regionalized rational method, stratified by cover type (Hewlett *et al* 1977), is the only practical possibility in ungauged small basins. The Flood Studies Report provides detailed information on selected major floods in the UK (see also Newson 1975; Rodda *et al* 1976). Although the causes of floods are largely climatic for headwater catchments, the conditions which tend to intensify floods are primarily related to characteristics of the catchment such as soil type, topography, degree of land drainage, and so on. Only in large catchments does the nature of the channel network exert a significant influence on the shape of the flood hydrograph (Ward 1978).

Runoff regime

The regime of a river may be defined as the seasonal variation in its runoff response and is usually portrayed by a curve based on monthly mean flow. Seasonal variations in the natural runoff of a drainage basin depend primarily on the relations between climate, vegetation, soils and rock structure, of which only the last can be strictly independent of climate (Beckinsale 1969). Beckinsale noted that there are such large areas of the world within which the annual pattern of runoff for small and moderately sized basins closely reflect the regional climatic rhythm that areal differentiation of hydrological regions is easily achieved by adapting Koppen's climatic divisions (Fig. 2.10). In Beckinsale's classification, Koppen's terms retain their climatic meaning:

A = tropical rainy climates; all months with mean over 18°C

B = dry climates with an excess of potential evaporation over precipitation

C = warm, temperate rainy climates

D = seasonally cold, snowy climate; mean temperature of the coldest month being below −3°C

Beckinsale applied the rainfall symbols of Koppen to provide the second capital letter in the code:

F = appreciable runoff all year

W = marked winter low flow

S = marked summer low flow

He added a further class to take into account regimes that occur in the snow and ice environments of high mountains outside the polar ice caps, codes EN and EG (HN and HG in Beckinsale's scheme) denoting nival and glacial regimes respectively. He also added a third category of small letters (not shown in Fig. 2.10) to make allowance for temperature regimes which have some relevance to hydrological regimes, such as the occurrence of high evaporation losses in summer.

Ward (1968) analysed river regimes for 37 British basins and showed that 29 were clearly characterized by winter maxima and summer minima. A reasonably uniform precipitation input combines with modest summer evaporation to reduce flows significantly, although not drastically, in that season. Thus, Beckinsale's code CF applies. However, even in a small country like the UK, variations in river regime do occur, some being climatic in origin and others relating to the physical characteristics of the basin itself. Figure 2.11 shows river regimes for several British basins; flow for each month is expressed as a dimension-

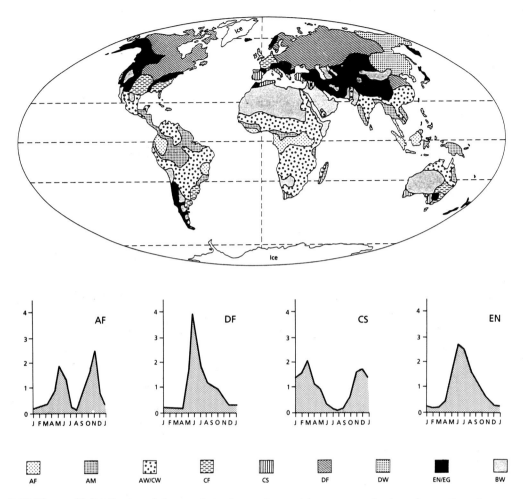

Fig. 2.10 The world distribution of characteristic river regimes with type examples from four regions. Letters are specified in the text. (Adapted and simplified from Beckinsale 1969.)

less index — the ratio of the monthly mean flow to the mean monthly flow. Regimes for Cheriton and Foston Beck are typical of catchments where groundwater dominates: monthly flows vary little from the mean monthly flow, suggesting both minimal quickflow inputs and sustained groundwater flow in summer; peak discharge is relatively late (January or February) indicating the delay involved during recharge. Regimes for Snaizeholme Beck and Blackfoss Beck are more extreme: lack of baseflow means that flows are very low in summer; winter flows are high and peak earlier because quickflow dominates the runoff response. The Slapton Wood regime is also

surprisingly extreme: though a catchment dominated by subsurface stormflow, it is evident that flows are quite low in summer after the soil has drained. In winter when much subsurface stormflow is generated, flows are very much higher than the mean flow. The regime for the River Falloch in Scotland is typical of a mountainous stream with little groundwater storage: summer flows are relatively low whilst winter flows are high and peak early in November in response to high rainfall at that time; a second peak in March (which seems to be typical of many upland streams in Britain) perhaps indicates the influence of snowmelt. Apart from the groundwater-fed

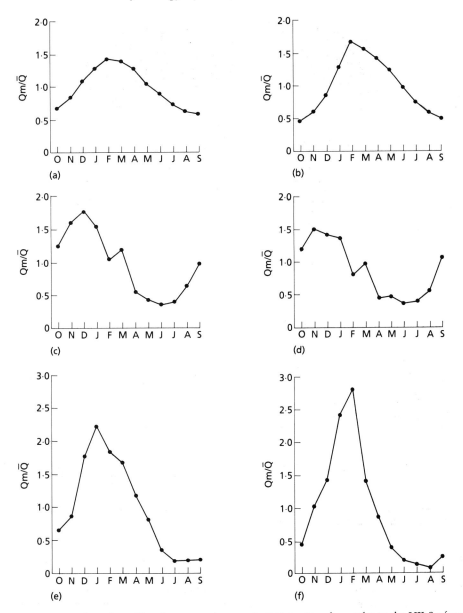

Fig. 2.11 Runoff regimes for selected headwater catchments in Britain. Numbers refer to the UK Surface Water Archive classification of gauging stations. $Qm/\bar{Q}$, monthly mean flow as a ratio of annual mean flow. (a) Cheriton (42008); (b) Foston Beck (26003); (c) Snaizeholme Beck (27047); (d) Falloch (85003); (e) Blackfoss Beck (27044); and (f) Slapton Wood.

streams, all regimes show the characteristic regime drought referred to in section 2.3 above: despite its humid climate, many British streams have seasonally recurrent low flows. Such low flows could provide the limiting conditions of the aquatic habitat, and so be of particular interest for managers of stream biota.

Water balance

Increasingly, headwater catchments are being used for water supply, so that, even here, aggregation of flow figures into discharge totals is important to show the total resource available. The water balance over a selected time period can be evaluated as follows:

$$P - Q - G - \Delta S - E = 0$$

where P is precipitation, Q is stream discharge, G is groundwater discharge, E is evaporation and ΔS is change in storage. In many cases it is assumed that the catchment is watertight and that no inflow or outflow of groundwater occurs; however, this may not be the case where there are aquifers. On an annual basis it is also often assumed that no change in storage takes place from year to year, although this may well not be the case. If a water balance is required for a shorter period, then changes in storage must be measured. Ferguson and Znamensky (1981) discussed the errors associated with water balance computations and showed that where measurements are inaccurate, the balance period needs to be longer. Probably most uncertainty has surrounded the estimation of basin-wide totals of precipitation and evaporation: these problems are reviewed fully by Rodda *et al* (1976) and Shaw (1988).

Figure 2.12 shows the monthly water balance for the Slapton Wood stream from October 1974 to September 1977, which includes a period of major drought. Potential evaporation was calculated using the Penman formula and these data were then used to calculate actual evaporation using Penman's 'root constant' concept. The full procedure is given in Shaw (1988, Chapter 11). It should be noted that such figures are now routinely available in the UK via the Meteorological Office's MORECS scheme. Particularly notable on Fig. 2.13 is the rainfall deficit in 1975 and 1976, especially during the winter months. Accordingly, very low runoff resulted in the winter of 1975–76. Soil moisture deficits were consistently high during the summers of 1975 and 1976, and soils were barely recharged during the intervening winter. In both summers soil moisture deficits were sufficiently high to preclude evaporation at the potential rate. By contrast, the period from September 1976 onwards was one of the wettest on record, and soil moisture and streamflow recovered quickly; high flow in the winter was associated with a major period of nitrate leaching (Burt *et al* 1988). Assuming an arbitrary initial storage of 500 mm at the end of September 1974, catchment storage had fallen to 368 mm by October 1975 and to 277 mm at the end of August 1976; nevertheless, by the end of February 1977 storage had fully recovered to 577 mm.

Flow frequency and duration

The contrast between baseflow- and quickflow-dominated basins may also be seen clearly using flow duration curves. These are prepared by grouping daily mean flows into selected discharge classes, starting with the lowest values. The cumulative frequency, expressed as a percentage of the total, is then the basis for the flow duration curve, which gives the percentage of time during which any selected discharge may be equalled or exceeded (Shaw 1988). Such cumulative frequency curves are most conveniently plotted on probability paper (on which normal distributions plot as a straight line). In Fig. 2.13 flow duration curves are plotted using a dimensionless flow axis (daily mean flow divided by mean daily flow) to allow comparison between basins. The shape of the flow duration curve gives a good indication of the catchment's runoff response to precipitation: a steeply sloping curve indicates very variable flow, usually from catchments with much quickflow and little baseflow; flow duration curves with a flat slope result from the dampening effects of high infiltration and groundwater storage. Thus, in Fig. 2.13, the Coln, Windrush and Foston Beck all indicate groundwater catchments with sustained, reliable baseflow (minimum flow is still a relatively high percentage of mean flow) and low flood flows (maximum flow is at most only 3–4 times the mean flow). By contrast, Snaizeholme Beck and Shiny Brook, both peat-covered catchments in the Pennine uplands, have very steep flow duration curves; this indicates minimal baseflow in summer, since there is virtually no groundwater and little drainage from the peat (Burt *et al* 1990), and very high flood

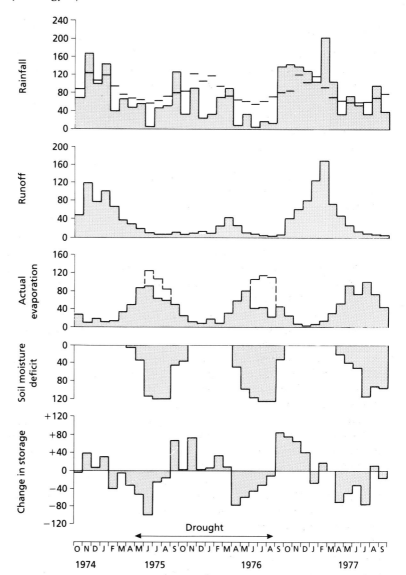

Fig. 2.12 Monthly water balance for the period from October 1974 to September 1977 for the Slapton Wood stream. Dashes on the rainfall histogram indicate mean monthly totals; dashed lines on the evaporation histogram indicate potential (Penman) evaporation. All units are in millimetres.

runoff, given the widespread production of surface flow in peat-covered catchments (Burt & Gardiner 1984). The Slapton Wood stream is intermediate to these two groups: relatively low flow in summer but high winter runoff due both to surface quickflow contributions from a variety of source areas (zones of permanent surface saturation, roads, tracks and fields of low infiltration capacity; Heathwaite *et al* 1990) and to delayed peaks in subsurface stormflow (Burt & Butcher 1985a, 1985b).

Low flow

Only recently has much attention been focused on low flows, with the result that standard definitions and methods of analysis are lacking. Unlike floods, low flows tend to be prolonged so that analyses based on mean flow over intervals ranging from a week to a month tend to be more useful than those based on daily mean flow; also, flow on a single day is too readily affected by abstractions. UK river agencies have often

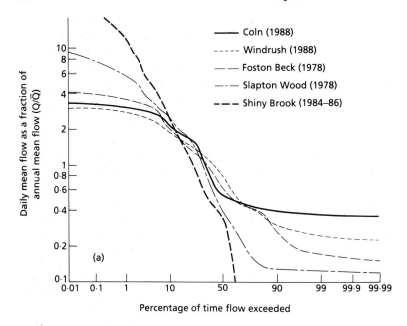

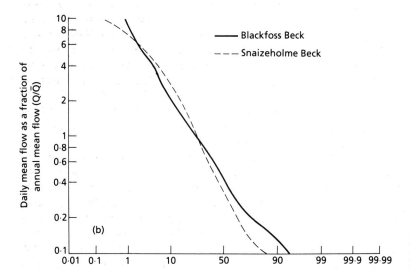

Fig. 2.13 Flow duration curves plotted as a dimensionless index for selected headwater catchments in Britain.

adopted the mean annual minimum 7-day flow as their index of dry weather flow; the Low Flow Studies Report (IH 1980) used the mean annual minimum 10-day flow. The Institute of Hydrology (IH 1989) noted that the daily mean flow exceeded for 95% of the time remains a useful low flow parameter for the assessment of river water quality consent conditions. Like floods, low flows may be analysed in terms of flow frequency or flow duration; for the latter, a flow duration curve of low slope indicates a groundwater basin with reliable low flows (*cf.* Fig. 2.13). Given its regularity, graphs of recession flow may be very useful for low-flow prediction purposes, although their use as indicators of the flow processes operating during recession is probably very limited (Anderson & Burt 1980). Recent droughts in the UK have been described by Doornkamp *et al* (1980) and by Marsh and Lees (1985).

Long-term variations in flow

Long-term variations in flow are caused by climatic variation and by human impact. Human influence on streamflow may be classified into two groups: *direct* impacts such as construction and operation of reservoirs, direct abstractions for domestic supply or irrigation, and diversions of streamflow. Urbanization changes permeable land surfaces to impermeable and probably has one of the most dramatic effects of any land-use change on hydrology, particularly on the volume and speed of storm runoff. Even in rural catchments, roads and villages may constitute a significant area of impermeable surface. In headwater catchments *indirect* impacts related to the condition of the land surface are probably more important, however. They are crucial because they can significantly change the conditions governing runoff formation in the basin. They include agricultural practices, deforestation, urbanization and drainage of wetlands. A quantitative evaluation of human impact on streamflow is complicated because numerous causal factors operate simultaneously. Moreover, deliberate human actions may overlap in the short or intermediate term with natural variations in flow, the amplitude of which may considerably exceed the magnitude of the cultural changes.

Given the intimate connection between runoff processes and streamflow in headwater catchments, it has proved possible to conduct hydrological experiments which allow the evaluation of the role and importance of each anthropogenic factor individually. Classically, such experiments have involved a 'paired catchment' approach, following the pioneering example of Bates and Henry (1928) at Wagon Wheel Gap, Colorado, USA. In this, two nearby or adjacent catchments are selected on the basis of their similarity in size, cover type, aspect and suitability for streamflow gauging. Runoff from the two basins is compared during a period of 'calibration' when both basins

remain unchanged. Then, one catchment is 'treated' while leaving the other unchanged as a control. The purpose of such experiments is to establish absolute differences in water yield between different basins, rather than to establish the absolute water balance for each basin (Hewlett 1982). Notable paired catchment studies in forest hydrology have been conducted at the Coweeta Hydrologic Laboratory in North Carolina, USA (Swank & Crossley 1988) and at the Institute of Hydrology's catchments at Plynlimon, mid-Wales (Kirby *et al* 1991). Plot studies, such as that of Law (1956), may provide useful evidence relating to process rates and the direction of change following a treatment, but basin scale experiments are preferable in that, depending on the basin size and the character of the manipulation, they reduce 'oasis' effects and also avoid the experimental difficulties associated with attempting to compute a full water balance. In some cases, specific studies are established on individual catchments in a process of standardization (or calibration of a catchment on itself; Hursh, quoted in Swank & Crossley 1988). However, this requires climate to remain constant during the treatment period if the effects of a treatment are to be evaluated with confidence. Techniques such as double-mass analysis (Dunne & Leopold 1978) may also be useful to indicate changes in flow where only the record from a single basin is available.

Space does not allow a full account of the effects of all possible human influences on streamflow; extensive reviews are given in Ward and Robinson (1990), for example. By way of brief illustration, some examples from forest hydrology will be described. Most paired catchment experiments have used water-balance methods to study changes in water yield at annual or monthly timescales (see review by Swank *et al* 1988). Changes in the size and shape of storm hydrographs have been examined by Swank *et al* (1982), and by Hewlett and Helvey (1970) who found that the main change following forest clearance was an increase in quickflow volume; peak discharge and time to peak were not significantly different. Hewlett and Helvey argued that forest clearance would produce moister soils; this would encourage the generation of increased volumes of sub-

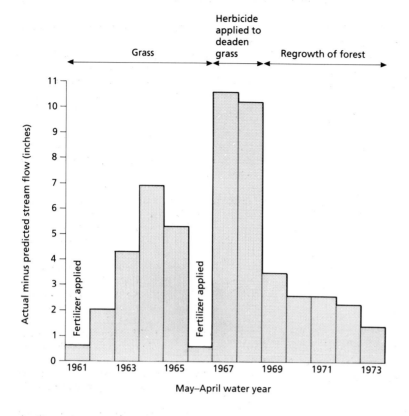

Fig. 2.14 Changes in annual water yield on watershed 6 at Coweeta, North Carolina, USA, in response to changes in land surface cover (see text for details).

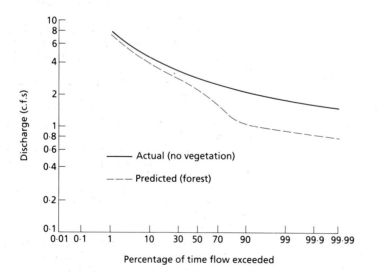

Fig. 2.15 Flow duration curves for mature forest and a surface of deadened grass; watershed 6, Coweeta, North Carolina, USA. c.f.s., cubic feet per second.

surface stormflow on the cleared catchment so that the mean increment to the hydrograph due to the treatment occurred on the falling limb of the storm hydrograph. Changes in flow duration have been examined (Burt & Swank 1992) for an experiment first reported by Hibbert (1969). Changes in annual water yield over the study period are shown in Fig. 2.14. Following forest clearance and its replacement by grass, water yields increase, the actual amount depending on the vigour of grass growth; as grass production declines, water yields rise. When herbicide is applied to deaden the grass, the water yield increases significantly. Thereafter, forest regrowth is allowed to progress naturally and water yields gradually decline towards the expected level. Figure 2.15 shows the predicted and actual flow duration curves for the 1967 water year. This shows that the increase in water yield at that time was associated particularly with increases in baseflow but that the discharge for a given exceedance probability was higher throughout the range of flows. Such results show that forest clearance provides an increase in water yield and that, although flood peaks may increase, the main effect is that of more baseflow. On this evidence, claims by some environmentalists that forest clearance causes a *decrease* in water yield would seem to be false.

ACKNOWLEDGEMENTS

I am grateful to Professor Mike Church for a typically thoughtful and thorough review of an earlier draft of this chapter. Figure 2.8 was kindly supplied by Dr Tony Edwards of the National Rivers Authority, Yorkshire Region, UK.

REFERENCES

Abdul AS, Gillham RW. (1984) Laboratory studies of the effects of the capillary fringe on streamflow generation. *Water Resources Research* **20**: 691–8. [2.2]

Anderson MG, Burt TP. (1977) A laboratory model to investigate the soil moisture conditions on a draining slope. *Journal of Hydrology* **33**: 383–90. [2.3]

Anderson MG, Burt TP. (1978) The role of topography in controlling throughflow generation. *Earth Surface Processes* **3**: 331–44. [2.2]

Anderson MG, Burt TP. (1980) Interpretation of re-cession curves. *Journal of Hydrology* **46**: 89–101. [2.4]

Anderson MG, Burt TP. (1985) *Hydrological Forecasting*. John Wiley & Sons, Chichester. [2.4]

Anderson MG, Burt TP. (1990) Subsurface runoff. In: Anderson MG, Burt TP (eds) *Process Studies in Hillslope Hydrology*, pp 365–400. John Wiley & Sons, Chichester. [2.2]

Atkinson TC. (1978) Techniques for measuring subsurface flow on hillslopes. In: Kirkby MJ (ed.) *Hillslope Hydrology*, pp 73–120. John Wiley & Sons, Chichester. [2.1]

Bates CG, Henry AJ. (1928) Forest and streamflow experiments at Wagon Wheel Gap, Colorado. *Monthly Weather Review* Supplement 30. [2.4]

Beckinsale RP. (1969) River regimes. In: Chorley RJ (ed.) *Water, Earth and Man*, pp 176–92. Methuen, London. [2.4]

Betson RP. (1964) What is watershed runoff? *Journal of Geophysical Research* **69**: 1541–52. [2.2]

Beven KJ. (1985) Distributed models. In: Anderson MG, Burt TP (eds) *Hydrological Forecasting*, pp 405–36. John Wiley & Sons, Chichester. [2.1]

Beven KJ, Germann PF. (1982) Macropores and water flow in soils. *Water Resources Research* **18** (5): 1311–25. [2.4]

British Standards Institution (1964) Methods of measurement of liquid flow in open channels. *BS 3680*. [2.1]

Burt TP. (1978) *Runoff processes in a small upland catchment with special reference to the role of hillslope hollows.* Unpublished PhD thesis, University of Bristol, UK. [2.2]

Burt TP. (1986) Runoff processes and solutional denudation rates on humid temperate hillslopes. In: Trudgill ST (ed.) *Solute Processes*, pp 193–249. John Wiley & Sons, Chichester. [2.2]

Burt TP, Arkell BP. (1986) Variable source areas of stream discharge and their relationship to point and non-point sources of nitrate pollution. *International Association of Hydrological Sciences Publication* **157**: 155–64. [2.2]

Burt TP, Butcher DP. (1985a) On the generation of delayed peaks in stream discharge. *Journal of Hydrology* **78**: 361–78. [2.2, 2.4]

Burt TP, Butcher DP. (1985b) Topographic controls of soil moisture distribution. *Journal of Soil Science* **36**: 469–76. [2.2, 2.4]

Burt TP, Gardiner AT. (1984) Runoff and sediment production in a small peat-covered catchment: some preliminary results. In: Burt TP, Walling DE (eds) *Catchment Experiments in Fluvial Geomorphology*, pp 133–52. GeoBooks, Norwich. [2.2, 2.4]

Burt TP, Swank WT. (1992) Flow frequency responses to hardwood-to-grass conversion and subsequent succession. *Hydrological Processes* **6**: (in press). [2.4]

Burt TP, Arkell BP, Trudgill ST, Walling DE. (1988) Stream nitrate in a small catchment in south west England over a period of 15 years (1970–1985). *Hydrological Processes* **2**: 267–84. [2.4]

Burt TP, Heathwaite AL, Labadz JC. (1990) Runoff production in peat-covered catchments. In: Anderson MG, Burt TP (eds) *Process Studies in Hillslope Hydrology*, pp 463–500. John Wiley & Sons, Chichester. [2.4]

Calder IR, Newson MD. (1979) Land-use and upland water resources in Britain—a strategic look. *Water Resources Bulletin* **15**: 1628–39. [2.4]

Church MA. (1988) Floods in cold climates. In: Baker VR, Kochel RC, Patton PC (eds) *Flood Geomorphology*, pp 205–29. Wiley Interscience, Chichester. [2.2, 2.3]

Church MA, Woo MK. (1990) Geography of surface runoff: some lessons for research. In: Anderson MG, Burt TP (eds) *Process Studies in Hillslope Hydrology*, pp 299–325. John Wiley & Sons, Chichester. [2.2]

Coles N, Trudgill ST. (1985) The movement of nitrate fertilizer from the soil surface to drainage waters by preferential flow in weakly structured soils, Slapton, south Devon. *Agriculture, Ecosystems and Environment* **13**: 241–59. [2.2]

Doornkamp JC, Gregory KJ, Burn AS. (1980) *Atlas of Drought in Britain 1975–76*. Institute of British Geographers, London. [2.4]

Dunne T. (1978) Field studies of hillslope flow processes. In: Kirkby MJ (ed.) *Hillslope Hydrology*, pp 227–93. John Wiley & Sons, Chichester. [2.2]

Dunne T, Black RD. (1970) An experimental investigation of runoff production in permeable soils. *Water Resources Research* **6**: 478–90. [2.2, 2.4]

Dunne T, Black RD. (1971) Runoff processes during snowmelt. *Water Resources Research* **10**: 119–23. [2.2]

Dunne T, Leopold LB. (1978) *Water in Environmental Planning*. Freeman, San Francisco. [2.2, 2.3, 2.4]

Dunne T, Price AG, Colbeck SC. (1976) The generation of runoff from subarctic snowpacks. *Water Resources Research* **12**: 677–85. [2.2]

Ferguson HL, Znamensky VA. (1981) Methods of computation of the water balance in large lakes and reservoirs. *UNESCO Studies and Reports in Hydrology 31*. UNESCO, Paris. [2.4]

Ford DC, Williams PW. (1989) *Karst Geomorphology and Hydrology*. Unwin Hyman, London. [2.3]

Germann PF. (1986) Rapid drainage response to precipitation. *Hydrological Processes* **1**: 3–14. [2.2]

Gilman K, Newson MD. (1980) *Soil pipes and pipeflow: a hydrological study in upland Wales*. British Geomorphological Research Group Research Monograph No. 1. GeoBooks, Norwich. [2.2]

Gregory KJ, Walling DE. (1973) *Drainage Basin Form and Process*. Edward Arnold, London. [2.1]

Goudie AS (ed.) (1990) *Geomorphological Techniques* 2nd edn. Unwin Hyman, London. [2.1]

Heathwaite AL, Burt TP, Trudgill ST. (1990) Land-use controls on sediment production in a lowland catchment, south-west England. In: Boardman J, Foster IDL, Dearing JA (eds) *Soil Erosion on Agricultural Land*, pp 69–86. John Wiley & Sons, Chichester. [2.2, 2.4]

Herschy RW (ed.) (1978) *Hydrometry, Principles and Practice*. Wiley Interscience, Chichester. [2.1]

Hewlett JD. (1961) Watershed management. In: *Report for 1961 Southeastern Forest Experiment Station*. pp. 62–66 US Forest Service, Asheville, North Carolina. [2.2]

Hewlett JD. (1982) *Principles of Forest Hydrology*. University of Georgia Press, Athens, USA. [2.4]

Hewlett JD, Helvey JD. (1970) Effects of forest clearfelling on the storm hydrograph. *Water Resources Research* **6**: 768–82. [2.4]

Hewlett JD, Hibbert AR. (1963) Moisture and energy conditions within a sloping soil mass during drainage. *Journal of Geophysical Research* **68**: 1080–7. [2.3]

Hewlett JD, Hibbert AR. (1967) Factors affecting the response of small watersheds to precipitation in humid areas. In: Sopper WE, Lull HW (eds) *Proceedings of the International Symposium of Forest Hydrology*, pp 275–90. Pergamon, Oxford. [2.1]

Hewlett JD, Cunningham GB, Troendle CA. (1977) Predicting stormflow and peakflow from small basins in humid areas by the R-index method. *Water Resources Bulletin* **13**: 231–54. [2.4]

Hibbert AR. (1969) Water yield changes after converting a forested catchment to grass. *Water Resources Research* **5**: 634–40. [2.4]

Horton RE. (1933) The role of infiltration in the hydrological cycle. *Transactions of the American Geophysical Union* **14**: 446–60. [2.1, 2.2]

Horton RE. (1940) An approach towards a physical interpretation of infiltration capacity. *Proceedings of the Soil Science Society of America* **4**: 399–417. [2.2]

Horton RE. (1945) Erosional development of streams and their drainage basins; hydrophysical approach to quantitative morphology. *Bulletin of the Geological Society of America* **56**: 275–370. [2.2]

Institute of Hydrology (1980) *Low Flow Studies Report*. Institute of Hydrology, Wallingford, UK. [2.4]

Institute of Hydrology (1989) *Hydrological Data UK, 1988*. Surface Water Archive, Institute of Hydrology, Wallingford, UK. [2.4]

Jones JAA. (1981) *The nature of soil piping: a review of research*. British Geomorphological Research Group Research Monograph No. 3. GeoBooks, Norwich. [2.2]

Kirby C, Newson MD, Gilman K. (1991) Plynlimon research: the first two decades. *Institute of Hydrology Report* **109**: pp 188. Wallingford, UK. [2.4]

Kirkby MJ. (1978) Implications for sediment transport.

In: Kirkby MJ (ed.) *Hillslope Hydrology*, pp 325–63. John Wiley & Sons, Chichester. [2.2]

Kirkby MJ. (1985) Hillslope hydrology. In: Anderson MG, Burt TP (eds) *Hydrological Forecasting*, pp 37–75. [2.2]

Kirkby MJ, Chorley RJ. (1967) Throughflow, overland flow and erosion. *Bulletin of the International Association of the Science of Hydrology* **12**: 5–21. [2.2]

Kneale WR. (1986) The hydrology of a sloping, structured clay soil at Wytham, near Oxford, England. *Journal of Hydrology* **85**: 1–14. [2.2]

Kneale WR, White RE. (1984) The movement of water through cores of a dry (cracked) clay-loam grassland topsoil. *Journal of Hydrology* **67**: 361–5. [2.2]

Langbein WB, Schumm SA. (1958) Yield of sediment in relation to mean annual precipitation. *Transactions of the American Geophysical Union* **39**: 1076–84. [2.2]

Law F. (1956) The effect of afforestation upon the yield of water catchment areas. *Journal of the British Waterworks Association* **38**: 489–94. [2.4]

Marsh T, Lees M. (1985) *Hydrological Data UK, The 1984 Drought*. Institute of Hydrology/British Geological Survey, Wallingford, UK. [2.4]

Mosley MP. (1979) Streamflow generation in a forested watershed, New Zealand. *Water Resources Research* **15**: 795–806. [2.2]

NERC (1975) *Flood Studies Report*. Natural Environment Research Council, London. [2.2, 2.4]

Newson MD. (1975) *Floods and Flood Hazard in the United Kingdom*. Oxford University Press, Oxford. [2.4]

Oakes DB, Pontin JMA. (1976) Mathematical modelling of a chalk aquifer. *Water Research Centre Technical Report* **24**: 37 pp. Water Research Centre, Medmenham, UK. [2.3]

O'Loughlin EM. (1981) Saturated regions in catchments and their relations to soil and topographic properties. *Journal of Hydrology* **53**: 229–46. [2.2]

O'Loughlin EM. (1986) Prediction of surface saturation zones in natural catchments by topographic analysis. *Water Resources Research* **22**: 794–804. [2.2]

Robinson M, Beven KJ. (1983) The effect of mole drainage on the hydrological response of a swelling clay soil. *Journal of Hydrology* **64**: 205–23. [2.2]

Rodda JC, Downing RA, Law FM. (1976) Systematic Hydrology. Butterworth, London. [2.1, 2.4]

Shaw EM. (1988) *Hydrology in Practice* 2nd edn. Van Nostrand Reinhold, London. [2.4]

Sherman LK. (1932) Streamflow from rainfall by the unit hydrograph method. *Engineering News Record*
108: 501–5. [2.1]

Stephenson GR, Freeze RA. (1974) Mathematical simulation of subsurface flow contributions to snowmelt runoff, Reynolds Creek, Idaho. *Water Resources Research* **10**: 284–94. [2.2]

Swank WT, Crossley DA. (1988) *Forest Hydrology and Ecology at Coweeta*. Ecological Studies 66. Springer-Verlag, New York. [2.1, 2.4]

Swank WT, Douglass JE, Cunningham GB. (1982) Changes in water yield and storm hydrographs following commercial clearcutting on a southern Appalachian catchment. In: *Proceedings of the Symposium on Hydrological Research Basins*, pp 583–94. Sonderh. Landeshydrologie, Bern. [2.4]

Swank WT, Swift LW, Douglass JE. (1988) Streamflow changes associated with forest cutting, species conversions, and natural disturbances. In: Swank WT, Crossley DA (eds) *Forest Hydrology and Ecology at Coweeta*, pp 297–312. Springer-Verlag, New York. [2.4]

Thorne CR, Zevenbergen LW, Burt TP, Butcher DP. (1987) Terrain analysis for quantitative description of zero order basins. In: *Proceedings of the International Symposium on Erosion and Sedimentation*, **165**: 121–30. Corvallis, Oregon. IAHS Publication. [2.2]

Troendle CA. (1985) Variable source area models. In: Anderson MG, Burt TP (eds) *Hydrological Forecasting*, pp 347–404. John Wiley & Sons, Chichester. [2.2]

Ward RC. (1968) Some runoff characteristics of British rivers. *Journal of Hydrology* **6**: 358–72. [2.4]

Ward RC. (1978) *Floods: A Geographical Perspective*. Macmillan, London. [2.4]

Ward RC, Robinson M. (1990) *Principles of Hydrology* 3rd edn. McGraw-Hill, London. [2.1, 2.4]

Weyman DR. (1973) Measurements of the downslope flow of water in a soil. *Journal of Hydrology* **20**: 267–84. [2.3]

Whipkey RZ. (1965) Subsurface stormflow from forested watersheds. *Bulletin of the International Association of Scientific Hydrology* **10**: 74–85. [2.2]

Whipkey RZ, Kirkby MJ. (1978) Flow within the soil. In: Kirkby MJ (ed.) *Hillslope Hydrology*, pp 121–44. John Wiley & Sons, Chichester. [2.2]

Yair A, Lavee H. (1985) Runoff generation in arid and semi-arid zones. In: Anderson MG, Burt TP (eds) *Hydrological Forecasting*, pp 183–220. John Wiley & Sons, Chichester. [2.2]

Zaslavsky D, Sinai G. (1981) Surface hydrology: 5 parts. *Journal of the Hydraulics Division, Proceedings of the American Society of Civil Engineers* **107**: 1–93. [2.2]

3: Analysis of River Regimes

A.GUSTARD

3.1 INTRODUCTION

The previous chapter described the hydrology of headwater catchments where the river regime can be closely related to catchment processes. Research at this scale has generally focused on small (less than 100 km²), topographically well-defined, impermeable upland valleys. However, in terms of source areas for larger river systems, flat agricultural areas and catchments with a significant groundwater component are of equal importance for runoff generation. With increasing scale and heterogeneity of catchment properties the link between process and hydrological response becomes obscured. Furthermore, processes such as channel, bank and floodplain storage, and the interaction between the river and the alluvial aquifer and the greater diversity of climate and hydrogeology, increase their significance in larger catchments. For example, a flood wave moving down a large river system will result in an increase in storage within each reach. With permeable alluvial deposits, river water will move from the river into bank storage and if the flood is of sufficient magnitude water will move across and be stored on the floodplain adjacent to the channel. These processes will influence the river regime by attenuating the flood wave, reducing both the magnitude of the flood peak and the velocity of the wave.

The regime of large rivers will be controlled by the diverse nature of upstream subcatchments which will have a range of climate, topography, geology, soils and land use. The Rhine, with a catchment area of 185 300 km², has a range of precipitation from 3000 mm per annum in the alpine headwaters to 600 mm where it enters the

North Sea (Friedrich & Müller 1984). In the Alpine region 50% of the precipitation falls as snow compared with less than 10% in the lower Rhine. This pattern results in an Alpine flow regime in the Rhine upstream of the confluence with the Main, with snowmelt producing a higher mean flow in the summer than the winter months. Further downstream below the confluence with the Mosel the mean flow in the winter is higher than in the summer as a result of the increased catchment area with a maritime climate with less snowfall. There is a similar diversity in land use in the Rhine catchment including forestry, viticulture, arable and pastoral agriculture, and urbanization. Although each land use will produce a distinct hydrological impact at the catchment scale of less than 100 km², in the large river system of the Rhine individual controls of particular land use (or topography, soils and hydrogeology) will be obscured. Furthermore the river regime will be influenced by a number of direct artificial influences, including abstractions and discharges, groundwater pumping, reservoir impoundments and flood protection schemes.

One approach to modelling these systems has been to develop relationships between the hydrological regime and the main catchment characteristics that control regime variability. The hydrological regime is defined by a number of flow statistics derived from daily flow data. These include the mean flow, the annual variability and seasonal pattern of runoff, the cumulative frequency distribution of daily flows and the distribution of annual minimum series of given durations. Flood frequency distributions are also derived from daily data for large catchments, but from hourly or shorter time-interval data for catchments with rapid response times. Catchment characteristics are single number indices of a particular physical property of the catchment. Thus, scale may be measured by topographic or groundwater catchment area or by the length of the main stream. Precipitation may be indexed by mean annual precipitation, the proportion of snowmelt or the depth of rainfall for a specified duration and return period. Catchment characteristics have been derived for a number of other variables including drainage density, channel slope, land use and soil type. Relationships have been developed between hydrological regimes and the controlling catchment characteristics using multiple regression. For example, the mean annual flood has been related to catchment characteristics which can be derived from readily available maps (NERC 1975). Such methods can then be used to estimate design floods at sites which do not have recorded flow data.

A knowledge of different river regimes and the techniques used to analyse them provides an important basis for understanding other environmental aspects of larger river systems. For example, flood frequency analysis has been found useful in understanding the spatial pattern of pollution in floodplain soils on the Meuse river (Rang & Schouten 1989). The time and location of geomorphological change on the Rhone river (Vivian 1989) have also been assisted by analysing the seasonal changes in river regimes from 1920. Thus, the impact of hydropower developments and abstractions for irrigation and water supply are related to the natural variability of river flows based on an analysis of a number of continuous flow series.

Where river flow data are available at the site of interest, they can be used to analyse aspects of the hydrological regime directly. However, there is frequently a need for estimating flow characteristics at ungauged sites. Use must then be made of physically based, or conceptual, models or of regional relationships between flow indices and catchment characteristics. The following sections describe a number of different ways of defining hydrological regimes, concluding with a summary of approaches used to estimate river regimes when recorded flow data are not available.

3.2 CLASSIFICATION OF HYDROLOGICAL REGIMES

Global classifications

A number of different procedures have been developed for the classification of hydrological regimes. At the global scale these have been based primarily on climate, using monthly rainfall and temperature, or based on mean values of the water or energy budget. Climatic classifications are, however, difficult to apply to larger river basins which cross climatic divides. This problem has been addressed by Parde (1955), who developed a classification based on the seasonal variation of river flow, and by Beckinsale (1969), who modified Koppen's climatic classification for hydrological regime definition.

A recent hydrological classification (Falkenmark & Chapman 1989) is based on a threefold subdivision of potential evaporation into cold, temperate and warm regions, and subdivisions into dry and humid according to the ratio of precipitation to potential evaporation (Unesco 1979). Although the importance of a number of hydrological processes in modifying climatic inputs was recognized, the need to provide a simple global classification resulted in using only two hydrological categories. These were 'areas with catchment response', with an organized natural drainage network typical of sloping land, and 'flatlands', defined as areas with a less organized natural drainage network.

Within the context of comparative hydrology, this classification defined temperate regions with a mean annual potential evaporation between 500 and 1000 mm, dry regions having a ratio of precipitation to potential evaporation of less than 0.75 and humid regions a ratio greater than 0.75. Table 3.1 gives examples of temperate regions typical of each category. The concept of hydrological regions has been used for regional analysis of both flood and low flow frequency with the objective of developing design methods for estimation at the ungauged site. The traditional approach was to combine flow statistics, for example the annual maximum flood from a number of gauging stations within a geographically defined area (NERC 1975). However, this

Table 3.1 Examples of temperate regions (Falkenmark & Chapman 1989)

Humid temperate sloping land	Dry temperate sloping land	Humid temperate flatlands	Dry temperate flatlands
Western Europe	Steppes of USSR	East Europe	Buenos Aires province of
South-west South	Mongolia	Asia and North America	Argentina, Caspian and
America	North China	centred approximately	Hungarian Plain
Pacific north-west coast	Patagonia in South	on latitude 50°	
of North America	America		
Japan	Coastal strip of Australia		
New Zealand	between Adelaide and		
Tasmania	Melbourne		

has resulted in some regions being defined that are not significantly different from one another (Wiltshire 1986) with very heterogeneous flood frequency characteristics being included within a single region. The search for a more objective definition of a homogeneous hydrological region has led to geographical regions being replaced by grouping basins that are physically similar to each other (Acreman & Wiltshire 1989), such as small, steep, wet catchments or large, flat, dry catchments. A problem with this (and the traditional) approach is that unambiguous assignment of the ungauged basins to one or other region creates discontinuities at the boundaries. This may be overcome by allowing fractional membership to more than one region. However, Acreman and Wiltshire (1989) have argued that once fractional membership is included in the procedure, regions become redundant. Each site can comprise its own region with estimation at the ungauged site being achieved by a weighted average of the observed flood frequency curves from a number of sites. This is a major departure from the traditional approach to defining regions, but with the ease of access to large hydrological and thematic databases it represents a very probable approach for design estimation in the future.

Hydrological regions, however they are defined, provide a basis for identifying areas where there is a similarity in hydrological response. Although this does not necessarily infer that there is a similarity in hydrological process, regions can be used to guide the extent to which hydrological models developed and calibrated in one location

can be used for hydrological prediction in another. At the global scale, regions can be used to ensure that hydrological results and methodologies are not transferred into areas where they are inappropriate. For relationships between hydrological regimes and catchment properties to be developed, statistical definitions of river regimes are required. These are normally derived from long-term observations of the daily mean flow hydrograph or peak flow statistics, and include a number of different measures which describe the mean, seasonal variability and extremes of flood and low flows, and other properties of the river hydrograph such as recession (defining the rate of decrease of discharge) and base flow (the component of river discharge with a slow response to precipitation which maintains river flow in dry periods) characteristics. Hydrological regions enable flow characteristics of different rivers to be compared and contrasted; they assist in evaluating historical changes in river regimes and provide a hydrological basis for water quality and ecological research.

Statistical definition of river regimes

This section summarizes different methods for defining river regimes, ranging from indices of the average annual regime to 'inter-annual' and 'intra-annual' variability. The analysis of the streamflow hydrograph to estimate extremes of both flood and drought are considered, together with methods for determining hydrograph recession and base flow characteristics.

Mean flow

The mean flow is the most fundamental variable for comparing the regime of different rivers, as well as for evaluating available water resources, for estimating changes in historic flow sequences and for determining the impact of human activity. By expressing the mean flow as an average depth over the topographic catchment area, comparisons can be made between catchments with different areas and between precipitation and runoff (assuming that there are no losses to groundwater and that topographic and groundwater catchment areas are the same).

Variations in mean flow are controlled primarily by variations in annual precipitation and annual evaporation, although differences in land use and groundwater flow will impose local differences on any regional trend. In humid temperate regions, annual precipitation increases with altitude and proximity to the sea, being, for example, in excess of 2000 mm in upland areas of western and central Europe and below 600 mm in continental humid temperate areas. Variation in annual potential evaporation is less marked, ranging between 500 and 700 mm with regional variations controlled by latitude, altitude and wind speed. Actual evaporation over most of humid temperate Europe is between 500 and 600 mm (Unesco 1978) with variations resulting from differences in potential evaporation, land use and soil type.

Mean runoff in humid temperate regions typically ranges from over 2000 mm in mountainous areas (e.g. western Britain) to approximately 100 mm (e.g. in north central France, eastern England and eastern Germany) where precipitation and evaporative losses are nearly in balance. However, these regional values obscure local variability. For example, the Val de Bonce catchment (topographic catchment area 203 km^2) gauged at Montboissier, south of Paris, France, has a mean annual runoff based on its topographic catchment area of only 8 mm. This observed value is much lower than the expected runoff of approximately 100 mm and results from a considerably reduced mean river discharge. This is primarily caused by a very high groundwater component of flow which flows through the chalk aquifer and thus 'bypasses' the gauging station. River regimes are considerably influenced by the hydrogeological characteristics of the underlying aquifer. This can result in the location of the topographic catchment boundary being different from the groundwater boundary which may give rise to differences in topographic and groundwater catchment areas, and between the observed and expected mean runoff. Local variability in runoff also arises from different losses by evaporation and transpiration as a result of land-use differences. For example, reductions in mean runoff equivalent to 290 mm for a fully forested coniferous catchment compared with a grassland catchment have been identified from studies of catchments of approximately 10 km^2 in central Wales (Kirkby & Newson 1991). As one moves downstream to larger catchments, the influence of a particular hydrogeological unit or land-use change is diminished as a result of the river system draining a more diversified region with a greater variability of hydrogeology and land use.

The annual variability of river runoff is a simple index of river regimes and is also an important variable for water resource assessment. The FREND (Flow Regimes from Experimental and Network Data) project analysed the coefficient of variation (CV) of annual runoff from over 500 catchments (Fig. 3.1) from 13 European countries (Gustard *et al* 1989) with catchment areas generally less than 500 km^2. The mean value was 0.28 and the spatial distribution over western and northern Europe indicated consistently low values of less than 0.20 over maritime western areas of the British Isles and south-western Norway and higher values between 0.40 and 0.60 in parts of eastern England, eastern France and central Germany. These differences were found to relate to the value of mean runoff with high annual variability in drier catchments, a result also found by McMahon (1979a). However, within more limited geographical regions, other variables may be significant. For example, Kovacs (1989), studying catchments in Hungary, found that the CV increased as the catchment area decreased.

The variability of annual runoff in this European region is compared with that of a sample of 126 'Unesco' rivers from different continents in Fig. 3.2. It can be seen that a wide range of

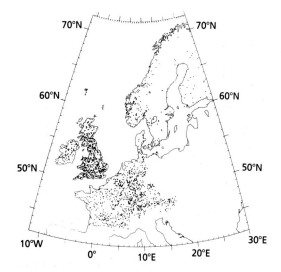

Fig. 3.1 Location of FREND gauging stations in Europe.

annual variability is exhibited by European rivers. Of course, the very stable rivers of the world typical of the humid tropics with CVs of less than 0.1 and the very variable rivers with CVs in excess of 0.7 that are typical of arid and semiarid environments are not represented in humid temperate Europe. Ward and Robinson (1990) make similar contrasts between that of the River

Thames, which for the period 1883–1986 was 0.29, and that of the River Darling in Australia for the period 1881–1959 which was 1.46. Figure 3.3 illustrates this greater range in the variability in the Australian flow series and the characteristic way in which there are sustained periods of low annual runoff interspersed with short periods of high runoff. This highlights the difficulties in estimating simple flow statistics such as the mean runoff even from long records in arid or semiarid regimes.

Seasonal variability in runoff

In discussing the definition of distinct seasonal flow regimes, Arnell (1989) highlights the difficulties of classifying a continuous process into discrete regions and presents examples of monthly runoff histograms from eight European rivers (Fig. 3.4, Table 3.2). Most of the examples are characterized by maximum flows in the winter and this is enhanced in western Europe where winter precipitation is highest. In continental areas where most precipitation falls in summer when evapotranspiration is highest, there is less difference between summer and winter flows. These patterns are modified in mountainous and continental areas where snowmelt is an important seasonal influence, as illustrated in the histo-

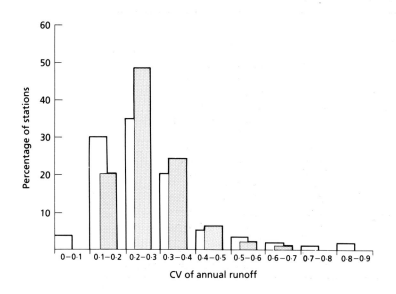

Fig. 3.2 Variability of annual runoff for 126 rivers of the world and European rivers (after Gustard *et al* 1989) ☐ 126 Unesco rivers (McMahon & Mein 1986); ☐ 577 European rivers.

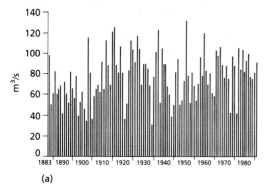
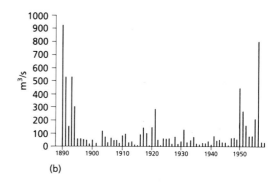

Fig. 3.3 Comparison of annual runoff between River Thames, UK and Darling River, Australia (after Ward & Robinson 1990). (a) Thames at Kingston (Teddington) 1883–1986; and (b) Darling at Menindee 1890–1958.

grams for the Lutschine, Grosser Regen and Versanan catchments.

The importance of hydrological processes in modifying hydrological regimes is clearly illustrated by comparing the histograms for the Kym and the Pang. Both catchments have similar maritime humid climates but they exhibit significantly different regimes. The Pang catchment which is underlain by permeable chalk has greater storage capacity enabling high flows to be maintained throughout the summer. This contrasts with the Kym catchment which is underlain by impermeable clay, and thus with limited storage the summer flows cannot be sustained. Larger catchments with areas in excess of 800 km^2 will be influenced by a number of different subcatchments, generally with a wide range of hydrogeology. The response of the larger catchments is thus the sum of a number of different regimes, each with their own seasonal variability in runoff.

Ward (1968) analysed 37 British flow records and demonstrated that 29 of them fell clearly into a simple summer minimum–winter maximum regime. Furthermore, regional patterns in the

Table 3.2 Location and area of example catchments (Arnell 1989)

River and gauging station	Area (km^2)	Location
Kym at Meagre Farm	137.5	East England
Pang at Pangbourne	170.9	South-east England
Severn at Plynlimon	8.8	Central Wales
Ammer at Oberammergau	114.0	Bavaria, Germany
Grosser Regen at Zweisel	177.0	Lower Danube, Germany
Reiche Ebrach at Herrnsdorf	169.0	Central Germany
Schussen at Magenhaus	246.0	Bavaria, Germany
Wurm at Randerath	305.0	North-west Germany
Zusam at Pfaffenhofen	505.0	Bavaria, Germany
Gardon de Mialet at Roucan	239.0	South France
Layon at St. George	250.0	West France
Maumont at La Chanourdie	162.0	South-west France
Orgeval at Le Theil	104.0	North central France
Versanan at Halaback	4.7	Southern Sweden
Lutschine at Gsteig	379.0	Central Switzerland
Yeongsan at Naju	2060.0	South-west Republic of Korea

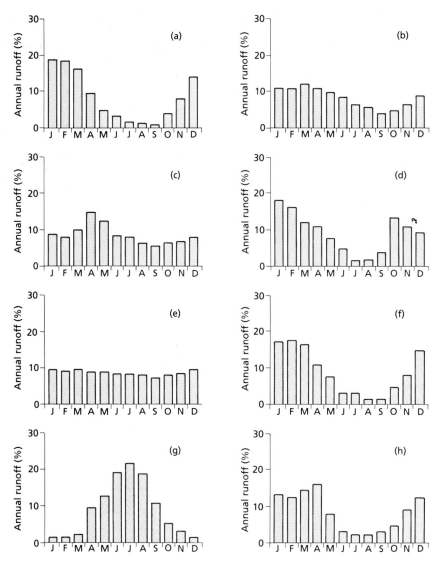

Fig. 3.4 Monthly runoff histograms, showing average monthly runoff as a percentage of average annual runoff for eight of the catchments listed in Table 3.2 (after Arnell 1989). (a) Kym at Meagre Farm; (b) Pang at Pangbourne; (c) Grosser Regen at Zweisel; (d) Gardon de Mialet at Roucan; (e) Wurm at Randerath; (f) Orgeval at le Thiel; (g) Lutschine at Gsteig; and (h) Versanan at Halaback.

month of maximum and minimum runoff were related to both variations in climatic factors and the presence of aquifers in the catchment. A study of more than 500 flow records in the UK (Institute of Hydrology 1980a) stratified the flow records into three groups defined by the 95th percentile discharge expressed as a percentage of the mean flow (ADF). An analysis was thus carried out on all 'impermeable catchments' (Q95 less than 15%), 'average catchments' (Q95 between 15% ADF and 30% ADF) and 'permeable catchments' (Q95 greater than 30% ADF). Having reduced the influence of catchment hydrogeology it was possible to use standard contouring routines to evaluate the spatial pattern of monthly runoff. Figure 3.5 illustrates the distribution of the mean runoff in January and August for impermeable catchments. The figures illustrate the clear regional variability due to climatic controls which is obscured when catchments with contrasting hydrogeologies are included in the analysis. The complete set of maps provides a useful procedure for estimating the seasonal distribution of monthly flows at ungauged sites.

Flow duration curve

The cumulative frequency distribution of daily mean flows (Fig. 3.6) shows the percentage of time during which specified discharges are equalled or exceeded during the period of record. The relationship is normally referred to as the flow duration curve, and although it does not convey information about the sequencing properties of flows it is one of the most informative methods of displaying the complete range of river discharges from low to flood flows. A normal probability scale is commonly used for the frequency axis and a logarithmic scale for the discharge axis. If the logarithms of the daily discharges are distributed normally then the plotted points will lie on a straight line, so aiding esti-

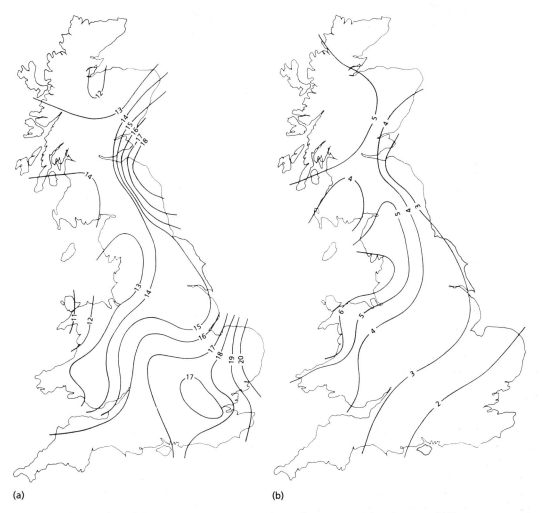

(a) (b)

Fig. 3.5 Spatial variability of the percentage mean annual runoff in January (a) and August (b) (from Institute of Hydrology 1980a).

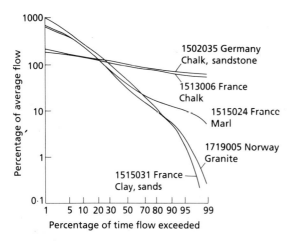

Fig. 3.6 One-day flow duration curves for catchments with contrasting geology (after Gustard *et al* 1989).

mation from the curve of discharges for different frequencies. Note that the discharge axis in Fig. 3.6 is standardized by the average flow. This facilitates comparison between catchments, because it reduces differences in the location of the curve caused by differences in the mean annual runoff.

Figure 3.6 illustrates the strong control of catchment geology in determining the distribution of daily flows. Rivers supported by aquifers have well-sustained low flows, whilst flood discharges are reduced as a result of the storage provided by permeable soils and geology. In contrast the gradients of curves from impermeable catchments are much steeper, reflecting the higher flood flows and lower low flows. It is also interesting to

note from Fig. 3.6 the similarity of the flow duration curves from the same lithology but from very different locations in Europe. Such clear contrasts are normally evident only in small catchments of less than 500 km^2 with homogeneous hydrogeology. This suggests that within an area that is broadly 'humid temperate', local catchment controls in catchments of less than 500 km^2 exert a greater influence on short-term runoff variability than regional climatic conditions. With increasing catchment area, the contributions of a number of diverse catchments lead to an 'averaging' of catchment response. As a result there is greater similarity of flow duration curves from large basins in a similar climatic area, particularly those basins in excess of 10 000 km^2.

Flow duration curves can be derived from individual months or groups of months (Institute of Hydrology 1980b) to provide a more detailed analysis of the seasonal distribution of river flows. These can be presented in the form of a flow duration surface where the discharge (or water level) is plotted on the vertical axis, the month on the horizontal axis and the flow duration curve as a parameter. The flow duration surface (Kovacs 1989) is a useful method for characterizing the seasonal variability of the full range of flows. Figure 3.7 illustrates the water level regime from two gauging stations 100 km apart on the two main rivers with very different regimes in the Carpathian basin. The extensive alpine areas above the permanent snow-line are the main source of water for the Danube at Budapest, and this gives rise to a prolonged snowmelt period

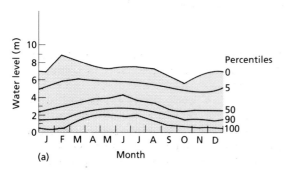

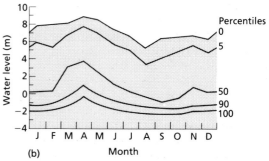

Fig. 3.7 Level duration surfaces characterizing different regimes (after Kovacs 1989). (a) River Danube at Budapest (1921–50); and (b) River Tisza at Szolnok (1921–50).

and a similar probability of flooding between March and August. The low flows are similarly influenced by this alpine region, with low flows being lower in January and February than in the drier months of September and October. In contrast, the headwaters of the Tisza river are much lower and are below the permanent snow-line, resulting in the high floods due to spring snow-melt being confined to April, which is also the time of the highest low flows.

Flow frequency curve

While the flow duration curve displays information about the proportion of time during which a flow is exceeded, the flow frequency curve shows the proportion of years when a flow is exceeded, or equivalently the average interval in years that the river falls below a given threshold discharge. Figure 3.8 illustrates the plot derived from an analysis of 10-day annual minimum discharges from six flow records in western Europe. The analysis can be derived from minima of other durations, for example 1, 30, 60, 90 and 180 days. The annual minima are standardized by the mean flow to enable frequency curves derived from both large and small catchments and catchments with high and low mean precipitation to be more easily compared. The figure is based on using a Weibull extreme value distribution which has been used in a number of low-flow investigations in the USA (Matalas 1963; Joseph 1970), in Malawi (Drayton *et al* 1980) and in the UK (Institute of Hydrology 1980b). Figure 3.8 illustrates the strong control that the geology of the catchment imposes on the slope and the position of the flow frequency curve. There are also close similarities between the curves from limestone lithologies from catchments in the UK and those from the karst area of Yugoslavia.

The similarity of river regimes defined by their flow frequency curve was investigated in the FREND project (Gustard *et al* 1989) by pooling flow frequency curves from 643 European rivers. Annual minima series were first grouped according to the value of their mean annual 10-day minima, MAM(10), which was expressed as a percentage of the mean flow. For example, all relatively impermeable catchments with a

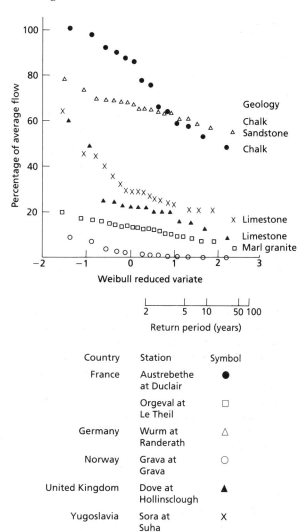

Fig. 3.8 Annual minima series for rivers with contrasting geology (after IAHS 1990).

MAM(10) between 10% and 20% of the mean flow were analysed and then all permeable catchments with MAM(10) between 30% and 40% of the mean flow. The analysis was carried out by dividing these groups of stations into smaller subsets from different geographical regions of Europe and deriving pooled flow frequency curves. The pooled curves were derived by plotting the mean value of the x and y co-ordinates for class intervals of the Weibull reduced variate, e.g.

0.0−0.05, 0.05−1.0 etc. Figure 3.9 illustrates the results of this analysis which demonstrates the similarity of flow frequency curves in different geographical regions. For example, the ratio of the 50- to the 2-year return period 10-day annual minima is very similar in the Seine basin to that of Norway for rivers with the same value of MAM(10). The full analysis showed that although there was great spatial variability in low flow statistics such as MAM(10) there was similarity in the relationships between frequent and extreme drought events across Europe.

Flood frequency curve

A similar procedure for estimating the frequency of low flows can be applied to annual maximum flood discharges to derive flood frequency curves. A number of different distributions including log-normal, Pearson, log-Pearson type 3, Wakeby and extreme value distributions have been fitted to

flood data (Cunnane 1988). In assessing which was the most appropriate distribution to use, Cunnane concluded that the choice of distribution is dependent upon the fitting technique, the goodness of fit test, and the assumptions made about the real world distribution.

Arnell (1989) used a general extreme-value distribution fitted by the method of probability-weighted moments to compare the flood regimes of eight European catchments. Figure 3.10 shows the relationship between annual maximum flood (standardized by the mean annual flood) and return period. Flood frequency curves typical of lowland regions with low annual rainfall are illustrated by the three steepest curves. These result from data series with a large number of years with low annual maxima but with some extreme events caused by summer thunderstorms. Lowland catchments with more maritime climates have flatter frequency curves as shown by the Ammer and Zusman data. The Zusman

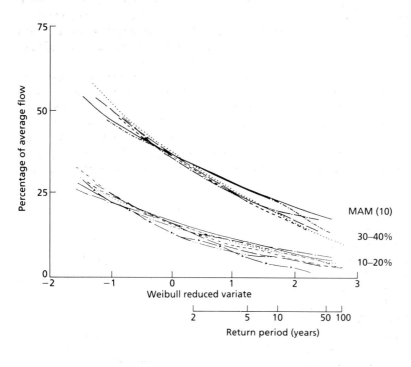

Fig. 3.9 Pooled 10-day mean annual minimum series expressed as a percentage of the average flow. Stations grouped by value of 10-day mean annual minima and hydrometric areas. (after Gustard & Gross 1989).

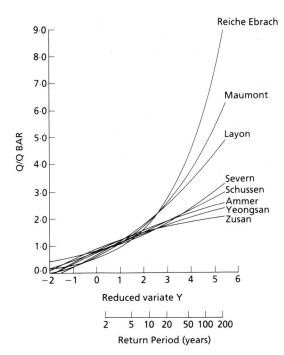

Fig. 3.10 Flood frequency curves for eight catchments listed in Table 3.2 (after Arnell 1989). Discharge standardized by mean annual flood (QBAR).

Table 3.3 Flow regimes in north-west Europe (from Gustard *et al* 1989)

Regime type	Description
Glacier	Dominant high flows in July and August, mostly caused by snowmelt in combination with precipitation
Mountain	Dominant spring flood in April–July, low flows in winter
Inland	Dominant spring flood in March–June, low flows in winter, but higher flows in the autumn
Transition	High runoff in spring and autumn, low-flow spells in both summer and winter
Maritime	High runoff in autumn and winter, with lowest flows in summer

(505 km^2) and Yeongsen (2060 km^2) are the catchments with the flattest frequency curves and the largest catchment areas, a trend found by Farquharson *et al* (1987) for catchments in excess of 100 000 km^2 in area. Steeper curves are generally found in more upland catchments (Severn), although flatter curves may be observed in upland catchments with lakes or more significant snowmelt (Schussen).

An analysis of more than 1500 flow records investigated the seasonal aspects of flood regimes in north-west Europe which were classified into one of five groups based on the number of distinct flood seasons and the relative magnitude of the spring and autumn floods (Gustard *et al* 1989). A summary of this classification, which is based on daily flow data, is presented in Table 3.3. The analysis of daily mean flow data is valuable for both flood and low flows on large river basins in excess of 500 km^2. However, details of catchment processes are often not apparent when average daily discharge is analysed, and Fig. 3.11 illus-

trates the rapid fluctuations in discharge which can occur even on large catchments. It illustrates the diurnal variation in runoff draining an area of 195 km^2 on the Aletschgletscher basin, of which 67% is glaciated. The hydrograph is from a typical melt period in July 1975 during a period of high temperatures and high radiation.

A detailed analysis of annual maximum data (Roald 1989) enabled flood frequency distributions to be subdivided into ten groups. These were based on a cluster analysis of station values of the specific mean annual flood and the CV of the annual maximum flood. The cluster analysis partitions the data into homogeneous groups,

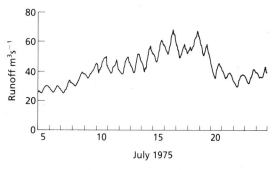

Fig. 3.11 Diurnal variations of runoff from the Aletschgletscher River (after Lang 1989).

for each of which a pooled frequency curve was derived. Although there was some spatial coherence in the regional distribution of the ten groups, it was not possible accurately to predict group membership from basin characteristics. However, the method was considered to have advantages over geographically defined regions which are not necessarily homogeneous in terms of flood frequency distributions.

Figure 3.12 shows a range of pooled flood frequency distributions based on a world flood study using data from 1121 gauging stations in 70 different countries (Farquharson *et al* 1987). They illustrate the lower variability of flood discharges from large river basins in the same region. For example, in Iran the average frequency curve was flatter for catchments with an area greater than 75 000 km^2 compared with smaller catchments.

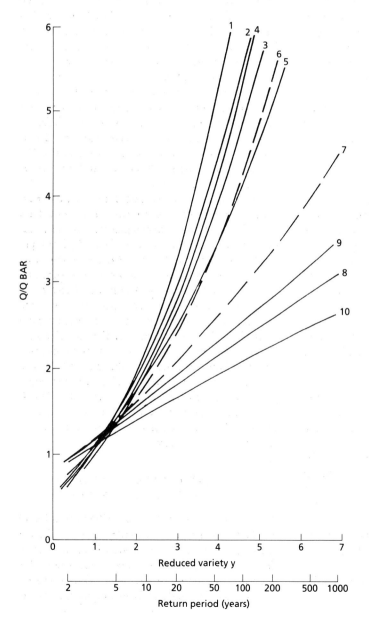

Fig. 3.12 Regional flood frequency distributions—Asia and Australasia; standardized by mean annual flood (General Extreme Value (GEV) fitted by Probability Weighted Moments (PWM)) (from Farquharson *et al* 1987). 1, Saudi Arabia and Yemen; 2, Jordan; 3, Iran (all); 4, Iran (area <7500 km^2); 5, Iran (area >7500 km^2); 6, Sri Lanka; 7, India, Kerala State; 8, India and Bangladesh (all); 9, India and Bangladesh (area <100 000 km^2); 10, India and Bangladesh (area >100 000 km^2).

This trend was also found in data from India and Bangladesh where catchments greater in area than 100 000 km^2 had the flattest frequency curve. The steep flood growth curves typical of arid and semiarid environments in Saudi Arabia, Yemen, Jordan and Iran are also illustrated.

Other measures of flow regime

There are a number of additional methods of describing the flow regime including, for example, the volume of flood response calculated using a separation between quick response flow and base flow (Hewlett & Hibbert 1967). A number of techniques have also been developed for estimating the base flow discharge using daily flow data. A method developed by Kille (1970) is based on the analysis of monthly minimum flows and has been applied in Germany and Poland. More recently, Demuth (1989) has automated this procedure and applied it to 102 small research basins in western and northern Europe.

The Low Flow Studies Report (Institute of Hydrology 1980b) recommends using the base flow index (BFI) derived from mean daily flow data for classifying the low-flow response of a catchment and for indexing the storage at the catchment scale. The index is the ratio of base flow (calculated from a hydrograph separation procedure) to total flow. The procedure for calculating the index is as follows:

1 Divide the mean daily flow data into non-overlapping blocks of 5 days and calculate the minima for each of these blocks, and let them be called $Q_1, Q_2, Q_3, \ldots, Q_n$.

2 Consider in turn (Q_1, Q_2, Q_3), (Q_2, Q_3, Q_4), $\ldots, (Q_{n-1}, Q_n, Q_{n+1})$, etc. In each case, if $0.9 \times$ central value is less than outer values, then the central value is a turning point for the baseflow line. Continue this procedure until all the data have been analysed to provide a derived set of baseflow ordinates $QB_1, QB_2, QB_3, \ldots QB_i$ which will have different time periods between them.

3 By linear interpolation between each QB_i value, estimate each daily value of $QB_1 \ldots QB_n$.

4 If $QB_i > Q_i$ then set $QB_i = Q_i$.

5 Calculate V_A, the volume beneath the recorded mean daily flows Q_n.

6 Calculate V_B, the volume beneath the baseflow line between the first and last baseflow turning points $Q_1 \ldots Q_n$.

7 The BFI is then V_B/V_A.

Relationships were derived relating BFI to low-flow statistics, and recommendations were made concerning the estimation of BFI at the ungauged site from catchment geology. The index has been applied in Canada (Pilon & Condie 1986), Zimbabwe and Malawi (Wright 1989), New Zealand (National Water and Soil Conservation Authority 1984) and Norway (Tallaksen 1986). These studies have indicated that some modification of the index and care in interpretation are required when applying the separation procedure in different river regimes. However, the use of a standard procedure has facilitated the comparison of catchment response from a number of different countries. The BFI has also been used for calibrating the hydrological response of soils in the UK (Boorman & Hollis 1990) and in mainland Europe (Gustard & Gross 1989).

Recession properties calculated either from short time-interval data for flood events or daily data for low-flow investigations are also an important property of the flow hydrograph. Toebes and Strang (1964) and Hall (1968) reviewed a number of different procedures for estimating recession characteristics. A number of analytical procedures have been developed for the rapid calculation of recession parameters based on daily flow data. These include studies by Demuth (1989) who related recession properties to basin characteristics in Europe, Tallaksen (1989) who investigated the temporal variability in recession properties in Norway, and Schwarze *et al* (1989) who used a computer-aided analysis of flow recessions in a water-balance investigation of 30 catchments in Germany.

3.3 ESTIMATING REGIMES AT UNGAUGED SITES

The previous sections have described a number of methods for defining particular aspects of river regimes based on the analysis of recorded flow data. There are, however, occasions when this approach is not appropriate. First, when there are no recorded flow data at or in the vicinity of the site of interest. Second, when estimates of the change in flow regime are required following a change in catchment land use, for example urbanization. Third, when a change in flow regime will occur as a result of water resource development, for example groundwater abstraction or reservoir impoundment.

A wide range of hydrological models has been developed for evaluating these problems. The most recent modelling development has been in the area of physically based distributed models. The component processes of the hydrological system, for example the unsaturated zone and the saturated zone, are described by partial differential equations representing the conservation of mass and momentum operating on a grid cell basis. Beven and O'Connell (1982) have reviewed the general concepts of physically based models, examples of which include IHDM (Beven *et al* 1987) and SHE (Abbott *et al* 1986a, 1986b). Although this family of models has demanding data requirements it provides a physically based simulation of catchment behaviour and is therefore able to model a number of different changes occurring simultaneously on a catchment.

There has been a wider use of simpler conceptual models. Such models may be lumped or distributed depending on whether or not the spatial distribution of hydrological variables within the catchment is considered. Figure 3.13 illustrates a typical structure of a conceptual model which has been used extensively in land-use change modelling (Blackie & Eeles 1985). Both physically based and conceptual models can be used to simulate a time series from which statistics of the flow regime can be calculated.

A number of regional studies have been carried out with the objective of estimating flow statistics at ungauged sites. These studies relate flow variables to catchment characteristics, for example mean annual precipitation, catchment area, slope, land use and soil type, using multiple regression analysis techniques. These statistical models are generally calibrated on all the available flow data within a region, and are thus based on a large number of station-years of data, which minimizes the errors in estimating hydrological extremes. However, because the models do not incorporate catchment processes they are not

appropriate for addressing problems such as land-use change or the impact of water resource development on downstream flow regimes. Examples of regional low flow studies include Simmers (1975) for New Zealand, Musiake *et al* (1975) for Japan, Knisel (1963), Mitchell (1957), Hines (1975) and Riggs (1973, 1990) in the USA, and McMahon (1969) for Australia. In Europe, studies have been carried out by Martin and Cunnane (1976) in Ireland, by Wittenberg (1989) in Germany, by Moltzau (1990) in Norway, and by the Institute of Hydrology (1980a, 1980b) in the UK. Similarly, a number of flood studies have been completed, including those in the UK (NERC 1975), New Zealand (Beable & McKerchar 1982) and North America (Chong & Moore 1983).

3.4 CONCLUSIONS

Recent advances in the analysis of river regimes have been in three main areas. The first of these has been the further development of methods for the definition of river regimes. For example, Nathan and McMahon (1990) have developed automated techniques for the estimation of recession and base flow which have been applied to 186 catchments in south-eastern Australia. Second, the definition of hydrological regions has developed from the concept of geographical regions defined by administrative or topographic boundaries to the 'region of influence approach' (Burn 1990). This concept considers that each site on a river defines its own unique region, and estimation of flow variables at an ungauged site should be based on a weighted average of observed flow variables at a number of other sites (Acreman & Wiltshire 1989). Third, the development of international databases has led to a classification of regimes over much wider areas. For example, the FREND project has produced a time series and thematic database for much of western and northern Europe (Gustard *et al* 1989). The analysis of these data has enabled regionalization techniques to be developed across national boundaries using consistent methods of analysis. Similarly, a world flood study (Farquharson *et al* 1987) has led to a greater understanding of the main global controls on flood frequency. Haines and colleagues (1988) have carried out a global classifi-

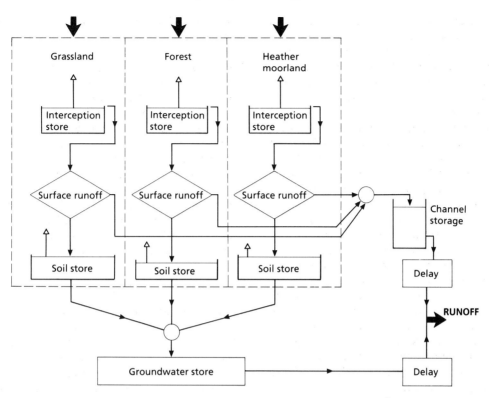

Fig. 3.13 The Institute of Hydrology integrated land-use model.→—Water movement; ↑ evaporation/transpiration; -○-areal summation; ↓ precipitation.

cation of regimes, using 32 000 station-years of monthly streamflow data from 969 stream-gauging stations. This study defined 15 seasonal flow regimes and presented the first world map of regime type based on hydrograph analysis alone. The availability of an increasing number of these national and international databases will enable continued advances to be made both in the definition of river regimes and in analysing their spatial distribution. This will be complemented by the application of geographical information systems which will enhance the speed, number and accuracy of catchment characteristic calculation, eventually leading to improvements in estimating flow regimes at the ungauged site.

GLOSSARY

Base flow index the ratio of base flow to total flow derived from a hydrograph separation of daily flows.

Coefficient of variation of annual runoff a measure of the year-to-year variability of runoff calculated by dividing the standard deviation of annual runoff by the mean.

Flood frequency curve shows the relationship between the average interval between years (return period) in which the river exceeds a given flood discharge. The curve is normally derived from hourly or shorter time-interval data but can be derived from daily flow data for large catchments. Comparison between curves from different catchments is assisted by standardizing the discharge ordinate by the mean of the annual maximum flood series.

Flow duration curve the cumulative frequency distribution of daily (or monthly) mean flows. The curve shows the relationship between discharge and the percentage of time for which a given discharge is exceeded. The curve can be derived for the period of record, individual years, monthly or groups of months. Comparisons between curves from different catchments are assisted by dividing the discharge ordinates by the mean flow or by the catchment area.

Low flow frequency curve the curve shows the average

interval between years (return period) in which the river falls below a given discharge. It can be derived from daily or monthly data and from one or D-day consecutive flows, for example the 10-day or 90-day annual minimum. Comparisons between curves from different catchments are assisted by standardizing the discharge ordinates by the mean flow or catchment area.

Q95 the 95th percentile discharge derived from the flow duration curve is the discharge exceeded for 95% of the days of the period analysed. The 95th percentile is equivalent to the discharge which is exceeded 18 days a year on average. Other percentiles can be derived from the flow duration curve such as Q5, a high discharge, or Q50, the median discharge.

Recession the rate of decrease in river flow normally derived from hourly (or shorter time-interval) data for flood events or daily data for low-flow recessions.

REFERENCES

Acreman MC, Wiltshire SE. (1989) The regions are dead. Long live the regions. Methods of identifying and dispensing with regions for flood frequency analysis. In: *FRIENDS in Hydrology* (Proceedings of the Bolkesjo Symposium, Norway). IAHS Publication No. 187, pp 175–88. Wallingford, UK. [3.2, 3.4]

Abbott MB, Bathurst JC, Cunge JA, O'Connell PE, Rasmussen J. (1986a) An introduction to the European Hydrological System—Système Hydrologique Européen 'SHE' (1): History and philosophy of a physically based, distributed modelling system. *Journal of Hydrology* 87: 45–59. [3.3]

Abbott MB, Bathurst JC, Cunge JA, O'Connell PE, Rasmussen J. (1986b) An introduction to the European Hydrological System—Système Hydrologique Européen 'SHE' (2): Structure of a physically based, distributed modelling system. *Journal of Hydrology* 87: 61–77. [3.3]

Arnell NW. (1989) Humid temperate sloping land. In: Falkenmark M, Chapman T (eds) *Comparative Hydrology*, pp 163–207. Unesco, Paris. [3.2]

Beable ME, McKerchar AI. (1982) Regional flood estimation in New Zealand. *Water and Soil Technical Publication* 20, National Water and Soil Conservation Organisation. Wellington, New Zealand. [3.3]

Beckinsale RP. (1969) River Regimes. In: Chorley RJ (ed.) *Water, Earth and Man*, pp 455–71. Methuen, London. [3.2]

Beven KJ, O'Connell PE. (1982) *On the role of physically based distributed modelling in hydrology*. IH Report No. 81. Institute of Hydrology, Wallingford, UK. [3.3]

Beven K, Calver A, Morris EM. (1987) *The Institute of Hydrology distributed model*. IH Report No. 98. Institute of Hydrology, Wallingford, UK. [3.3]

Blackie JR, Eeles CWO. (1985) Lumped catchment models. In: Anderson MG, Burt TP (eds) *Hydrological Forecasting*, pp 311–45. John Wiley and Sons, Chichester. [3.3]

Boorman DB, Hollis JM. (1990) *Hydrology of soil types: a hydrologically-based classification of the soils of England and Wales*. Ministry of Agriculture, Fisheries and Food. Conference of River and Coastal Engineers, Loughborough University, July 1990. Wallingford, UK. [3.2]

Burn DH. (1990) Evaluation of regional flood frequency analysis with a region of influence approach. *Water Resources Research* 26: 2257–65. [3.4]

Chong SK, Moore SM. (1983) Flood frequency analysis for small watersheds in southern Illinois. *Water Resources Bulletin* 19 (2): 277–82. [3.3]

Cunnane C. (1988) Methods and merits of regional flood frequency analysis. *Journal of Hydrology* 100: 269–90. [3.2]

Demuth S. (1989) The application of the West German IHP Recommendations for the analysis of data from small research basins. In: *FRIENDS in Hydrology* (Proceedings of the Bolkesjo Symposium, Norway). IAHS Publication No. 187, pp 46–60. Wallingford, UK. [3.2]

Drayton RS, Kidd CHR, Mandeville AN, Miller JB. (1980) *A regional analysis of river floods and low flows in Malawi*. IH Report No. 72. Institute of Hydrology, Wallingford, UK. [3.2]

Falkenmark M, Chapman T. (1989) *Comparative Hydrology*. Unesco, Paris. [3.1]

Farquharson FAK, Green CS, Meigh JR, Sutcliffe JV. (1987) Comparison of flood frequency curves for many different regions of the world. In: Singh VP (ed.) *Regional Flood Frequency Analysis*, pp 223–56. Reidel, Dordrecht, Holland. [3.2, 3.4]

Friedrich G, Müller D. (1984) Rhine. In: Whitton BA (ed.) *Ecology of European Rivers*, pp 263–315. Blackwell Scientific Publications, Oxford. [3.1]

Gustard A, Gross R. (1989) Low flow regimes of northern and western Europe. In: *FRIENDS in Hydrology* (Proceedings of the Bolkesjo Symposium, Norway). IAHS Publication No. 187, pp 205–12. Wallingford, UK. [3.2]

Gustard A, Roald LA, Demuth S, Lumadjeng HS, Gross R. (1989) *Flow Regimes from Experimental and Network Data* (FREND). Institute of Hydrology, Wallingford, UK. [3.2, 3.4]

Haines AT, Finlayson BL, McMahon TA. (1988) A global classification of river regimes. *Applied Geography* 8: 255–72. [3.4]

Hall FR. (1968) Base flow recessions—a review. *Water Resources Research* 4 (5): 973–98. [3.2]

Hewlett JD, Hibbert AR. (1967) Factors affecting the response of small watersheds to precipitation in humid areas. In: Sopper WE, Lull HW (eds) *Forest*

Hydrology, pp 275–90. Pergamon, Oxford. [3.2]

Hines MS. (1975) Flow-duration and low-flow frequency determinations of selected Arkansas streams. *United States Geological Survey, Water Resources Circular No. 12*. Government Printing Office, Washington, DC. [3.3]

Institute of Hydrology (1980a) *Seasonal flow duration curve estimation manual*. Low Flow Studies Report No. 2.4, Institute of Hydrology, Wallingford, UK. [3.2, 3.3]

Institute of Hydrology (1980b) *Research report*. Low Flow Studies Report No. 1, Institute of Hydrology, Wallingford, UK. [3.2, 3.3]

IAHS (1990) *Front cover regionalisation in hydrology* (Proceedings of the Ljubljana Symposium, Yugoslavia). IAHS Publication No. 191. Wallingford, UK. [3.2]

Joseph ES. (1970) Probability distribution of annual droughts. *Journal of the Irrigation Division ASCE* **96** (IR): 461–74. [3.2]

Kille K. (1970) Das Verfahren MoNMQ, ein Beitrag zur Berechnung der mittleren Langjahrigen Grundwasser–neubildung mit Hilfe der monatlichen Niedrigwasser–abflusse. *Zeitschrift der deutschen geologischen Gesellschaft*, Sonderheft Hydrologie und Hydrochemie, pp 89–95. [3.2]

Kirkby C, Newson MD. (1991) *Plynlimon research: the first two decades*. IH Report No. 109. Institute of Hydrology, Wallingford, UK. [3.2]

Knisel WG. (1963) Baseflow recession analyses for comparison of drainage basins and geology. *Journal of Geophysical Research* **68** (12): 3649–53. [3.3]

Kovacs K. (1989) Measurement and estimation of hydrological processes. In: Falkenmark M, Chapman T (eds) *Comparative Hydrology*. 75–104 Unesco, Paris. [3.2]

Lang H. (1989) Sloping land with snow and ice. In: Falkenmark M, Chapman T (eds) *Comparative Hydrology*. Unesco, Paris. pp 146–62. [3.2]

McMahon TA. (1969) Water resources research: aspects of a regional study in the Hunter Valley, New South Wales. *Journal of Hydrology* **7**: 14–38. [3.3]

McMahon TA. (1979a) Hydrological characteristics of arid zones. In: *The Hydrology of Areas of Low Precipitation* (Proceedings of the Canberra Symposium, Australia). IAHS Publication No. 128, pp 105–23. Wallingford, UK. [3.2]

McMahon TA, Mein RG. (1986) *River and Reservoir Yield*. Water Resources Publication, Colorado. [3.2]

Martin JV, Cunnane C. (1976) *Analysis and prediction of low-flow and drought volumes for selected Irish rivers*. Institution of Engineers of Ireland. Dublin. [3.3]

Matalas NC. (1963) *Probability distribution of low flows*. United States Geological Survey Professional Paper 434–A. Washington, DC. [3.2]

Mitchell WD. (1957) *Flow duration of Illinois streams*. Department of Public Works and Buildings, State of Illinois. [3.3]

Moltzau B. (1990) *Low flow analysis: a regional approach for low flow calculation in Norway*. Department of Geography Report No. 23, University of Oslo. [3.3]

Musiake K, Inokuti S, Takahasi Y. (1975) Dependence of low flow characteristics on basin geology in mountainous areas of Japan. In: *The Hydrological Characteristics of River Basin* (Proceedings of the Tokyo Symposium, Japan). IAHS Publication No. 117, pp 147–56. Wallingford, UK. [3.3]

Nathan RJ, McMahon TA. (1990) Evaluation of automated techniques for base flow and recession analysis. *Water Resources Research* **26** (7):1465–73. [3.4]

National Water and Soil Conservation Authority (1984) *An index for base flows*. Streamland 24, Water and Soil Directorate, Ministry of Works and Development, Wellington, New Zealand. [3.2]

NERC (1975) *Flood Studies Report*. Natural Environment Research Council, London. [3.1, 3.2, 3.3]

Parde, M. (1955) *Fleuves et Rivières* 3rd edn. Armand Colin, Paris. [3.2]

Pilon PJ, Condie R. (1986) Median drought flows at ungauged sites in Southern Ontario. *Canadian Hydrology Symposium* (CHS86), National Research Council of Canada, Regina. [3.2]

Rang MC, Schouten CJ. (1989) Evidence for historical heavy metal pollution in floodplain soils: the Meuse. In: Petts GE (ed.) *Historical Change of Large Alluvial Rivers*, pp 127–42. John Wiley and Sons, Chichester. [3.1]

Riggs HC. (1973) Regional analyses of streamflow techniques. *Techniques of Water Research Investigations*, Book 4, Chapter B3. USGS Washington, DC. [3.3]

Riggs HC. (1990) Estimating flow characteristics at ungauged sites. In: *Regionalization in Hydrology* (Proceedings of the Ljubljana Symposium, Yugoslavia). IAHS Publication No. 191, pp 159–70. Wallingford, UK. [3.3]

Roald LA. (1989) Application of regional flood frequency analysis to basins in North West Europe. In: *FRIENDS in Hydrology* (Proceedings of the Bolkesjo Symposium, Norway). IAHS Publication No. 187, pp 163–74. Wallingford, UK. [3.2]

Schwarze R, Grünewald U, Becker A, Frülich W. (1989) Computer-aided analysis of flow recessions and coupled basin water balance investigations. In: *FRIENDS in Hydrology* (Proceedings of the Bolkesjo Symposium, Norway). IAHS Publication No. 157, pp 75–84. Wallingford, UK. [3.2]

Simmers I. (1975) The use of regional hydrology concepts for spatial translation of stream data. In: *The Hydrological Characteristics of River Basins* (Proceedings of the Tokyo Symposium, Japan). IAHS Publication

No. 117, pp 109–18. Wallingford, UK. [3.3]

Tallaksen L. (1986) *An evaluation of the base flow index (BFI).* Department of Geography, University of Oslo. [3.2]

Tallaksen L. (1989) Analysis of time variability in recession. In: *FRIENDS in Hydrology* (Proceedings of the Bolkesjo Symposium, Norway). IAHS Publication No. 187, pp 85–96. Wallingford, UK. [3.2]

Toebes C, Strang DD. (1964) On recession curves. 1 – Recession equations. *Journal of Hydrology, New Zealand* **3** (2): 2–15. [3.2]

Unesco (1978) World water balance and water resources of the earth. *Studies and Reports in Hydrology 25.* Unesco, Paris. [3.2]

Unesco (1979) Map of the world distribution of arid regions. *MAB Technical Note 7.* Unesco, Paris. [3.2]

Vivian H. (1989) Hydrological changes of the Rhône river. In: Petts GE (ed.) *Historical Changes of Large Alluvial Rivers*, pp 57–78. John Wiley and Sons, Chichester. [3.1]

Ward RC. (1968) Some runoff characteristics of British rivers. *Journal of Hydrology* **vi** (4): 358–72. [3.2]

Ward RC, Robinson M. (1990) *Principles of Hydrology.* McGraw-Hill, London. [3.2]

Wiltshire SE. (1986) Regional flood frequency analysis II: Multivariate classification of drainage basins in Britain. *Hydrological Sciences Journal* **31**: 334–46. [3.2]

Wittenberg H. (1989) Regional analysis of flow duration curves, case studies on catchments in north-west Germany. In: *FRIENDS in Hydrology* (Proceedings of the Bolkesjo Symposium, Norway). IAHS Publication No. 187, pp 213–20. Wallingford, UK. [3.3]

Wright EP. (1989) The basement aquifer research project 1984–1989: final report to the Overseas Development Administration. *British Geological Survey Technical Report WD/89/15.* Wallingford, UK. [3.2]

4: Modelling Hydrological Processes for River Management

N. P. FAWTHROP

4.1 INTRODUCTION

Hydrological modelling can be of use in many aspects of river management. It is an essential 'tool of the trade'. This chapter discusses the various types of models and gives examples to illustrate the sort of analyses which are possible. The chapter concludes with a section of practical advice about how to avoid some of the pitfalls and make the most of hydrological modelling at whatever level it is approached.

4.2 HYDROLOGICAL MODELLING: A CONSTITUENT OF RIVER MANAGEMENT

Water industry professionals together with their consultants and advisors are well aware of the conflicts of interest which often arise in river management (Cook *et al* 1992). In recent years that awareness has extended rapidly to the general public. There has been a dramatic increase in the prominence of environmental issues. Rivers, streams and all aspects of the aquatic environment are receiving more attention than ever before.

The science of hydrology is basic to almost every aspect of river management. It deals with the occurrence and movement of water in, on and over the land surface (details are given in Calow & Petts 1994). Without quantitative knowledge of hydrological processes most of the analysis carried out within the disciplines of Water Resources Planning, Water Quality Planning and River Engineering would not be possible.

Hydrological analysis is required for studies of all kinds. At one extreme may be the prediction of flows for the design of a channel improvement scheme, or assessment of why a small stream has dried up. At the other may be the design of a region-wide water transfer scheme involving water supply, irrigation and navigation. The two are very much connected, as the consequences of one may affect the other. Techniques appropriate to the complexity of the problem need to be applied across the whole spectrum. What techniques might be appropriate? How can all those interrelated components of the hydrological cycle be disentangled? The answer, more often than not, is by some form of modelling.

For the purposes of this chapter modelling is defined as: *Computer software which quantitatively represents the response of a water resource system (natural and/or artificial) to a sample of information.* Some of the words used here are significant. First, a model *represents* reality; it is conceptually similar to the real world, but simplifying assumptions are invariably involved. Secondly, any historical record against which a model has been calibrated is only one of many possible sequences of events; it is thus a *sample*, in time and in space. Care must be taken to ensure that this is representative of the period and location for which predictions are to be made. Thirdly, the *information* which triggers the response of the system could be virtually anything. Models are commonly used to predict the hydrological performance of a catchment consequent upon some change which could be either natural (e.g. rainfall, channel morphology) or artificial (e.g. abstraction, reservoir release). A model should never be extended too far or used to examine scenarios for which it was not designed.

These are the most important of many issues

relating to the use of models in practice, which are discussed further in Section 4.6.

Selecting the right tool for the job

The scientific literature abounds with descriptions of hydrological models to investigate every conceivable aspect of the subject, but only a small proportion are of any practical use to the applied hydrologist. The following extract is from a paper by J.C. Dooge;

> The subject of mathematical modelling in hydrology is characterised by: (1) A proliferation of approaches and techniques. (2) A further proliferation of models based on any one of these particular approaches. (3) A failure to develop adequate techniques for the evaluation of specific models and for choice between models in a given situation. (4) A widening gulf between research techniques and operational methods.

In many respects this is as true today as when it was written some 20 years ago (Dooge 1972).

It is understandable and healthy that there has been a proliferation of approaches and techniques amongst researchers and other specialists. This chapter, however, is for practising engineers and scientists, for whom the approach to modelling has to be different. Models are 'tools of the trade', and like tools in any profession they have to be robust, effective and easy to use. A large choice is unnecessary and leads only to confusion. Proof of this is found in the ready acceptance of packaged models, particularly if they form part of a 'cookbook' approach within standard procedures.

However, standard packages are not always the answer. Unless extensively developed and properly tested they often fail to deliver what is required of them. A lack of flexibility is their biggest disadvantage and there are fewer opportunities to use standard models than might be imagined. There is a preference for models which can be tailored to a particular application.

Examples of both packaged and tailored models are used in the following sections to illustrate the types of models which are available for a range of applications. They are described in broad categories of: (i) catchment simulation; (ii) groundwater simulation; and (iii) river channel models.

The emphasis is upon models which are in use now, with the objective of illustrating a little of the background theory, but more importantly how they have been implemented in practice.

4.3 CATCHMENT SIMULATION

Models are needed to simulate the behaviour of catchments for a variety of purposes including:

1 water balance and water resources assessment work;

2 making forecasts of river flow;

3 quality control and infilling/extending flow records;

4 testing conceptual understanding of hydrological processes.

The objective is to simulate the land phase of the hydrological cycle, with the model output usually being a prediction of river flow. Interest in flows, and hence the emphasis on modelling, is often at one or other of the extremes of the flow range. For example, ecological studies of 'in river needs' would be mostly concerned with an analysis of low flows. Conservation studies in the floodplain would be more concerned with high flows.

Some of the terms used to describe different types of model are not always self-explanatory, so a brief summary may help. The processes which link rainfall with river flow are governed by physical principles, but solutions based upon the direct application of the laws of physics are impractical at the catchment scale. *Conceptual* models are distinguished from *empirical* models according to whether they involve consideration of physical processes, or are simply mathematical relationships based on observation and analysis. *Stochastic* models incorporate statistical chance or probability, whereas *deterministic* models do not.

Conceptual models are usually deterministic in that they seek to uniquely represent our understanding of physical processes, but their mathematical algorithms incorporate parameters which need to be calibrated. At the most simple level are 'lumped' models which employ relatively simple and spatially averaged mathematical representations of input and output to stores (or 'tanks') in order to simulate the passage of water through a catchment. At the other extreme are physically based 'distributed' models which represent processes in far more detail and with a corresponding increase in computational effort.

Lumped conceptual models

The term 'lumped' is frequently used in relation to hydrological modelling (Blackie & Eeles 1985). It indicates spatial averaging so that, for example, one store may represent soil water storage throughout the whole of a catchment covering several hundreds of square kilometres. The area covered could equally be as small as a few hectares. Flows into and out of the stores are controlled by mathematical equations. Parameters define the size of the stores and the rate of flow between them. Lumped models must be used with caution and at a level of spatial definition appropriate to the problem being investigated.

Figure 4.1 illustrates one example of this type of model in schematic form. The HYRROM model (NERC 1989) is a commercially available conceptual rainfall runoff model developed for use on personal computers (PCs). The stores fill and empty to provide flows which depend upon the volume of water they contain. Each store affects the timing and magnitude of the flow at the catchment outlet. There are nine parameters:
- one which defines the size of the interception store;
- one which controls the evaporation losses;
- one which controls the proportion of water entering the runoff store (the remainder enters the soil store);

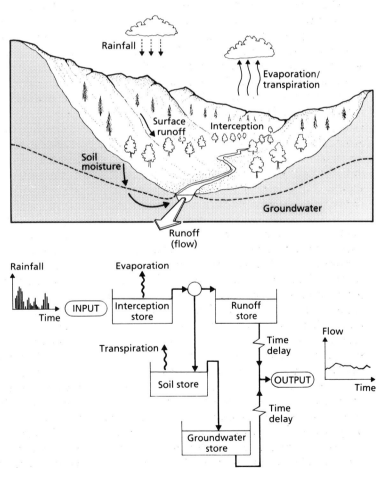

Fig. 4.1 The simulation of rainfall–streamflow processes by the HYRROM model.

- three which control the contribution of the runoff store to catchment flow;
- three which control the contribution of the groundwater store to catchment flow.

The process of model calibration involves adjusting the parameter values until acceptable agreement is obtained between the model predictions and a sample of observed flows. In this case there is an option for automatic parameter optimization. This is a useful facility, as are the user-friendly data entry screens and graphical output. These are the advantages of a packaged system, but as is often the case there are disadvantages due to the lack of flexibility. For instance, in this version of the model there is no allowance for abstractions or discharges, which limits its use to catchments with minimal artificial influences.

There are few commercially available 'packaged' lumped models available in the UK. An alternative to HYRROM which has recently been updated to run on PCs is HYSIM (Manley 1978, 1993). HYSIM has 17 parameters, but many are based on measurable catchment characteristics. Flows from separately modelled sub-catchments can be passed downstream using kinematic wave routing (see Section 4.5).

The reason that there are not more general purpose models is that they nearly always have limitations of one sort or another. As many hydrologists have computer programming skills one basic model is often tailored to particular requirements. This is more likely to happen with relatively simple lumped models than it is with anything more complex and most hydrologists will have access to a catchment model of this type. Lumped models can be incorporated in to particular applications, as illustrated by the following examples.

The London Reservoirs
Drought Management System

One of the main uses of hydrological models is to provide input to, or form a component of, water resources system models. Such models simulate the operation of regional schemes and may be used for yield assessment, system design, short-term operational planning or long-term strategic planning. Moore *et al* (1989) describes a decision support system which is used for the management of London's pumped storage reservoirs on the River Thames. Hydrological models form only a part of the overall system, which also includes simulation of the reservoir operations, a risk assessment procedure and a decision support user interface. A conceptual rainfall-runoff model is utilized, which had been previously developed for the River Thames (Greenfield 1984). The key stages of the decision support system are:

1 The rainfall-runoff model is run over recent time to produce a simulated flow sequence up to the current day. The results are compared with observed flow, and the conceptual stores of the model adjusted so that the modelled and observed flows agree. The conceptual model then encapsulates information on the current hydrological status of the catchment.

2 If the current day was (say) 1 July 1993 and an assessment of the next 6 months was required, the 6-monthly sequences of daily rainfall starting on 1 July would be extracted from a database for every year of record – in this case 103 going back to 1890. Each of these sequences would be run through the rainfall-runoff model to produce 103 equi-probable sequences of river flow.

3 The river flow scenarios are then used as input to the water resources system model to derive equi-probable scenarios of reservoir levels. These are then analysed statistically to provide statements about the probability of certain actions being required.

4 The method can be refined by incorporating long-range weather forecasts.

The Thames Model is now being used to simulate catchment response to climate change and land-use scenarios (Wilby *et al* 1993).

The Great Ouse Resource Model

The Great Ouse is an 8580-km^2 catchment in Eastern England. River flow records are available from some 50 gauging stations. The nature of the flow hydrographs varies considerably, reflecting the changing geology. There are three aquifers, the most important of which is the Chalk, as well as large areas of impermeable clay. This is the driest region in England and the water resources are utilized for agricultural, industrial and public water supplies. The spatial and temporal distribution of river flows are much affected by these abstractions.

Effective catchment management demanded that hydrologists were able to describe the flow regime in quantitative terms at any location. Extrapolation of data from the gauging station network was possible, but often time consuming and inconsistent. The prediction of future flow regimes consequent upon changes in abstraction and discharges was even more difficult. It was decided to develop a model of the catchment to provide the sort of information which was regularly required.

The Great Ouse Resource Model (Oakes & Keay 1990) is a semi-distributed model of the hydrological processes within the catchment. The river system is represented by 392 river reaches, linked together at node points (Fig. 4.2). The inflows and outflows for each reach are calculated, then accumulated downstream so that flows can be derived at every node in the model. This is carried out at weekly timesteps for a 26-year period.

The representation of a reach is shown schematically in Fig. 4.3. There are, in effect, 392 very simple lumped models linked together to form a distributed model. The input requirements of the program are:
- weekly values of effective rainfall, calculated using hydrometeorological data and a soil moisture accounting model (Thompson *et al* 1981);
- monthly surface water abstractions and discharges, and groundwater abstractions for each of 64 sub-catchments. These were derived from aggregated statistics based on returns from abstracters.

The abstraction and discharge data can be modified to account for anticipated future changes, and the model run for any conceivable scenario. The main outputs are flow duration statistics, which integrate the effects of all up-

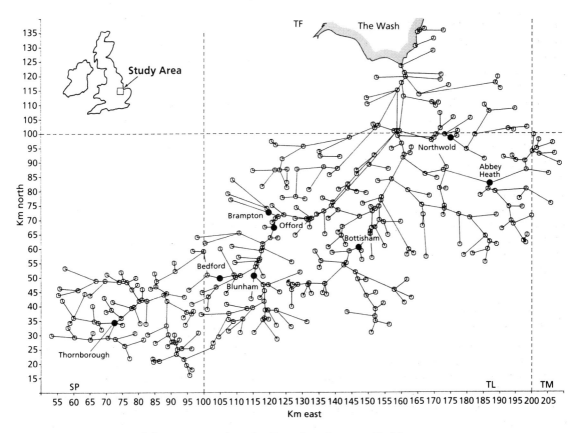

Fig. 4.2 Representation of the river network in the Great Ouse Resource Model.

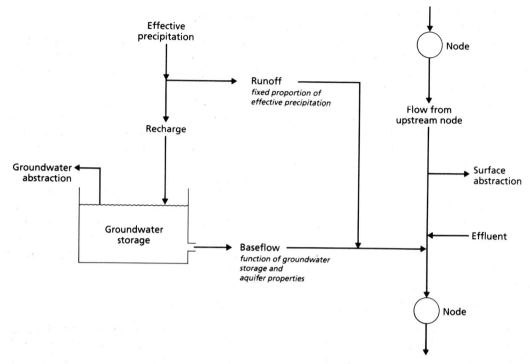

Fig. 4.3 Representation of reaches in the Great Ouse Resource Model (see also Fig. 4.2).

stream changes and summarize modifications to the flow regime. Such statistics may be used, for example, to evaluate the impact of applications to abstract water or to estimate the flows which will be available for diluting the effluent from sewage works.

The facility to calculate flow statistics so easily for so many locations is extremely useful. A user-friendly interface and graphical presentation of the results means that the model can be run by non-specialists, but the documentation draws the attention of all users to its limitations and to the degree of precision which can realistically be achieved. If greater precision is needed a more complex model will be required. An example of a model for one aquifer unit in the Great Ouse catchment is described in Section 4.4.

Physically based distributed models

A simplistic view of hydrological modelling is that there are basically two types of application. The first is to test our conceptual theories of hydrological processes and the second is to predict catchment behaviour of practical relevance. Despite significant advances in recent years the application of physically-based distributed models is still largely restricted to the former. Nevertheless, they are of interest to anyone involved in hydrological modelling, and may in the future move along the spectrum of practicality.

Physically-based distributed models usually represent surface and subsurface flows using a grid (or net) of interconnected nodes spaced only tens or hundreds of metres apart. Flows vertically and in two dimensions between the nodes are calculated using equations based on physical laws (e.g. Darcy's law). The parameters are *in theory* physically definable and measurable. The SHE (Système Hydrologique Européen) model (Abbott *et al* 1986a,b) is the most ambitious project undertaken to date to develop a physically based distributed model for commercial use. The IHDM (Institute of Hydrology Distributed Model) (Beven *et al* 1987) is the other one most often referred to in the UK.

The theoretical advantages of a distributed model are, however, often outweighed by the penalty of extremely demanding data requirements. The number of grid nodes, and hence parameters, are very high. Even on intensively monitored research catchments the number of parameters is often too high to be able to determine values in the field, and a degree of spatial aggregation is inevitable. Beven (1989) clearly spells out the problems of using these models in practice, and they are well summarized by Eeles *et al* (1990). Attention is drawn to the problems associated with using physical parameters determined at a point in a heterogeneous system and applied to a grid square. On the other hand at least one version of SHE (Danish Hydraulics Institute 1991) is commercially available and aims to provide a flexible model which can be applied at varying levels of complexity.

The manipulation of spatial data is aided by geographical information systems. These assist greatly in the storage and manipulation of the kind of data required by distributed models. Large databases of, for example, soil types or land use at grid scales less than $1 \, km^2$, are now commercially available. Similarly, topography may be incorporated into distributed models through the use of digital terrain maps. In all fields the use of remotely-sensed data from satellites and aircraft is increasing.

A simplification of the distributed modelling approach is made in TOPMODEL (Quinn *et al* 1989). This uses a digital terrain map to determine a topographic index which is used to predict the dynamics of variable source areas. The number of parameters is kept deliberately low and is usually around three.

A yet more pragmatic approach is to utilize semi-distributed models, which are in effect lots of lumped models linked together in some way, perhaps by some form of river routing. The London Reservoirs Drought Management System and Great Ouse Resource Model, described above, in fact fall into this category. Another good example is the RORB runoff routing model (Lowing & Mein 1981; Laurenson & Mein 1983; Rylands & Lee 1992), which uses a catchment subdivision based on isochrones.

Empirical models

This section considers how models and the processing power of computers can be used to define empirically the nature of hydrological systems.

Empirical models offer little insight to physical processes. They are sometimes referred to as 'black box' models, because inputs are converted to outputs based only on statistical correspondence. In many situations, if there is a sufficiently strong and consistent relationship between input and output an understanding of the intervening catchment processes is not necessary. Nevertheless, the empirical relationships can sometimes be interpreted in terms of implicit catchment response. This is best illustrated by means of examples.

The Unit Hydrograph Model

A milestone in the development of hydrological science was the introduction of the concept of the unit hydrograph by Sherman in 1932. This model is still in widespread use today. Figure 4.4(a) illustrates the idea of how a T hour unit hydrograph (TUH) results from a unit depth of *effective* rainfall falling in *T* hours over the catchment. The 'unit' is usually 1 mm and in this case the ordinates of the unit hydrograph are $m^3 s^{-1}$ per millimetre of rain. The volume underneath the curve of the hydrograph is equivalent to 1 mm depth of effective rainfall over the entire catchment area. The dimensions of the unit hydrograph will vary between catchments, depending on those characteristics which affect catchment response such as area, geology, slope, shape and land use. The ordinates of a unit hydrograph can be derived from synchronous records of river flow and rainfall. This is a purely mathematical procedure. The unit hydrograph is simply a mathematical expression of the catchment response, but relationships can be established with physical parameters of the catchment, so in that sense it does have physical meaning.

The unit hydrograph concept involves several assumptions which simplify its application to catchment scale rainfall–runoff modelling.

1 Runoff is directly proportional to effective rain-

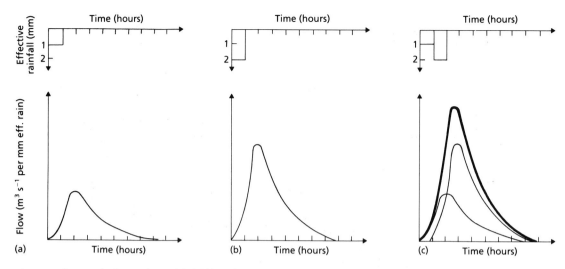

Fig. 4.4 The unit hydrograph model: (a) the 1-hour UH for 1 mm of effective rainfall; (b) superposition, the runoff hydrograph for 2 mm of effective rainfall in 1 h; (c) convolution, the runoff hydrograph for 1 mm followed by 2 mm of effective rainfall.

fall. Thus two units of effective rainfall falling in T hours would produce a runoff hydrograph with ordinates twice that of the unit hydrograph (Fig. 4.4b).

2 Effective rainfall in consecutive T-hour periods results in a runoff hydrograph which is the sum of the lagged component hydrographs for each hour (Fig. 4.4c).

3 The shape of the unit hydrograph is time invariant.

These assumptions are never entirely valid, and there are other weaknesses such as the assumption of uniform catchment–wide rainfall (which is common to most lumped models). The topic of baseflow separation and the analysis of what proportion of the total rainfall is effective rainfall would need many pages to explain fully. Nevertheless, the unit hydrograph is a sound technique which has the great advantage of simplicity, and it is central to many procedures for modelling the rainfall–runoff process. This is particularly the case in engineering hydrology where the Flood Studies Report (NERC 1975) recommends it as one of the standard approaches to flow estimation throughout the UK. As such it is likely to be used in the design of many channel improvement schemes.

An example of the use of the unit hydrograph model is given for the River Nene in the English East Midlands (Fig. 4.5). Unit hydrographs were derived from gauged flow records for each of the major catchments, and from relationships with catchment characteristics for all the ungauged sub-catchments (Cole *et al* 1981). They were then combined with hypothetical rainfalls to generate design flows for input to a mathematical hydraulic model of the main river (see Section 4.5 for a more complete description). This calculation (as simplified in Fig. 4.4c) is referred to as the 'convolution' of a time series of rainfall and unit hydrograph ordinates. Data preparation is often time-consuming, so recent developments in unit hydrograph theory as described in the next section are very welcome.

The IHACRES Model

IHACRES is an acronym for 'Identification of unit Hydrographs And Component flows from Rainfall, Evaporation and Streamflow data' (Jakeman *et al* 1990). The model used in IHACRES is a development of unit hydrograph theory, which overcomes many of the practical difficulties of that approach by utilizing an approximation based

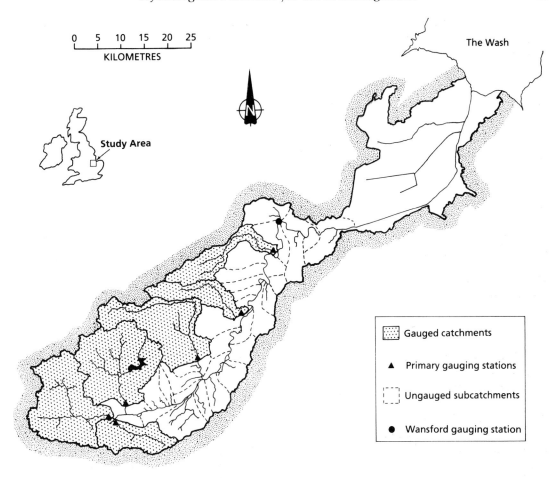

Fig. 4.5 Sub-catchments of the River Nene.

on 'transfer functions'. A PC version may soon become available commercially.

Transfer function models predict flows based on a limited number of previous rainfalls and flows. It is possible to demonstrate their mathematical equivalence to unit hydrographs, but they have numerous advantages; in IHACRES these are:

• there are fewer parameters, so it is possible to adopt a more effective optimizing approach to parameter estimation;

• predictions are made of total streamflow and the quickflow/slowflow components separately;

• data inputted are time series of *total* rainfall and river flow, not effective rainfall and rapid response runoff. Hydrograph separation is an integral part of model identification and not, as in conventional unit hydrograph derivation, a necessary prerequisite;

• continuous time series of data can be utilized rather than the discrete 'well behaved' events which are necessary to avoid instability in unit hydrographs.

Figure 4.6 illustrates one version of IHACRES in schematic form. It can be applied to catchments varying greatly in size. Jakeman *et al* (1991) give examples varying from a 743-km^2 catchment with high baseflow modelled with a daily timestep to a 0.016-km^2 flashy catchment modelled with an hourly timestep. Figure 4.7

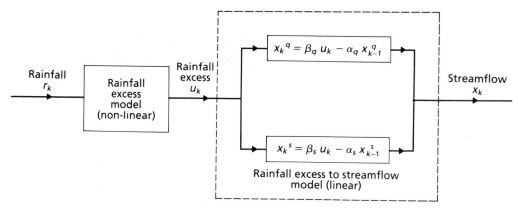

Fig. 4.6 IHACRES model.

shows the model output from the two catchments, illustrating the prediction of both total flow and the baseflow component. Further examples of use of the model, including an interesting application to baseflow separation, are given in Littlewood and Jakeman (1993).

Although the parameter values of the transfer function models have some physical interpretation it is important to realize the difference between empirical models such as this and the conceptual models described earlier. The parameter values of IHACRES summarize the catchment as it was during the calibration period, so this model would not be appropriate for predicting changes in riverflow due to changes in specific components of the water balance (such as abstractions). On the other hand IHACRES would be well suited to many other modelling applications where either characterization or comparison of catchments is required. It could be used wherever a prediction of catchment response due to rainfall is needed. For example:
- detecting the effect of land use change;
- predicting the impact of climate change;
- infilling or extending river flow records.

Models for operational flood forecasting are not considered in this chapter, but it is worth mentioning that empirical models, and particularly transfer function models, are well suited to this type of application. They need few parameters, are computationally efficient and have minimal data requirements (Cluckie & Tilford 1989).

4.4 GROUNDWATER SIMULATION

Groundwater models as described here may be thought of as catchment simulation models which focus on flows underground rather than in surface watercourses. They are useful for:
- initial evaluation and appraisal studies preceding field investigations;
- interpreting the results of a field investigation;
- prediction and forecasting to estimate future field behaviour.

For groundwater flow the physical laws which govern cause and effect are generally well understood and a deterministic approach is usually possible, albeit with numerous simplifications and assumptions. For contaminant transport there are more unknowns and stochastic techniques may form an integral part of the model in order to quantify uncertainties.

Groundwater flow models

Due to the complexity of groundwater flow mechanisms a lumped approach is usually inadequate. In order to understand aquifer behaviour it is necessary to consider aspects such as:
- properties of the aquifer such as the hydraulic conductivities and storage coefficients. These often change markedly over short distances both laterally and vertically;
- significant features such as fissures and preferential flow horizons;

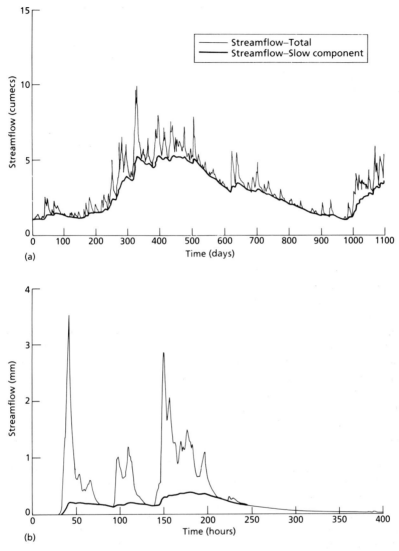

Fig. 4.7 Examples of flow simulation by IHACRES: (a) modelled flows for a catchment of 743 km²; (b) modelled flows for a catchment of 0.016 km².

● recharge processes. These are many and varied, such as infiltration recharge through soils, recharge into swallow holes, recharge from leaking water supply pipes in urban areas and recharge through river beds;
● the location of features such as abstractions and springlines which can exert a controlling influence on flows.

Spatial detail must be incorporated. This is achieved by subdividing the aquifer into many 'cells' in the form of a grid or net overlaying the study area. Given estimates of the aquifer parameters, mathematical equations based on Darcy's Law define the relationships between water levels and flows between the cells. When combined with expressions for the conservation of mass and (for time-variant problems) the effect of changing storage, simultaneous equations are defined which may be solved using numerical solutions. Different numerical procedures may utilize either a rectangular grid (finite difference) or a polygonal grid (integrated finite difference or finite element). The grid need not be uniform. More detail may be required in areas around features of interest.

Modelling may be used to examine a small area rather than the whole aquifer. For example, radial flow models are used to investigate aquifer characteristics around a borehole in more detail than is possible using analytical solutions to pumping test data (Rushton & Redshaw 1979).

Models may be for steady-state conditions or be time variant. They may be one-, two- or three-dimensional and consider any number and combination of confined, unconfined and leaky aquifers. The skill required for model development reflects the complexity of the situation. The *use* of such models, however, need not be the province of specialists (Spink & Hughes 1992).

The type of model which river-management professionals are most likely to encounter is a regional aquifer model being used to predict changes in aquifer behaviour consequent upon additional or relocated abstractions. In the UK these have been in use since the early 1970s, gradually increasing in refinement as knowledge, experience and the available computing resources have all improved. Rushton (1986) reviews the approach and gives some practical considerations and examples. A review giving more emphasis to American non-proprietary models (which are increasingly being used in the UK) is provided by Anderson and Woessner (1992). The main objective is always to represent adequately the features of groundwater flow in the aquifer. The success of a model is measured by its ability to translate the results back to physical meaning and its effectiveness in answering the questions which motivated the model study. The following sections describe several such models in some detail.

Lodes–Granta Model

The Chalk northeast of Cambridge, England, is a major source of water for public supply. Ab-

stractions have been increasing and in recent years there has been growing concern about the effect upon the watercourses in the area. The National Rivers Authority has a duty to protect the water environment whilst ensuring that water resources are utilized effectively. There is a conflict of interests in the area, with on the one hand two water companies keen to maximize their use of cheap and high-quality groundwater, and on the other local pressure groups and conservation interests very anxious to ensure that there is no further deterioration in low flows.

The maintenance of streamflows using pumped groundwater is quite common in England. It was decided to initiate a feasibility study into the possibility of this option to compensate the effect of additional groundwater abstractions. It involved detailed hydrogeological investigations and the development of a mathematical model (Rushton & Fawthrop 1991). The model is a distributed regional groundwater model based on the solution of finite difference approximations to the groundwater flow equations. It has a regular 1-km grid and uses a 7-day timestep.

It was very important to have a conceptual understanding of the hydrogeology of the aquifer before proceeding with the design of the river support scheme. The significant features in so far as they needed to be incorporated into the model are summarized below.

1 Springlines were associated with the outcrops to two bands of hard, well-fissured chalk with relatively high hydraulic conductivities. Ninety-eight springs and locations where river–aquifer interaction took place were included. Their discharge characteristics were defined by coefficients so that the modelled flows could be calibrated against periodic current meter measurements.

2 These two bands also provided important flow paths to pumped boreholes. There was no attempt to explicitly model the layering, but there was a feedback facility to restrict borehole yields if groundwater levels fell below threshold levels related to the depth of the major fissures.

3 In most Chalk aquifers, the zone of fluctuation of the water table defines a zone of high hydraulic conductivity. Consequently the transmissivity reduces rapidly as the water table falls. This was incorporated into the model by defining a relationship between hydraulic conductivity and groundwater level. It proved to be essential in order to simulate correctly the magnitude of the seasonal changes in flow towards the springs and streams.

The model was calibrated against observed groundwater levels and streamflows for the 17-year period from 1961 to 1987. The fit was suf-ficiently good to provide confidence that the understanding of the aquifer flow mechanisms and dynamic water balance were adequately represented.

The model was used to test whether the new boreholes for both river support and public supply could provide water reliably and without affecting existing boreholes, wells and protected wetlands. Output from the model was used to predict

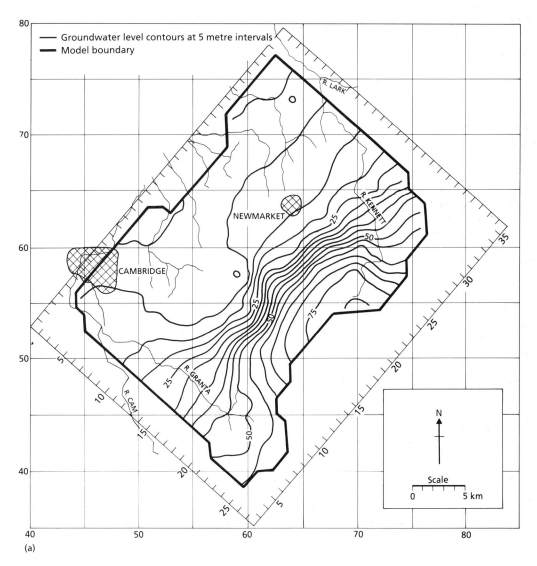

(a)

Fig. 4.8 Examples of output from the Lodes—Granta Model: (a) groundwater head distribution October 1974 (current abstraction); (*over page*) (b) a comparison of heads at one location showing the effect of increased abstraction; (c) a comparison of flows from a single spring showing the effect of increased abstraction.

Chapter 4

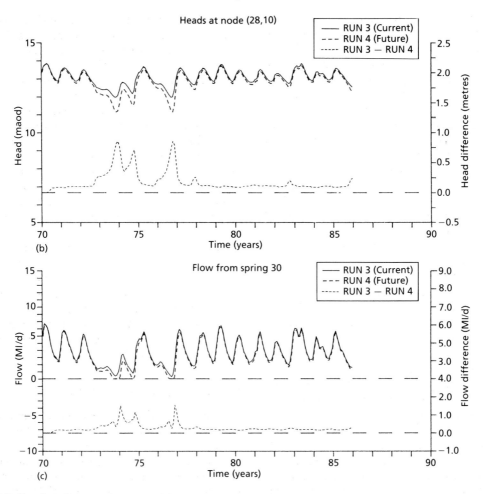

Fig. 4.8 *Continued*

the effect upon water levels and springflows at particular locations. It was invaluable in both the design of the groundwater development scheme and in illustrating its consequences during the consultation phase. Figure 4.8 shows examples of the sort of output used.

Low-flow investigations using an integrated model

Some of Britain's most pressing environmental concerns are about low flows in certain over-utilized rivers. A recent review by the National Rivers Authority identified the 40 catchments where problems are most apparent and remedial action is necessary. Although lack of rainfall causes low flows in rivers, the problem is sometimes exacerbated by over-abstraction or changing land use. It would not be possible to isolate the various components without the utilization of hydrological models.

Many of the most seriously affected rivers are in southern England. Three examples are:
1 The River Darent in Kent (Brown *et al* 1991).
2 The River Allen in Dorset (Newman & Symonds 1991).

3 The Wallop Brook and Bourne Rivulet (tributaries of the River Test) in Hampshire (Midgley & Jones 1991).

In all three cases low flows are sustained by groundwater discharges from the chalk aquifer. Hydrological investigations had to quantify the dynamic water balance, involving the assessment of recharge, runoff to the rivers, groundwater flow, springflow/seepage from groundwater to the rivers, leakage from the river back to groundwater and the impact on each of these of groundwater abstraction. The examples listed above were selected because they were all modelled using the same *Integrated Catchment Model* (Mott MacDonald 1993).

The model contains three main simulation modules, together with graphical pre- and post-processing facilities. The three modules are:

1 A lumped model to simulate the spatial and temporal distibution of recharge and runoff.
2 A model to derive the flow-level relationships for river reaches.
3 An integrated groundwater and river flow routing model.

The recharge model is a development of the Stanford Watershed Model (Crawford & Linsley 1966). This is a well known lumped model in which the water balance takes account of variations in storage and the rate of water flow both above and below the surface. Recharge estimates made in this way could be verified in the context of the total water balance for the catchment. Recharge was the primary input to the groundwater simulation model.

The groundwater model is fully distributed and can simulate complex, heterogeneous, multi layer aquifer systems. It is based on the Integrated Finite Difference Method (IFDM) which allows for the use of irregularly shaped model polygons. This permitted the more detailed investigation of selected areas (Fig. 4.9).

A distinguishing feature of the IFDM model is the representation of interaction between the river and underlying aquifers. The calculation of flow between the river and the aquifer and the routing of flow down the river network is part of the iterative computation which solves the equations governing flow between the model polygons. River–aquifer interaction depends on

their relative water levels. Groundwater levels come from the groundwater flow equations. River levels come from a hydrodynamic model which relates river flow to level using open channel hydraulic computations (see Section 4.5). Flow accretion profiles (Fig. 4.10) were produced to show locations where the river gains and loses water under various water level conditions.

The benefit of using an integrated model is apparent when the relationships between the various hydrological processes need to be examined. Comparison, mapping and plotting of hydrological variables is done in the knowledge that all interdependencies have been accounted for.

There is a general trend in groundwater modelling towards a more integrated approach. The SHE model (Section 4.3) carries this concept through to its ultimate conclusion, but others have adopted a more pragmatic approach. MODFLOW (McDonald & Harbaugh 1988) is a widely used groundwater model which has several add-on modules.

The South Humberside Chalk Model

Groundwater models are usually developed at the design stage of a scheme, satisfactory calibration is achieved and predictions made. It is less common for models to be updated in order to verify performance under the new abstraction regime and even rarer for models to be sufficiently well developed to use for short-term operational planning. One exception is a model of the South Humberside Chalk (Spink & Hughes 1992), which is used routinely in the short-term local management of the aquifer.

During the 1989–92 drought in eastern England this model was run at 3-monthly intervals to establish the amounts of water which could safely be abstracted from various locations. Careful management was necessary in order to maximize drinking-water supplies from boreholes inland, whilst maintaining groundwater flow towards the sea so that saline water did not encroach upon industrial abstractions on the coast. The control of groundwater abstraction in response to hydrological conditions can prevent damage to the environment.

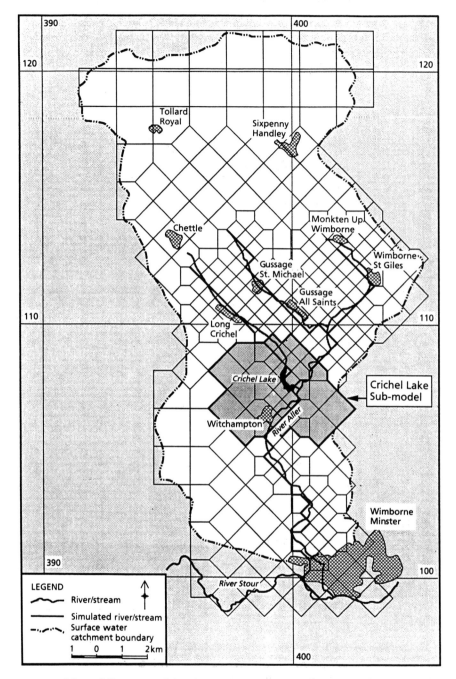

Fig. 4.9   An integrated finite difference model applied to the River Allen, UK.

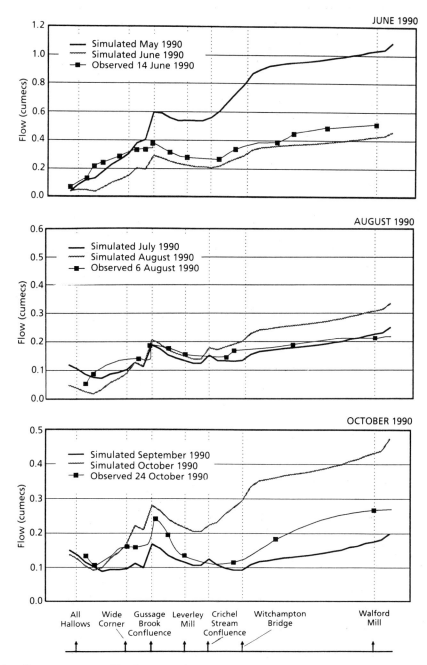

Fig. 4.10 River flow accretion profiles for 3 months in 1990 on the River Allen, UK.

Contaminant transport models

Attempts to include predictions of changing water quality adds another level of complexity. Contaminant transport models must be underpinned by reliable modelling of the regional flow mechanisms, to which are added equations representing the advection, dispersion and chemical reaction of the determinand being modelled. They have been used, for example, to estimate the rates of movement of hazardous waste from landfills, to forecast the effects of pollution incidents and to predict the intrusion of saltwater in coastal regions. Modelling the movement of solutes or contaminants by advection alone (movement with groundwater flow) is well proven through use of the 'particle tracking' technique, but groundwater quality modelling is a field where significant advances are still being made. Given the sometimes limited knowledge of chemical processes and reactions within the aquifer, limited data on the source and strengths of pollutants, and the intrinsic difficulties of modelling groundwater flow, problems concerning contaminant transport often extend mathematical models to the limit. Examples where models have been used successfully are described in Anderson and Woessner (1992) and Anderson *et al* (1992).

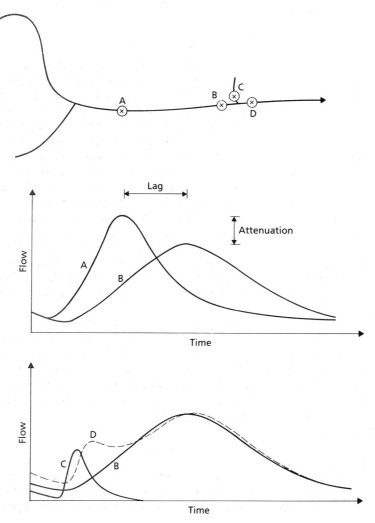

Fig. 4.11 River routing.

4.5 RIVER CHANNEL MODELS

Hydrologists are frequently asked for estimates of river flow or level along a reach of river for which no data exist. It is not always a straightforward matter to extrapolate information from a nearby site, and once again it is often necessary to employ modelling techniques of varying complexity.

Earlier chapters have referred to the characteristics of river hydrographs generated from different sizes and types of catchments. Flow 'routing' is a technique used to quantify the changing shapes of hydrographs at different points on a river system. Following a rainfall event flows increase rapidly in the smaller tributary streams. The water moves downstream at a rate controlled by the size and shape of the channel and its floodplain, the gradient of the river and the presence of any obstructions, which may be either natural or artificial. The shape of the hydrograph is different at every point on the river.

The sort of changes which occur are illustrated in Fig. 4.11. Between A and B where there are no lateral inflows the hydrograph has been modified in two ways.

1 The peak at B occurs later than it does at A. This lagging effect is known as 'translation' of the peak.

2 The peak flow is lower at B than it is at A, due to the effects of channel storage and friction. The flattening of the hydrograph is known as 'attenuation'.

As tributaries join the additional flows are superimposed upon the attenuation. The combination of B and C results in a larger hydrograph (D) which will itself be modified as it 'moves' downstream. This apparent movement of a hydrograph should not be confused with the actual movement of water. Hydrographs record the variation in flow or level through time at a point. The lag between the peaks of hydrographs at two points such as A and B is not the same thing as the time it would take a notional particle of water to travel between A and B. The translation lag is always less than the travel time.

Figure 4.11 refers only to flows. Conversion to levels is where difficulties often arise. For any point the relationship between level and flow may not be unique, for instance due to the seasonal effect of weedgrowth. The effects of weeds or any other form of obstruction are felt for some distance upstream. These are known as 'backwater' effects, and backwater analysis is commonly employed by river engineers to calculate levels under constant flow conditions.

When the additional dimension of time is introduced more complex modelling is required. Time variant modelling is needed whenever the effects of changing storage within the system need to be considered or when the 'boundary' conditions (e.g. tributary inflows or downstream levels) are changing so rapidly that constant flow assumptions are invalid. There are two broad categories of river routing models, often referred to as hydrological and hydraulic models.

Hydrological routing models

Hydrological routing models relate the inflow and outflow of a river reach using the principle of continuity and a relationship between flow and storage within the reach. Most hydrological textbooks give worked examples of this type of approach (the Muskingum method) and demonstrate how the translation and attenuation shown in Fig. 4.11 can be predicted (Shaw 1983; Wilson 1983). Such models are relatively simple and reasonably accurate, but there are limitations. They can only be used in situations where the relationship between level and flow is unique. Thus changing backwater effects from tides, confluences or moving structures cannot be incorporated, and they are not appropriate to low-gradient rivers where looped rating curves may exist.

The same limitation applies to the use of simplified hydraulic routing using the principal of kinematic waves (well described in Dingman 1984). Wave speed is derived from the flow, the dimensions of the channel and an estimate of channel roughness. More complex hydraulic models are described below.

Hydraulic routing models

In a manner very similar to developments in groundwater modelling the reduced cost of increasingly powerful computers has led to significant progress in the application of more sophisticated hydraulic models. These are based on numerical solutions to the equations describing the conservation of mass and momentum. As with groundwater modelling there are a number of mathematical approaches to the solution of the equations but most of them involve finite difference approximation and iterative solutions.

The user of results from such models will be more interested in how the river system is re-

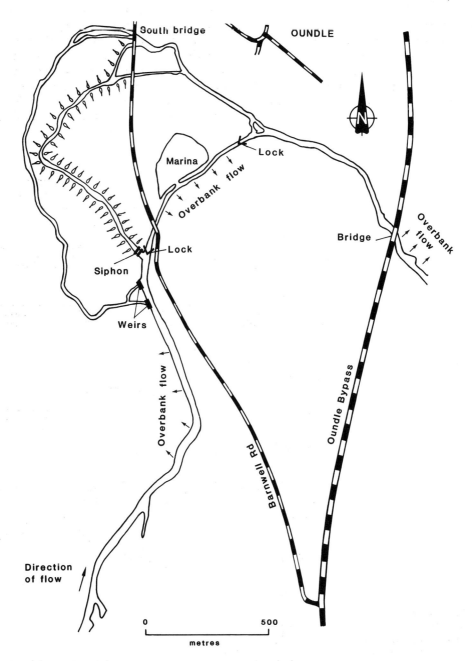

Fig. 4.12 The River Nene south of Oundle.

presented and what sort of things such models can do. The following examples illustrate some typical features.

The River Nene Model

The sort of detail that can be examined using a mathematical hydraulic model is illustrated in Fig. 4.12, which shows part of the River Nene in Northamptonshire. The river flows from south to north. During flooding the left bank of the river south of Barnwell Road becomes inundated and acts as a side weir. The water then enters the most westerly overflow channel to join with flow from over the weir control structures. The other overflow channel is fed from the main river by water passing through a set of air regulated siphons. This channel has raised banks to prevent flooding on to the surrounding land. The two overflow channels join to pass through the old arched Oundle South Bridge. The main river continues through a lock into a lower reach whose right-hand bank can also act as a side weir and then through a further lock structure before joining the now combined overflow channels. The river then continues on through a new road bridge.

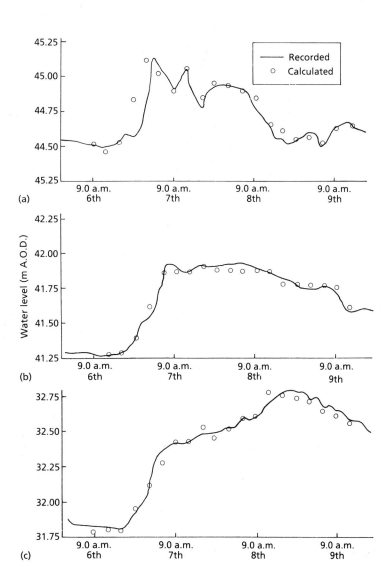

Fig. 4.13 Example output from the River Nene Model: (a) upstream of Doddington lock; (b) upstream of Wellingborough lock; (c) downstream of Upper Ringstead lock.

Time variant flows and levels can be modelled at points spaced only a few hundred metres apart, as shown in Fig. 4.13 (for elsewhere on the river). This is only 5 km out of 144 km, which includes some 120 structures and 34 bridges (Williams & Fawthrop 1988).

Such detail is necessary when analysis is required of the effects of river improvements upon other parts of the river. For example, a scheme of river widening and straightening at a particular point may reduce the degree of flooding there only to transfer the problem to a location farther downstream. Mathematical models are an effective way of examining options. This is particularly useful when it is important to look at the effectiveness and cost of alternatives, including those which minimize damage to the environment.

The Lower Colne Study

The Lower Colne Study (Gardiner *et al* 1987) is a good example of how mathematical modelling should be incorporated into clearly structured catchment management. The Colne lies to the northwest of London and forms a complex network of river channels, gravel pits and extensive floodplain (Fig. 4.14). Much of the catchment is heavily developed and major road and rail arteries cross the floodplain. It was the proposed construction of the M25 motorway on an embankment within the floodplain which initiated a comprehensive study into the river management objectives for the catchment. The impact of flood alleviation works would be of great concern to many organizations with environmental interests and it was apparent that the implementation of flood alleviation measures would be facilitated with their support. The use of a mathematical model allowed more rapid and complete evaluation of alternative options than would otherwise have been possible.

The proposals were formulated in two phases. Firstly, the model provided information on the frequency of flood risk and the extent, duration and depth of flooding in order to estimate the costs and benefits based on a preliminary scheme. At the same time an environmental database was being constructed by specialists in 12 disciplines covering all in-river and riparian interests. In the second phase the river engineers used the model to investigate options to minimize the environmental concerns raised by the specialists. The alternative engineering solutions for particular areas usually included:

- the rehabilitation or replacement of weirs and sluices;
- embankments or flood walls;
- the construction of new flood channels and associated control structures;
- improved maintenance programmes;
- localized channel enlargement.

The process of consultation was iterative until a preferred option was identified.

The model used was ONDA, a general purpose program developed by Halcrows, the consultants on this project. Other consultants have similar models. Open channels were represented by their cross-sections and roughness coefficients, and flows calculated using the St Venant equations for the conservation of mass and momentum. Structures and other constrictions such as low bridges were represented by standard hydraulic discharge equations. Although the modelling was only in sufficient detail to support the consultation phase, an assessment of the model accuracy based on the calibration runs showed that predictions were within 100 mm of the observed values in 68% of cases. The differences included errors quantifying tributary inputs, gaugeboard levelling and reading errors, and errors in representing the operation of river regulation structures. The subsequent design stage involved additional site surveys and more detailed modelling of localized areas.

Mathematical models have proved to be essential aids in catchment planning throughout the Thames Region (Gardiner 1990). Beyond the design phase of a land drainage improvement scheme they continue to play an ongoing role in adaptive river management.

Several mathematical hydraulic models are commercially available. ONDA, HYDRO, SALMON-F and MIKE 11 are used extensively in the UK. However, the magnitude of the task in setting up, calibrating and maintaining a model on the scale of either the Nene or Lower Colne examples should not be underestimated. It is a

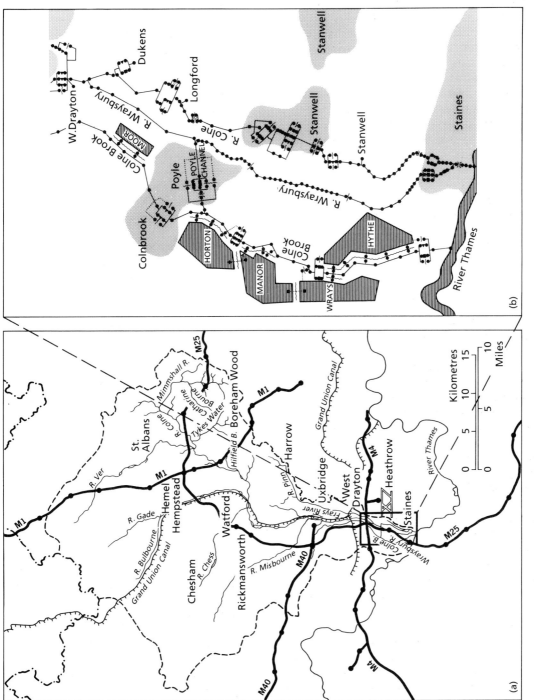

Fig. 4.14 The Lower Colne Study: (a) the Colne catchment; (b) schematic of part of the hydraulic model.

major undertaking which must be adequately resourced. Recent examples of the application of these models *by their developers* are as follow.

• The application of HYDRO to the Ouse Washes, part of the River Great Ouse in Cambridgeshire, England (Balfour *et al* 1989).

• The application of LORIS (the computational model within SALMON-F) also to the River Great Ouse in Cambridgeshire, UK (Slade & Samuels 1990).

• The application of MIKE 11 to the Kalani River catchment in Sri Lanka, in order to examine options for flood protection schemes and their impact on sediment transport (Van Kalken & Havno 1992).

4.6 MODELLING IN PRACTICE

With the growth of available databases over the past three decades, the number of applications where hydrological modelling can be useful has increased rapidly. Increasingly, hydrological models are being used to provide information for multi-disciplinary studies. Truly interdisciplinary modelling is still comparatively rare, but there are numerous examples of hydrological input to water quality or ecological models. Identification of the interface is the key issue. One common interface is the use of flow duration statistics as input to water quality planning models or models of river ecology for quantifying in-river needs, e.g. PHABSIM (Bullock & Johnson 1991).

It would be all too easy to allow impressive examples of modelling such as those illustrated in the previous sections to create the illusion that models can be used to solve any problem. There is a danger that the jargon and technology could create a misleading and superficial gloss. The reliability of model results depends upon a great many contributing factors. These include: the soundness of the underlying theory; the skills of the modeller in translating that theory to a computer program; the quality of the input data; and the applicability of the technique to the issue in hand. If models are decision-making tools then the users of the results are, *de facto*, decision-makers. Inappropriate use of the results is often caused by ignorance of the model's limitations and its inherent assumptions and simplifications.

The important thing for users of models to remember is that you must be prepared to ask questions. As with any discipline, different hydrologists will have their own favourite techniques of analysis and modelling. These will not always be the most up to date, or the most appropriate or the best for the given problem. You should always ask why a particular modelling technique was used. There is a danger that modellers get so engrossed in the methodology and mathematics that they fail to explain the basics.

Hydrological studies are but one component in successful river management. This chapter has described the wide range of hydrological model types which are available and given examples to illustrate how they are used. An appropriate model can be an invaluable tool in helping to explain the complexities of catchment hydrology, as long as it is used correctly. It is hoped that engineers and scientists in related disciplines will now be able to better utilize the specialist hydrological advice at their disposal.

REFERENCES

Abbott MB, Bathurst JC, Cunge JA, O'Connell PE, Rasmussen J. (1986a) An introduction to the European Hydrological System – Système Hydrologique Européen 'SHE' 1. History and philosophy of a physically-based distributed modelling system. *Journal of Hydrology* 87: 45–59. [4.3]

Abbott MB, Bathurst JC, Cunge JA, O'Connell PE, Rasmussen J. (1986b) An introduction to the European Hydrological System – Système Hydrologique Européen 'SHE' 2. Structure of a physically-based distributed modelling system. *Journal of Hydrology* 87: 61–77. [4.3]

Anderson MP, Woessner WW. (1992) *Applied Groundwater Modeling*. Academic Press, London. [4.4]

Anderson PF, Ward DS, Lappala EG, Prickett TA. (1992) Computer models for subsurface water. In: Maidment D (ed) *Handbook of Hydrology*. McGraw-Hill, New York. [4.4]

Balfour N, Guganesharajah K, Halifax P. (1989) Mathematical modelling of the Ouse Washes. In: *Proceedings of the Second National Hydrology Symposium*. British Hydrological Society/Institute of Hydrology, Wallingford, UK. [4.5]

Beven K. (1989) Changing ideas in hydrology – The case of physically based models. *Journal of Hydrology* 105: 157–72. [4.3]

Beven KJ, Calver A, Morris EM. (1987) *The Institute of*

Hydrology Distributed Model. Institute of Hydrology Report No. 98, Wallingford, UK. [4.3]

Blackie JR, Eeles CWO. (1985) Lumped catchment models. In: Anderson MG, Burt TP (eds) *Hydrological Forecasting*, pp. 311–45. John Wiley, Chichester. [4.3]

Brown RPC, Ironside N, Johnson S. (1991) Defining an environmentally acceptable flow regime for the River Darent, Kent. In: *Proceedings of the Third National Hydrology Symposium*, pp. 2.59–2.66. British Hydrological Society/Institute of Hydrology, Wallingford, UK. [4.4]

Bullock A, Johnson I. (1991) Towards the setting of ecologically acceptable flow regimes with IFIM. In: *Proceedings of the Third National Hydrology Symposium*, pp. 2.67–2.72. British Hydrological Society/Institute of Hydrology, Wallingford, UK. [4.6]

Calow P, Petts GE. (eds) (1994) *The Rivers Handbook*, vol 1. Blackwell Scientific Publications, Oxford. [4.2]

Cluckie ID, Tilford KA. (1989) *Transfer Function Models for Flood Forecasting in the National Rivers Authority Anglian Region.* Anglian Radar Information Project Report No. 3. Department of Civil Engineering, Salford University, England. [4.3]

Cole GA, Gustard A, Beran MA. (1981) *Flood Hydrology of the River Nene.* Institute of Hydrology Report for Anglian Water Authority, Wallingford, UK. [4.3]

Cook R, Evans D, Barham P. (1992) Abstractive V's non abstractive uses for water or 'A better future for Anglian rivers'. In: *Proceedings of the Brighton Symposium.* Institution of Water and Environmental Management, London. [4.2]

Crawford NH, Linsley RK. (1966) *Digital Simulation in Hydrology: Stanford Watershed Model IV.* Technical Report 39. Department of Civil Engineering, Stanford University, USA. [4.4]

Danish Hydraulics Institute (1991) *SHE: Système Hydrologique Européen/European Hydrogical System.* Groundwater Modelling Report. DHI, Hørsholm, Denmark. [4.3]

Dingman SC. (1984) *Fluvial Hydrology.* WH Freeman & Co, New York. [4.5]

Dooge JCI. (1972) Mathematical models of hydrologic systems. In: Biswas AK (ed) *Modelling of Water Resources Systems 1*, pp. 170–80. Harvest House, Montreal. [4.2]

Eeles CWO, Robinson M, Ward RC. (1990) Experimental basins and environmental models. In: Hooghart JC *et al* (eds) *Proceedings of the International Conference on Hydrological Research Basins and the Environment, Wageningen*, pp. 3–12. Netherlands Organisation for Applied Scientific Research, The Hague. [4.3]

Gardiner JL. (1990) The River Thames Strategic Flood Defence Initiative Planning, a model influence. In: White WR (ed) *International Conference on River Flood Hydraulics*, pp. 447–58. John Wiley, Chichester. [4.5]

Gardiner JL, Dearsley AF, Woolnough JR. (1987) The appraisal of environmentally sensitive options for flood alleviation using mathematical modelling. *Journal of the Institution of Water and Environmental Management* 1(2): 171–84. [4.5]

Greenfield BJ. (1984) *The Thames Catchment Model.* Internal Report, Thames Water Authority, Reading, UK. [4.3]

Jakeman AJ, Littlewood IJ, Whitehead PG. (1990) Computation of the instantaneous unit hydrograph and identifiable component flows with application to two small upland catchments. *Journal of Hydrology* 117: 275–300. [4.3]

Jakeman AJ, Littlewood IJ, Symons HD. (1991) Features and applications of IHACRES: A PC program for identification of unit hydrograph and component flows from rainfall, evapotranspiration and streamflow data. In: *13th IMACS World Congress on Computation and Applied Mathematics, Dublin.* [4.3]

Laurenson EM, Mein RG. (1983) *RORB version 3. Runoff Routing Program User Manual.* Department of Civil Engineering, Monash University, Australia. [4.3]

Littlewood IG, Jakeman AJ. (1993) Characterisation of quick and slow streamflow components by unit hydrographs for single and multi-basin studies. In: Robinson M (ed) *Proceedings of a Conference of Methods of Hydrologic Basin Comparison, Oxford, 1992*, pp. 94–105. Institute of Hydrology, Wallingford, UK. [4.3]

Lowing MJ, Mein RG. (1981) Flood event modelling – a study of two methods. *Water Resources Bulletin* 17(4): 599–606. [4.3]

McDonald MG, Harbaugh AW. (1988) *A Modular Three-Dimensional Finite-Difference Groundwater Flow Model.* Techniques of Water Resources Investigations 06-A1, United States Geological Survey. [4.4]

Manley RE. (1978) Simulation of flows in ungauged basins. *Hydrological Sciences Bulletin* 23(1): 85–101. [4.3]

Manley RE. (1993) *HYSIM Information Pack.* RE Manley Consultancy, Cambridge, UK. [4.3]

Midgley P, Jones CRC. (1991) Rehabilitation of the Wallop Brook and Bourne Rivulet. In: *Proceedings of the Third National Hydrology Symposium*, pp. 2.37–2.42. British Hydrological Society/Institute of Hydrology, Wallingford, UK. [4.4]

Moore RJ, Jones DA, Black KB. (1989) Risk assessment and drought management in the Thames Basin. *Hydrological Sciences Journal* 34(6): 705–17. [4.3]

Mott MacDonald. (1993) *The Integrated Catchment Model. User's Manual.* Mott MacDonald, Cambridge, UK. [4.4]

NERC (Natural Environmental Research Council).

(1975) *Flood Studies Report*: 5 vols. NERC, London. [4.3]

NERC (Natural Environmental Research Council). (1989) *HYRROM. User's Manaual.* NERC, London. [4.3]

Newman AT, Symonds RD. (1991) The River Allen – A case study. In: *Proceedings of the Third National Hydrology Symposium*, pp. 2.53–2.58. British Hydrological Society/Institute of Hydrology, Wallingford, UK. [4.4]

Oakes DB, Keay D. (1990) *A Resource Model of the Great Ouse River System.* Report No. CO2504-M. Water Research Centre, Medmenham, UK. [4.3]

Quinn P, Beven K, Morris D, Moore R. (1989) The use of digital terrain data in modelling the response of hillslopes and headwaters. In: *Proceedings of the 2nd National Hydrology Symposium*, pp. 1.37–1.47. British Hydrological Society. [4.3]

Rushton KR. (1986) Groundwater models. In: *Groundwater Occurrence, Development and Protection.* Institution of Water Engineers and Scientists, London. [4.4]

Rushton KR, Fawthrop NP. (1991) Groundwater support of streamflows in the Cambridge area, UK. In: Nachtebel HP, Kovar K (eds) *Hydrological Basis of Ecologically Sound Management of Soil and Groundwater*, pp. 367–76. IAHS Publication No. 202, Institute of Hydrology, Wallingford, UK. [4.4]

Rushton KR, Redshaw SC. (1979) *Seepage and Groundwater Flow.* John Wiley, Chichester. [4.4]

Rylands WD, Lee JK. (1992) The possibility of an integrated approach to the provision of storage in the Thames Region of the National Rivers Authority. In: Saul AJ (ed) *Floods and Flood Management*, pp. 169–85. Kluwer. [4.3]

Shaw EM. (1983) *Hydrology in Practice.* Van Nostrand Reinhold (UK), Wokingham, England. [4.5]

Slade JE, Samuels PG. (1990) Modelling complex river networks. In: White WR (ed) *International Conference on River Flood Hydraulics*, pp. 351–8. John Wiley, Chichester. [4.5]

Spink AEF, Hughes AG. (1992) Development of a graphical user interface for use in groundwater management. In: Zannetti P (ed) *Proceedings of the 4th International Conference on Computer Techniques in Environmental Studies*, pp. 575–88. Elsevier Applied Science, Amsterdam. [4.3, 4.4]

Thompson N, Barrie IA, Ayles M. (1981) *The Meteorological Office Rainfall and Evaporation Calculation System: MORECS.* Meteorological Office, Bracknell, England. [4.3]

Van Kalken T, Havno K. (1992) Multipurpose mathematical model for flood management studies and real time control. In: Saul AJ (ed) *Floods and Flood Management*, pp. 169–85. Kluwer, Dordrecht. [4.5]

Wilby R, Greenfield B, Glenny C. (1993) *A Coupled Synoptic-Hydrological Model for Climate Change Impact Assessment.* NRA Research Paper, Loughborough University, UK. [4.3]

Williams JJR, Fawthrop NP. (1988) A mathematical hydraulic model of the River Nene – A canalised and heavily controlled river. *Regulated Rivers: Research and Management* **2**: 517–33. [4.5]

Wilson EM. (1983) *Engineering Hydrology.* Macmillan, London. [4.5]

5: Water Quality
I. Physical Characteristics

D.E.WALLING AND B.W.WEBB

A review of existing manuals and texts covering the field of water quality indicates that the *physical* characteristics of water flowing in rivers and streams have been variously defined to embrace a wide range of parameters extending from suspended sediment concentration, turbidity and the presence of foam, through colour, taste, odour and dissolved oxygen content, to electrical conductivity, temperature and radioactivity. Some blurring of the distinction between physical and chemical properties clearly exists. For example, electrical conductivity primarily reflects the total dissolved solids content, which is a chemical parameter. Likewise, the content of dissolved gases could be viewed as a chemical characteristic. In this section attention will be limited to four parameters that are physical in nature and that have important implications for river ecology and for water use. These are suspended sediment concentration, colour, temperature and dissolved oxygen content. Furthermore, although all of these aspects of physical water quality may be strongly influenced by river pollution, the present chapter focuses on uncontaminated watercourses. The influence of pollution on physical water quality is considered in more detail in Calow and Petts 1994.

5.1 SUSPENDED SEDIMENT

The presence of suspended sediment or solids in river water is an important physical characteristic. Such sediment can have both a direct effect on aquatic life through damage to organisms and their habitat (e.g. Ritchie 1972; Muncy *et al* 1979) and an indirect effect through its influence on turbidity and light penetration. Decreased light penetration reduces primary production and hinders the growth of benthic macrophytes (e.g. US Environmental Protection Agency 1979). Increased sediment concentrations will also necessitate increased treatment of domestic and industrial water supplies (e.g. Castorina 1980).

Definitions, sources and budgets

The suspended sediment load of a river represents the fine-grained material transported in suspension, with its weight supported by the upward component of fluid turbulence. It is conventionally separated from material in solution by filtration through a 0.45-μm filter and may therefore include some colloidal material. Particles are commonly less than 0.2 mm in diameter and in most rivers the suspended load will be dominated by clay- and silt-sized particles (i.e <0.062 mm in diameter). A distinction is often made between the *wash load*, which comprises the finer material washed in from the catchment slopes and which commonly remains in suspension as it is transported downstream, and the *suspended bed material* which represents the coarser, sand-sized particles, mobilized from within the channel. The latter particles frequently move in and out of suspension and interact with the bedload component of sediment transport, according to the local hydraulic conditions within the channel.

The suspended sediment load is dominated by material eroded from a variety of sources within the upstream basin, and the precise nature and location of these will exert an important influence on both the character and the behaviour of the load. Potential sources include channel and bank erosion, gully erosion, erosion of tracks and unmetalled roads, sheet and rill erosion on the catchment slopes and the removal of sediment delivered to the channel by mass movement. In agricultural areas, soil loss from cultivated fields often accounts for a major proportion of the suspended sediment load. Other sources of sediment of ecological significance include mining, quarrying, construction work and effluent discharges.

When attempting to link the suspended sediment load of a drainage basin to the erosion processes and sources operating within it, it is important to recognize that only a proportion (probably a small one) of the sediment mobilized by erosion will find its way to the basin outlet. Much will be deposited within the system. The *sediment delivery ratio*, which expresses the sediment yield at the basin outlet as a proportion of the gross erosion and sediment mobilization within the basin, provides a simple basis for representing this link, although the many uncertainties involved in estimating and in interpreting this parameter should be recognized (*cf.* Walling 1983). Existing work indicates that sediment delivery ratios decline as drainage basin size increases, in response to the increased opportunities for deposition, and values are frequently as low as 10% in medium-sized basins. This trend is shown clearly by the relationship presented in Fig. 5.1 (a), although the range of delivery ratio values evident for a given basin size emphasizes the importance of other controls.

The sediment delivery ratio provides only a simple lumped, black-box representation of the linkages interposed between the erosion processes operating within a drainage basin and the downstream suspended sediment yield. An alternative approach to elucidating and quantifying these linkages is the establishment of a *sediment budget* (*cf.* Fig. 5.1(b)). With this, an attempt is made to quantify the various sediment sources and sinks, in order to provide an improved understanding of the linkages involved. The sediment budget approach advocated by Dietrich and Dunne (1978) and developed by Lehre (1982) and Swanson *et al* (1982) is important in focusing attention on the processes governing the suspended sediment yield at the basin outlet, but it should be seen as an essentially conceptual approach, since it is difficult to assemble precise information on the rates and fluxes involved for anything but a relatively small drainage basin. Recent advances in the use of caesium-137 for investigating rates and patterns of erosion and deposition within a drainage basin (*cf.* Walling & Bradley 1988; Walling 1990) and the use of specific sediment properties to fingerprint major sources (*cf.* Wall and Wilding 1976; Peart & Walling 1986) would, however, appear to offer considerable potential in this field.

Most of the sediment sinks or stores depicted in the examples of sediment budgets presented in Fig. 5.1(b) can be viewed as essentially permanent in the context of the short-term operation of the sediment conveyance system. Remobilization may, however, occur in the longer term as slope, valley and floodplain deposits are reworked (*cf.* Trimble 1976, 1981). The potential for short-term channel storage to attenuate the transmission of sediment through the channel system must nevertheless be considered (*cf.* Duijsings 1985, 1986), since this could introduce a distinct phase difference between sediment supply and sediment output, with sediment supplied to the channel during one season appearing at the basin outlet several months later.

The nature of suspended sediment

Traditionally, investigations of the suspended sediment loads of streams have focused on collection of gross data concerning both concentrations and loads. Increasing awareness of the wider environmental significance of the fine sediment transported by streams has, however, emphasized the need for information on the physical and chemical properties of the sediment. In this context it is important to note the distinction which is sometimes made between *suspended solids* and *suspended sediment*, the former referring to the total material in suspension and the latter to the mineral sediment or inorganic fraction. The term suspended sediment is, however, widely used to refer to both the inorganic and the organic components, and this interpretation has been adopted here. The properties of suspended sediment reflect clearly those of the source material from which it was derived, but any attempt to link the two must take account of the effects of

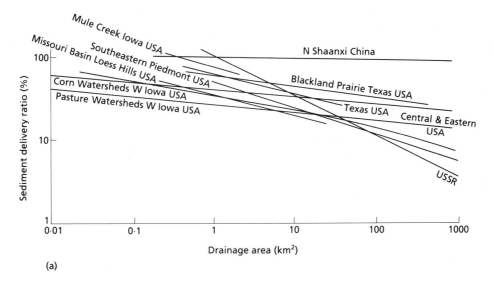

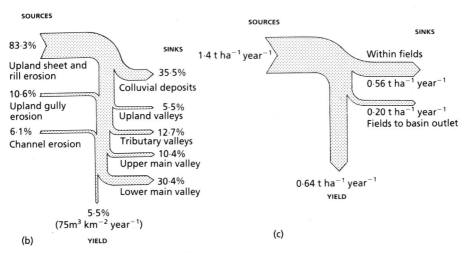

Fig. 5.1 (a) Relationships between sediment delivery ratio and drainage basin area proposed for various areas of the world (based on Walling 1983). Examples of sediment budgets established for (b) Coon Creek, Wisconsin, USA (360 km²) by Trimble (1981) and (c) for Jackmoor Brook, Devon, UK (9.8 km²) by Walling (1990).

selective erosion and deposition mechanisms operating within the sediment delivery system. Such mechanisms frequently increase the content of both fines and organic matter in the suspended sediment, relative to the source material.

Information on the organic matter content of suspended sediment is relatively sparse, but, as might be expected, existing data exhibit substantial spatial variation in response to the sources

involved and the nature of the source material itself. For example, Burton *et al* (1977) cited organic matter percentages of 30.6, 17.2 and 14.8, respectively, for sediment from forested-agricultural, suburban and urban drainage basins in northern Florida; Walling and Kane (1982) reported mean organic matter contents of between 8% and 14% for suspended sediment from four rivers in Devon, UK; whilst Skvortsov (1959) documented much lower values of between 1.5% and 3.1% for the Rion River draining to the Black Sea in the USSR.

Particle size analyses of suspended sediment are traditionally undertaken on the mineral fraction after chemical treatment to remove organic material. Available data demonstrate consider-able variation in the particle size characteristics of sediment from different rivers and streams in response to variations in source material and other physiographic controls (*cf.* Walling & Moorehead 1989). The potential range of particle size distributions associated with suspended sediment is illustrated in Fig. 5.2(a) which presents data for a

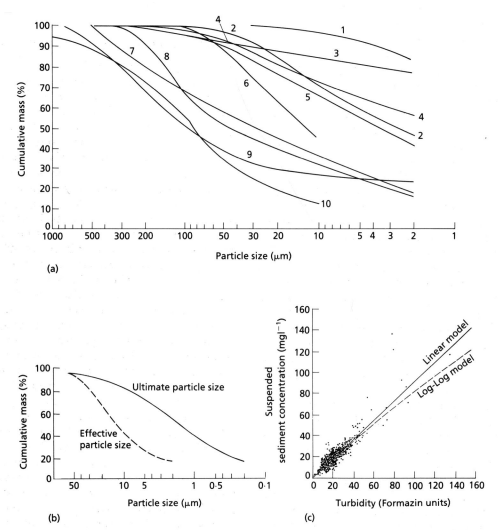

Fig. 5.2 (a) Examples of characteristic grain-size distributions of suspended sediment for various world rivers. (b) Comparison of typical ultimate and effective particle-size distributions of suspended sediment from the River Exe, Devon, UK. (c) Relationship between turbidity and suspended sediment concentration established for Geebing Creek, New South Wales, Australia. (Based (a) on Walling & Moorehead 1989; and (c) on Gippel 1989). 1 Barwon River, New South Wales, Australia; 2 Citarum River at Nanjung, Indonesia; 3 Sanaga River at Nachtigal, Cameroon; 4 River Nile at Cairo, Egypt (1925–6); 5 Amazon River at Obidos, Brazil; 6 Sao Francisco at S. Romao, Brazil; 7 Chulitna River near Talkeetna, Alaska; 8 Colorado River at Lees Ferry Arizona, USA; 9 Limpopo River, Zimbabwe; and 10 Huangfu River, China.

number of world rivers. Thus, for example, the situation in the Barwon River in Australia, where the majority of the suspended sediment is clay sized and <0.002 mm in diameter, may be contrasted with the Middle Yellow River in China where up to 60% of the suspended load may be composed of sand-sized particles (>0.062 mm) and clay-sized material represents less than 10% of the total load. Particle size in turn exerts an important influence on the basic mineralogy and geochemistry of fine sediment. The <0.002-mm fraction will, for example, be composed primarily of secondary silicate minerals, whereas quartz will dominate in the larger fractions. The specific surface area of sediment, which is a major control on its surface chemistry, also increases markedly with decreasing particle size, such that typical values for clay (20–800 m^2 g^{-1}) are several orders of magnitude greater than those for silt and sand.

Traditional laboratory determinations of particle size, involving chemical dispersion of the mineral fraction, may be unrepresentative of the *in situ* particle-size characteristics of suspended sediment transported by a river. This is likely to include a substantial proportion of aggregates comprising smaller particles (cf. Droppo & Ongley 1989). The processes responsible for producing such aggregates are poorly understood but are likely to involve a variety of mechanisms including electrochemical attraction and bonding associated with organic substances and bacteria. Figure 5.2(b) provides an example from a river in Devon, UK of the potential contrast between the *ultimate* particle-size distribution provided by traditional laboratory analysis and the *effective* particle-size distribution of *in situ* sediment. In this case the median particle sizes of the two distributions differ by almost an order of magnitude. Measurement of the effective size distribution inevitably involves significant problems and uncertainties, and in this example the measurements were made in the field using a sedimentation tube, immediately after collecting the sample.

Although the measured turbidity of river water commonly exhibits a close positive relationship with suspended sediment concentration (e.g. Fig. 5.2(c)), the precise form of the relationship will vary both spatially and temporally in response to variations in sediment composition and water colour. Particle size, shape and composition can all be expected to influence light attenuation and turbidity, and any attempt to use turbidity measurements as a surrogate for direct determinations of suspended sediment concentration should take careful account of such factors (cf. Gippel 1989).

Spatial variation of suspended sediment transport

In many applications it is the concentrations of suspended sediment occurring in a river (i.e. mg l^{-1}) rather than the specific sediment yield (i.e. t km^{-2} $year^{-1}$) that is of prime interest. However, the two are closely linked, since rivers with high suspended sediment yields are likely to exhibit high sediment concentrations. The average concentration in a river is by definition the mean annual sediment yield divided by the mean annual runoff volume. In Britain, for example, where suspended sediment yields are typically in the range 50–100 t km^{-2} $year^{-1}$ (cf. Walling 1990), maximum concentrations rarely exceed 5000 mg l^{-1} and average concentrations are of the order of 50–100 mg l^{-1}, whereas in the USA where sediment yields can exceed 1000 t km^{-2} $year^{-1}$, average concentrations are frequently in excess of 2000 mg l^{-1}. Most existing work on the spatial variability of suspended sediment transport has, however, been undertaken on annual suspended sediment yields. At the global scale, the sediment yields of small river basins (i.e. about 100 km^2) are known to range from less than 2 t km^{-2} $year^{-1}$ to in excess of 10 000 t km^{-2} $year^{-1}$ (Walling 1987), and the generalized global pattern described by Walling and Webb (1983) reflects control by a number of factors, including climate, rock type, relief, tectonic activity and land use. Similar factors control the pattern of average suspended sediment concentrations in US rivers mapped by Meade and Parker (1985) and presented in Fig. 5.3(a). Here the effect of climate is clearly evident, with average sediment concentrations showing a general increase across the country, in response to increasing aridity. At the more local scale, factors such as geology, soil type and land use assume increasing importance in controlling spatial patterns, and Fig. 5.3(b), based on the

Chapter 5

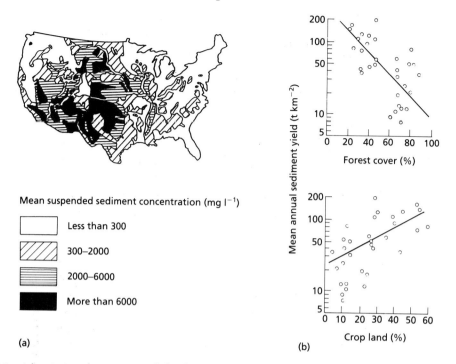

(a)

(b)

Fig. 5.3 (a) Spatial variation of mean suspended sediment concentrations within the conterminous USA (after Meade & Parker 1985). (b) Relationships between mean annual suspended sediment yield and land use established for catchments within the Potomac River basin by Wark and Keller (1963).

classic work of Wark and Keller (1963) in the basin of the Potomac River in eastern USA, highlights the role of land use in influencing soil loss and sediment yield.

Temporal variation in suspended sediment transport

General characteristics

Summary statistics such as annual suspended sediment yields and average concentrations can fail to convey a sense of the extreme temporal variability of suspended sediment transport in most rivers. For much of the time rivers may transport very little suspended sediment and the water will be essentially clear. A large proportion of the suspended sediment transport occurs during storm events when rainfall and storm runoff mobilize sediment from the upstream watershed and channel network. Figure 5.4(a), based on a 7.75-year programme of detailed monitoring undertaken by the authors in the 262 km² drainage basin of the River Creedy in Devon, UK, characterizes this variability by means of concentration–duration and load–duration curves. These indicate that most of the total sediment load is transported during only 5% of the time

Fig. 5.4 (*Opposite*) Characteristics of suspended sediment transport in the River Creedy, Devon, UK, 1972–80. (a) Cumulative frequency curves of suspended sediment concentration and suspended sediment yield. (b) The effectiveness of different flow classes for transporting suspended sediment. (c) A typical record of the variation of suspended sediment concentration during a sequence of storm runoff events. (d) A scatterplot for the relationship between suspended sediment concentration and discharge subdivided according to season and stage condition. (Winter: ● rising stage; ○ falling stage. Summer: ▲ rising stage; △ falling stage.) (e) Straight-line logarithmic relationships fitted to various subsets of the data plotted in (d); (1 All data; 2 winter; 3 summer; 4 rising stage; 5 falling stage; 6 winter rising; 7 winter falling; 8 summer rising; and 9 summer falling.)

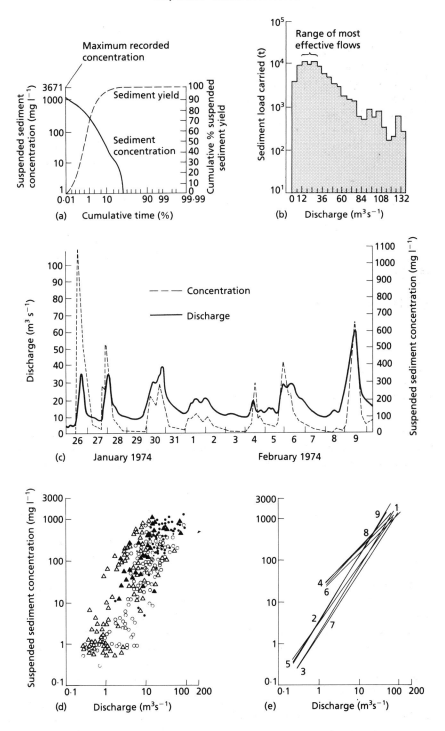

(a) Cumulative time (%)

(b) Discharge (m³s⁻¹)

(c) January 1974   February 1974

(d) Discharge (m³s⁻¹)

(e) Discharge (m³s⁻¹)

when sediment concentrations exceed about 100 mg l^{-1}. Equally, if suspended sediment concentrations below 20 mg l^{-1} are considered to be minimal, the concentration–duration curve indicates that significant concentrations occur only for 10% of the time. Most rivers can be expected to exhibit similar behaviour.

Maximum suspended sediment concentrations and loads are commonly transported during flood events and geomorphologists have frequently attempted to assess the relative importance of high-magnitude, low-frequency flood events and smaller more frequent events to the long-term sediment load. Lack of long-term records and the difficulties of measuring sediment transport during rare high-magnitude events severely hampers such analysis. The authors have, however, analysed the 7.75-year record of suspended sediment transport for the River Creedy in Devon, UK, referred to above, in an attempt to evaluate the relative importance of events of varying frequency (*cf.* Webb & Walling 1982). Figure 5.4(b) indicates the total load carried during this 7.75-year period by 23 equal discharge classes spanning the total flow range. Three flow classes between $12 \text{ m}^3 \text{ s}^{-1}$ and $30 \text{ m}^3 \text{ s}^{-1}$ each have loads in excess of $10\,000$ tonnes and represent the most effective discharge range for the transport of suspended solids. It appears that the more extreme flows experienced by this basin are less effective in removing suspended sediment. This conclusion accords with that advanced by Wolman and Miller (1960) on the basis of a study of the suspended sediment loads of several rivers in the USA, but the true significance of extreme catastrophic events, for which records are rarely available, still remains somewhat uncertain. For example, Meade and Parker (1985) referred to the impact of an extreme storm which struck northwestern California in December 1964. In 3 days the Eel River, which drains an 8063-km² drainage basin, transported more sediment than it had carried in the previous 7 years, and in 10 days it transported a total sediment load equivalent to that for 10 average years.

Short-term variations

In considering the major factors governing short-term temporal variations in suspended sediment concentrations, most workers have emphasized the dominant role of water discharge or flow magnitude. High suspended sediment concentrations are in most cases associated with periods of high discharge, as illustrated by Fig. 5.4(c), but closer inspection of this example indicates that suspended sediment concentration is not a simple function of water discharge and that other factors such as the time elapsed since a previous storm event are also important. Simple positive straight-line logarithmic relationships between suspended sediment concentration (C) and water discharge (Q) have nevertheless been established for many rivers (*cf.* Fig. 5.4(d)). These relationships can be described by the equation $C = aQ^b$, with the exponent b typically falling in the range $1-2$. The existence of a relationship between C and Q should not, however, be interpreted as a simple transport function, whereby increases in water discharge are associated with increased shear velocities and turbulence and therefore increased capacity to erode and transport sediment. In most cases the amount of suspended sediment carried by a river is several orders of magnitude less than its maximum transport capacity and the dominant control on suspended sediment concentration is the *supply* of material to the river. The existence of a relationship between concentration and discharge is rather a reflection of the fact that sediment supply increases during periods of storm rainfall and storm runoff and these periods are generally characterized by high discharges. Suspended sediment loads must therefore be treated as *non-capacity loads* and the logarithmic plot of concentration versus discharge will typically exhibit very considerable scatter (*cf.* Fig. 5.4(d)). Concentrations associated with a given level of discharge will frequently range over several orders of magnitude in response to variations in sediment supply.

There have been many attempts to account for the scatter apparent in sediment concentration/discharge relationships such as that portrayed in Fig. 5.4(d). Seasonal differentiation of the plot is often apparent and a distinction can also frequently be made between samples collected during periods of rising and falling stage (Fig. 5.4(e)). Hysteresis has also been widely identified in relationships between sediment concentration and discharge established for individual events, and may reflect both phase differences between the water discharge and sediment concentration response or a tendency for concentrations to be higher on the rising stage than on the falling stage or *vice versa*. Williams (1989) has attempted to classify the form of such hysteretic relationships into five types which are shown schematically in Fig. 5.5(a).

Longer-term trends

Erosion processes and sediment yields are particularly sensitive to changes in land use, and long-term sediment records will frequently exhibit marked fluctuations in response to land clearance and intensification of land use. Table 5.1 provides several examples of the magnitude of the impact of land-use change on sediment yields from small

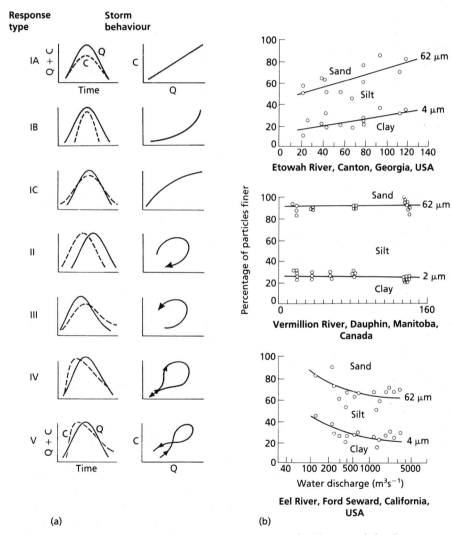

Fig. 5.5 (a) Illustrates the five characteristic response types associated with suspended sediment concentration/discharge relationships proposed by Williams (1989). (b) Presents contrasting examples of the response of the particle-size composition of suspended sediment to changing discharge (after Walling & Moorehead 1989).

Table 5.1 Results from experimental basin studies of the impact of land-use change on sediment yield

Region	Land-use change	Increase in sediment yield	Reference
Westland, New Zealand	Clearfelling	×8	O'Loughlin et al (1980)
Oregon, USA	Clearfelling	×39	Fredriksen (1970)
Texas, USA	Forest clearance and cultivation	×310	Chang et al (1982)
Maryland, USA	Building construction	×126–375	Wolman & Schick (1967)

drainage basins. These examples relate to increases, but other land-use changes, particularly the implementation of conservation measures, can lead to reduced sediment yields (*cf*. Lal 1982).

Temporal variation in sediment properties

Suspended sediment properties can also be expected to vary temporally in response to variations in sediment sources and the operation of erosion and sediment delivery processes. Such variations in properties are likely to be particularly marked in streams experiencing floods generated by both spring melt and summer rainstorms, since different sediment sources and erosion processes are likely to be associated with these events. Considering more short-term variations in response to fluctuations in discharge, the organic matter content of suspended sediment commonly exhibits an inverse relationship with discharge (e.g. Walling & Kane 1982). The behaviour of particle-size composition is, however, more complex and Fig. 5.5(b)) illustrates situations where particle size increases, decreases and remains essentially constant with increasing discharge. The tendency for suspended sediment to become coarser as flow increases may be accounted for in terms of increasing transport capacity and shear stress within the channel, whereas the reverse case is generally explained in terms of increased supply of fine sediment eroded from the slopes of the watershed and reflecting the interaction of runoff dynamics and sediment sources.

5.2 NATURAL COLOUR

Measurements of water colour conventionally distinguish between 'true colour' which represents that due to dissolved matter, and 'apparent colour' which also includes the effects of suspended solids. Measurements of true colour are made on water samples filtered through a 0.45-μm membrane filter and are the most widely cited (*cf*. Department of the Environment 1981). Colour can be seen as an important physical property of water primarily because of its implications for water supply and the need in some areas to reduce it to acceptable levels by water treatment. Recent increases in the colour of water in reservoirs in the Pennines, UK have, for example, resulted in increases in treatment costs by up to 10 times (*cf*. Naden & McDonald 1987). Colour also has a more indirect significance because of the complexing and adsorptive behaviour of dissolved organic colour with heavy metals (*cf*. Reuter & Perdue 1977; Malcolm 1985), which frequently results in a positive relationship between colour and the concentration of toxic metals such as mercury.

Causes

Colour in natural water usually results from the leaching of organic materials and is primarily a result of dissolved and colloidal humic substances, principally humic and fulvic acids. Highly coloured water occurs in many different environments where decaying vegetation is plentiful, as in swamps and bogs, and is frequently associated with acid conditions. In the UK, for example, it is primarily associated with upland moorland areas and there is evidence that colour increases in areas where the peat cover is being actively eroded or disturbed by burning or deep ploughing and ditching for afforestation.

Temporal variation

Little information currently exists on the detailed temporal behaviour of colour in streams, but recent concern for the increasing discoloration of water obtained from reservoirs fed by runoff from the peat moorlands of Yorkshire has focused attention on the need for detailed investigations of this parameter (*cf.* Edwards *et al* 1987). Available records indicate a pattern of annual variation, with highest levels of colour occurring during the autumn 'flush', when the peat is wetted, and accumulated soluble organic material is leached into the streams. The supply of readily soluble coloured organics may be increased after periods of extreme drought which promote aeration of the peat and its aerobic decomposition. The onset of enhanced colour levels usually occurs in October, although in wet summers it may begin as early as July, and it generally lasts for about 3 months. The decline in colour during the winter may reflect both the exhaustion of available supply or the influence of temperature or snowfall in limiting microbial activity. An analysis of monthly colour data for several catchments in Nidderdale reported by Naden and McDonald (1987) also demonstrated a positive relationship with monthly rainfall (Fig. 5.6(a)), indicating that maximum levels of colour are associated with periods of increased leaching and runoff. The same authors also reported a significant relationship with an antecedent moisture index representing the moisture deficit 3 months earlier (Fig. 5.6(b)). In this case samples collected following dry periods were seen to exhibit substantially higher levels of colour than those collected after wet periods.

In the longer term, evidence collected from the peat moorlands of Yorkshire also indicates that colour levels may demonstrate significant upward shifts through time. Figure 5.6(c), based on the work of McDonald *et al* (1989) presents monthly mean values of colour for water from Thornton Moor reservoir during the period 1968–88. The influence of drought conditions is clearly seen, since the severe droughts of 1976 and 1984 are both followed by significant upward shifts in colour. This is probably a result of the lowering of the water-table and the desiccation of the peat, which promotes aerobic decomposition and increases the availability of readily leached organic substances responsible for the discoloration once the moisture content has been replenished. Elsewhere, land-use changes, particularly moorland drainage, have resulted in more highly coloured runoff and there have also been suggestions that changes in precipitation chemistry, particularly increased acidity, have also been responsible for increased coloration of moorland streams.

5.3 WATER TEMPERATURE

Temperature represents one of the most important physical characteristics of river water. It affects other physical properties of rivers, such as dissolved oxygen and suspended solids content, and it influences the chemical and biochemical reactions which take place in lotic systems. The evolution, distribution and ecology of aquatic organisms is also fundamentally affected by river temperature (Rose 1967), and an enormous volume of research has been undertaken to investigate the thermobiology of fishes and invertebrates (e.g. Brett 1960; Fry 1964; Langford 1972; Ward & Stanford 1982; Crisp 1988a).

Temperature is a significant consideration in the domestic utilization of water since it has a bearing on the toxicity of contaminants, the efficacy of water treatment and the presence of tastes and odours (Everts 1963). Furthermore, temperature characteristics can determine the quality of water supplies for industrial and agricultural purposes, and are of particular economic importance in the provision of cooling water for the electricity generation industry (Langford 1983) and other industrial processes, and in the formation of ice in navigable waterways (Smith 1972). Human activities in utilizing water can also radically alter the thermal behaviour of a water-

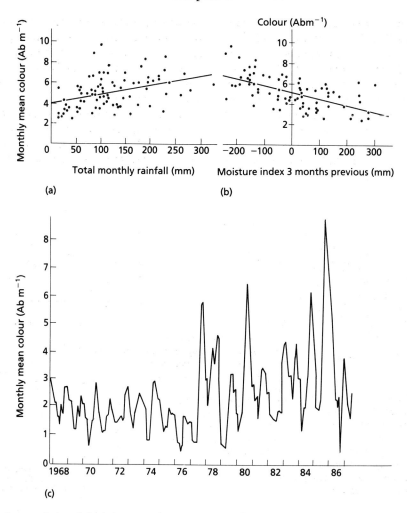

Fig. 5.6 Some characteristics of the behaviour of stream-water colour in upland catchments in Yorkshire, UK. The relationships between monthly mean colour and monthly rainfall (a) and catchment moisture status (b) are based on Naden and McDonald (1987), and the long-term trend in the colour of water from Thornton Moor Reservoir (c) is after McDonald *et al* (1989).

course, and direct impacts due to discharge of heated effluents (e.g. Parker & Krenkel 1969) and indirect modification associated with impoundment (e.g. Petts 1984) and land-use changes (e.g. Holtby 1988) have been extensively documented.

Temperature is a very easily and widely measured parameter of water quality, and further details of the instruments available, field applications and procedures, and problems of data collection and processing can be found in standard reference works (e.g. Stevens *et al* 1975).

Concepts

Energy budget

The temperature of a river is fundamentally determined by the transfers of heat energy which affect the water-course (Pluhowski 1970). Inputs of heat energy to a particular river reach arise as short-wave solar radiation and long-wave atmospheric and forest radiation, from condensation and precipitation, and through advection of heat from groundwater, from upstream and from tributary inflows (Fig. 5.7(a)). Heat energy is lost from the water-body through reflection of solar, atmospheric and forest radiation, by back radiation from the water surface itself, in the process of evaporation, and as the heat content of streamflow leaving a reach. Some terms in the energy budget, including convection and conduction between the water-body and its environment, can represent gains or losses depending on the thermal status of the channel margin, the water and the overlying air. The net heat exchange affecting a river, together with the volume of water, will determine the direction and extent of temperature change, although detailed studies (e.g. Brown 1969; Comer *et al* 1976; Troxler & Thackston 1977) indicate that the energy budget components can be highly variable in time, for example over a daily period, and in space, for example between forested and open watersheds (Fig. 5.7(b)). In

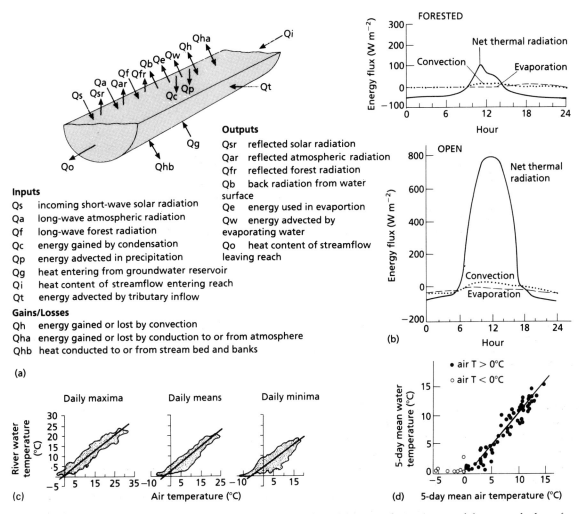

Inputs

Qs   incoming short-wave solar radiation
Qa   long-wave atmospheric radiation
Qf   long-wave forest radiation
Qc   energy gained by condensation
Qp   energy advected in precipitation
Qg   heat entering from groundwater reservoir
Qi   heat content of streamflow entering reach
Qt   energy advected by tributary inflow

Gains/Losses

Qh   energy gained or lost by convection
Qha   energy gained or lost by conduction to or from atmosphere
Qhb   heat conducted to or from stream bed and banks

Outputs

Qsr   reflected solar radiation
Qar   reflected atmospheric radiation
Qfr   reflected forest radiation
Qb   back radiation from water surface
Qe   energy used in evaportion
Qw   energy advected by evaporating water
Qo   heat content of streamflow leaving reach

Fig. 5.7 Energy budgets and air–water temperature relationships. (a) Principal components of the energy budget of a water-course. (b) Temporal variation in heat budget components of Deer Creek, USA (forested) and Berry Creek, USA (open catchment) (after Brown 1969). (c) Water–air temperature relationships in the River Clyst catchment, Devon, UK and (d) in a Pennine stream, Mattergill Sike, UK (after Crisp & Howson 1982).

general, however, evaporative, convective and conductive fluxes are found to be less important than the solar, atmospheric and back radiation terms in the energy budget (Brown 1969).

Air/water temperature relationship

Fluctuations in meteorological conditions together with the relatively high heat capacity of water, which damps the response of river temperature to changing energy gains and losses, ensure that the water surface is constantly striving to attain an equilibrium temperature. The latter is defined as the temperature at which there is no net energy exchange with the atmosphere (Edinger *et al* 1968; Dingman 1972). Air temperature is commonly taken as an approximation of the equilibrium temperature and is frequently used as a surrogate for the complexities of the energy budget components in predicting river water temperatures (Smith 1975; Walker & Lawson 1977). Strong linear relationships can often be established between air and river water temperatures (Smith 1981), although analysis of 2192 observations of daily maxima, means and minima in the River Clyst in Devon, UK (Webb 1987) has indicated that correlations are weakest for daily minimum values (Fig. 5.7(c)). Furthermore, departure from a linear relationship between air and water temperatures can arise in streams subject to winter freezing. This phenomenon has been demonstrated from an analysis of 5-day means of air and water temperatures during 1977–8 for Mattergill Sike in the northern Pennines, UK (Fig. 5.7(d)), and has been ascribed to latent heat effects associated with ice formation (Crisp & Howson 1982). Several studies have demonstrated the similarities of air and water temperature behaviour (e.g. Johnson 1971; Calandro 1973), although the latter often fluctuates over a much narrower range and tends to lag in its response behind the former (Smith 1968; Fowles 1975).

Temporal behaviour

The nature of water temperature behaviour varies significantly according to the temporal perspec-

tive under consideration, and river temperatures tend to exhibit more dramatic fluctuations over short rather than long time periods.

Longer-term trends

Although water temperature data have been collected in some Austrian and Russian rivers for more than 100 years, the availability of long-term and high-quality data series, on which to base an analysis of trends in river water temperatures, is generally limited. An increase in the temperature of the Mississippi River over a period of 40 years, however, has been inferred by a comparison of annual records collected in 1923 (Collins 1925) with those taken in 1962 (Blakey 1966). Mean annual water temperature was 1.6°C higher in the latter year despite the fact that mean air temperature was 2.2°C lower. The increase was thought to reflect the growing impact of human activity over the time period, although it was also recognized that these differences might be partly ascribed to the climatic vagaries of the individual years involved and to contrasts in measurement techniques (Blakey 1966). Smaller rivers, draining rural environments, may not evidence significant longer-term changes in temperature characteristics. Continuous monitoring at three stations within the Exe Basin, Devon, UK over a decade (1974–83) revealed the coefficient of variation for annual mean water temperatures to be in the range of 3–5% but did not indicate any significant upward or downward trend over the period (Webb & Walling 1985a).

Collection of records from a substantial run of years is necessary to define properly extreme temperature conditions and to establish adequately duration characteristics and other temperature statistics of ecological importance, such as accumulated degree hours above threshold temperatures (e.g. Macan 1958; Crisp & Le Cren 1970). An example of a temperature duration curve is shown in Fig. 5.8(a) for the River Exe at Thorverton, Devon, UK, which drains a rural catchment of 601 km^2. It is evident from this curve, which is based on 10 years of continuous data, that low water temperatures (<1°C) occur very infrequently (<0.5% of the time) in this

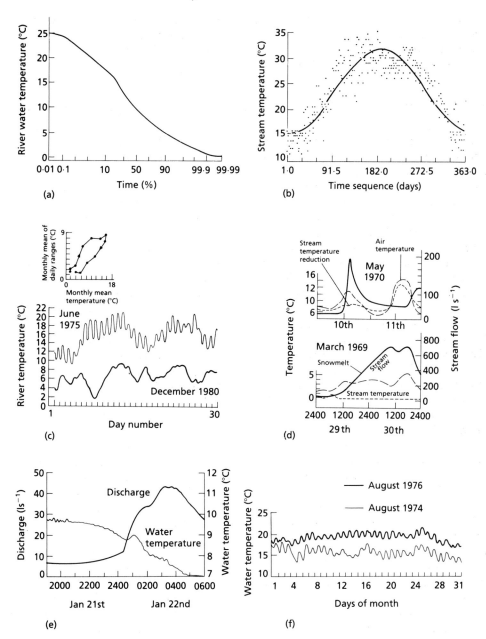

Fig. 5.8 Temporal fluctuations in water temperature. (a) A temperature duration curve for a Devon river; River Exe, Thorverton, UK (1974–83). (b) A multiyear harmonic to define annual temperature regime for Soan River, Dhok Pathan, Pakistan (after Steele 1982). (c) Daily and diurnal fluctuations in temperature of a Devon river; River Barle, Brushford, UK (inset depicts annual hysteresis in monthly average daily range in water temperature of a Pennine stream (Great Egglesthorpe Beck, UK, 1984), (after Crisp 1988b). (d) Effects of rain- and snowmelt-generated floods on water temperature in Upper Weardale, UK (after Smith & Lavis 1975). (e) Water temperature fluctuations in response to changing sources of storm runoff in a Georgia headwater stream; Lower Gage, Panola Mountain, USA (after Shanley & Peters 1988). (f) The impact of the 1976 drought on August water temperatures in the River Exe, Thorverton, UK (after Walling & Carter 1980).

temperate system, although a considerably greater occurrence of freezing conditions would be expected in smaller streams at higher altitudes and latitudes.

Annual regime

Most rivers exhibit a marked seasonal variation in water temperature, although the amplitude of the annual regime tends to be less in water-courses which receive a large groundwater flow component (Hopkins 1971; Crisp *et al* 1982). A simple harmonic curve of the form:

$$T(x) = A \left(\sin(bx + C)\right) + M \qquad (1)$$

where:

$T(x)$ = water temperature on day x of year (°C)
   A = amplitude of the harmonic (°C)
   b = 0.0172 radians per day ($2\pi/365$ days)
   x = number of days since beginning of year
   C = phase angle of harmonic (radians)
   M = mean of the harmonic (°C)

has often been used to characterize the seasonal cycle of river temperature variation (e.g. Collings 1969; Johnson 1971; Mosley 1982). This approach has been employed to define the seasonal regime in individual years (Ward 1963) and to derive a general annual cycle from observations made intermittently over a series of years (Steele *et al* 1974). An example of a multiyear harmonic function fitted to data collected between 1964 and 1972 for the Soan River at Dhok Pathan, Pakistan (Steele 1982) is given in Fig. 5.8(b). In some cases where the annual temperature regime is interrupted by a period of freezing (MacKichan 1967) or where the seasonal cycle is asymmetrical in nature (Moore 1967), it may be more appropriate to fit a higher order harmonic in order to characterize the annual march in temperature.

Daily and diurnal fluctuations

Superimposed upon the annual temperature regime of rivers are variations which take place from day to day and within a 24-hour period. In winter months, water temperature often fluctuates in response to changes in weather conditions, such as the passage of fronts, whereas in the summer period a clear diurnal cycle is generally

evident. This contrast in behaviour is apparent in continuous records of water temperature collected from the River Barle, Devon, UK (Webb & Walling 1985b) during June 1975 and December 1980 (Fig. 5.8(c)). A sine curve can again be employed to characterize the diurnal cycle of water temperature when it follows a regular and symmetrical pattern (Smith 1981). The magnitude of the diel range may exhibit hysteresis in the pattern of its variation throughout the year. Studies in the Great Egglesthorpe Beck and other streams of northern England (Crisp 1988b), for example, have shown that the monthly mean of daily range is higher relative to monthly mean temperature during periods of rising water temperatures in spring and early summer than during periods of falling water temperatures in late summer and autumn (Fig. 5.8(c)).

Storm and other events

Water temperatures also respond to changing flow conditions in the river channel. Investigations in a small Pennine stream in northern England (Smith & Lavis 1975) have shown that temperatures can be significantly reduced in summer months by rainfall which increases river discharge and in turn increases the thermal capacity of the watercourse. In the winter period, more prolonged depression of temperature is produced by hydrographs which are generated by snowmelt (Fig. 5.8(d)). Detailed monitoring of storm events at Panola Mountain, Georgia, USA (Shanley & Peters 1988) also indicates that stream temperatures may exhibit complex patterns during the passage of a hydrograph (Fig. 5.8(e)) which reflect the balance of surface and subsurface contributions to flow and the magnitude of difference in temperature between these two sources.

Extreme meteorological conditions also affect water temperature. The occurrence of prolonged drought with attendant high air temperatures and low volumes of flow can encourage unusually high water-temperature maxima and increase the range of diurnal fluctuation, especially in smaller rock-floored channels. In larger rivers, however, diel variation in water temperature may be reduced under drought conditions as a result of extended residence times and slow downstream transfer of channel flow (Walling & Carter 1980). This effect is evident when records from the River Exe at Thorverton, Devon, UK for August of the 1976 drought year are compared with those for August 1974 which was not affected by a severe drought (Fig. 5.8(f)).

Anomalously low water temperatures can also be generated in small streams through the chilling effects of heavy snowfall and wind-driven ice floes (Pluhowski 1972).

Spatial variation and controls

The thermal behaviour of rivers also varies considerably in space as well as in time. A variety of controlling factors determine the extent of variation and these controls, in turn, differ according to the spatial scale under consideration.

Macro-scale

Significant contrasts in the temperature characteristics of rivers are evident when differences over large geographical areas, encompassing global, continental, national and regional variations, are considered. Ward (1985) has suggested that water-courses in the southern hemisphere might exhibit more extreme temperatures, a lower incidence of freezing conditions and more unpredictable thermal behaviour than those in the northern hemisphere in response to the greater incidence of intermittent streams, to the occurrence of higher silt loads and more variable flows, and to the different distribution of land masses south of the Equator. Within the southern hemisphere (Fig. 5.9(a)), equatorial rivers are generally characterized by a restricted annual range in temperature, tropical rivers by very high maximum temperatures and intermediate annual ranges, and temperate rivers by large annual variability and in a few cases the occurrence of freezing (Ward 1985).

Macro-scale variations in water temperature behaviour are strongly controlled by latitude, altitude and continentality. These factors exert their influence by governing climatic conditions and especially air temperature characteristics. The effects of latitude on average annual water temperature are well seen at a continental scale

in the case of the conterminous USA (Blakey 1966), at a national scale for New Zealand (Mosley 1982), and at a regional scale for the eastern coast of the USA (Steele 1983), where the mean and amplitude of the annual water temperature harmonic has a strong correlation with latitudinal position (Fig. 5.9(b)). An example of a significant relationship between water temperature and altitude on a countrywide scale is shown for Pakistani rivers (Steele 1982) in Fig. 5.9(c), although in some regions a clearcut relationship between these variables is obscured by other factors. A thermal profile of Bolivian rivers (Wasson *et al* 1989), for example, shows that annual average and maximum temperatures vary considerably at given altitudes between the western and eastern slopes of the Cordillera Oriental (Fig. 5.9(d)).

Meso-scale

Temperature behaviour also varies between streams in a local area and within the tributaries and mainstream of individual river systems. Mean annual temperature can be related to altitude for a single catchment (Walker & Lawson 1977), although data from the Exe Basin, Devon, UK indicate that significant positive and negative residuals from this relationship can occur (Fig. 5.9(e)). The former are related to mainstream sites in large open valleys, whereas the latter are caused by vegetational shading and the influence of groundwater (Webb & Walling 1986). The moderating effects on thermal regime of subsurface flows are particularly apparent in catchments underlain by chalk lithologies (Crisp *et al* 1982). The temperature behaviour of a particular river catchment is often a function of several controls rather than a single factor. For example, the diurnal range of river temperature in the Nagara River, Japan has been related to drainage area, stream slope, rice-field area, mean altitude and relief, basin shape and catchment orientation (Miyazawa *et al* 1982).

Within an individual river system, a contrast can often be drawn between the thermal behaviour of the mainstream river and that of tributary streams (Boon & Shires 1976; Smith 1979). The latter tend to exhibit greater annual and diurnal

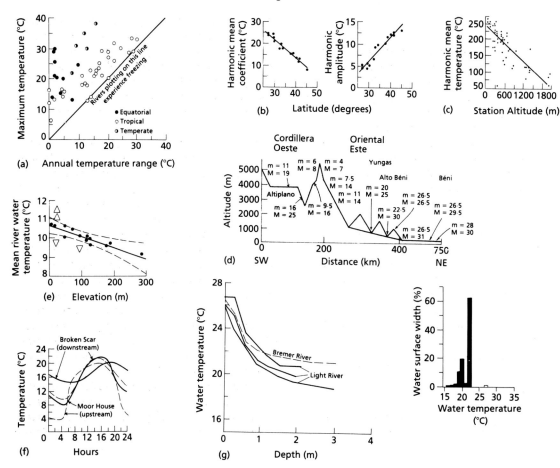

Fig. 5.9 Spatial variability in water temperature. (a) The relationship between maximum temperature and annual temperature range for water-courses in the southern hemisphere (after Ward 1985). (b) The relationship of harmonic mean temperature and amplitude of the temperature harmonic to latitude for streams of the East Coast, USA (after Steele 1983). (c) The relationship of harmonic mean temperature and altitude in Pakistani rivers (after Steele 1982). (d) Thermal profile of Bolivian rivers (after Wasson *et al* 1989); m = annual average temperature (°C), M = maximum temperature (°C). (e) The relationship of mean annual water temperature and elevation in the Exe Basin, UK. (f) Contrasts in diurnal temperature fluctuations at upstream and downstream locations in the River Tees, UK (after Smith 1979) (mean hourly variation 7–13 June 1969. —— River; – – – air). (g) Variation of temperature and depth in Southern Australian water-courses (after Morrissy 1971) (inset depicts lateral variability in water temperature in the Ashley River, New Zealand 16.12.82) (after Mosley 1983).

variation because their smaller volume of flow and lower thermal capacity make them more responsive to heat exchange processes. An increase in flow volume from river source to mouth also promotes systematic downstream trends in water temperature characteristics (e.g. Schmitz 1954; Smith 1972). In particular, diurnal fluctuations in water temperatures become less pronounced and lag increasingly behind air temperature fluctuations in the downstream direction, as results from the River Tees in northern England (Smith 1979) clearly demonstrate (Fig. 5.9(f)).

Micro-scale

Temperature characteristics may also vary along individual river reaches in response to factors such as bank-side vegetation (Weatherley & Ormerod 1990), groundwater seepage (Smith & Lavis 1975), channel depth (Sioli 1964), shape (Macan 1958) and orientation (Brown *et al* 1971), substrate conditions (Geijskes 1942) and silt content of the water (Reid & Wood 1976).

Within individual river reaches, there may be significant vertical and lateral variations in water temperature. Studies of pools in several rivers of South Australia (Fig. 5.9(g)) reveal an average temperature decrease with increasing depth of 1.77°C m^{-1} (Morrissy 1971), whereas an investigation of temperature in the distributaries of the braided Ashley River, New Zealand has revealed considerable lateral contrasts (Fig. 5.9(g)) between the main channel and smaller side channels kept cooler by seepage of groundwater (Mosley 1983).

5.4 DISSOLVED OXYGEN

Controls

Several gases, including nitrogen and carbon dioxide, may be dissolved in river water but of these oxygen is arguably the most significant because of its vital importance to aquatic organisms. The dissolved oxygen content of a watercourse is subject to physical, chemical and biological controls, and at any one time the oxygen concentration reflects a balance between the various sources and sinks for this gas in a stream system.

The ultimate source of oxygen in water is the atmosphere, although the solubility of oxygen in river water is a function of temperature and pressure conditions (Hem 1970). At a temperature of 0°C and normal atmospheric pressure at sea level (760 mmHg), the solubility of oxygen is 14.6 mg l^{-1}, assuming equilibrium is achieved between the water-body and overlying air, but this value falls to 7.63 mg l^{-1} at a temperature of 30°C (Fig. 5.10(a)). Reduction in the atmospheric partial pressure of oxygen with increasing altitude above sea level will also reduce the equilibrium concentration of O_2, as the relationships between the equilibrium solubility of oxygen and water temperature for sea level and 1000 m above sea level clearly demonstrate (Fig. 5.10(a)).

The attainment of oxygen equilibrium or saturation conditions in a river is often strongly mediated by biological processes within aquatic ecosystems. Autotrophic plants, including phyto-

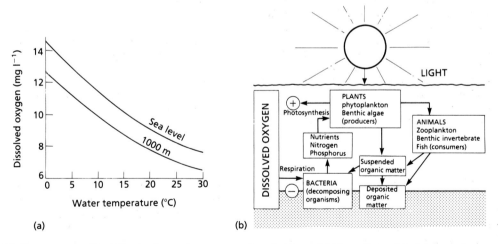

Fig. 5.10 Factors controlling dissolved oxygen concentrations in rivers. (a) The relationship between the equilibrium solubility of oxygen and water temperature (atmospheric pressure at sea level = 760 mm). (b) Interactions affecting dissolved oxygen in aquatic ecosystems (after Gras *et al* 1983).

plankton and fixed plants, produce oxygen in a water-body by photosynthesis. Photosynthetic activity by macrophytes and algae may cause an oxygen supersaturation during daylight hours. Butcher and colleagues (1930) indicated that supersaturation is more likely to be generated by algae, which produce oxygen that goes straight into solution, than by the higher plants, which produce bubbles of the gas, although daily oxygen contributions of up to 2.27 g^{-2} day^{-1} have been recorded for plants on weed beds in small British rivers (Owens & Edwards 1961). In contrast to photosynthesis, oxygen is consumed in a river by plant and bacterial respiration. It has been found, for example, that the respiration of a thick carpet of the benthic filamentous green alga *Cladophora glomerata* on the bed of the River Tees, UK had the effect of reducing the oxygen saturation to nearly 50% at night (Butcher *et al* 1937). Bacterial decomposition of organic matter, such as decaying water plants and dead leaves blown into the channel, and microbially accelerated oxidation of NH_3 to NH_4^+ also depletes dissolved oxygen concentrations. The influence of plants and bacteria on dissolved oxygen levels in rivers is complexly interrelated with grazing and carnivorous aquatic animals (heterotrophic consumers), and with suspended and deposited organic matter and dissolved nutrients in the river system. The main interactions in aquatic ecosystems (Fig. 5.10(b)) have been summarized by Gras *et al* (1983).

Variations

Dissolved oxygen concentrations in rivers may also vary markedly in time and over space. The annual regime of dissolved oxygen is inversely related to the annual cycle of water temperature. Concentrations are lowest in the summer months because of the decreased equilibrium solubility of oxygen, as results from the Red Cedar River, Michigan, USA (Ball & Bahr 1975) have clearly demonstrated (Fig. 5.11(a)). The impact of photosynthesis and respiration may cause a marked diurnal cycle in stream oxygen levels. This phenomenon is especially pronounced in rivers where high nutrient concentrations have stimulated excessive plant growth (Simonsen & Harremoës 1978; Gower 1980) and photosyn-

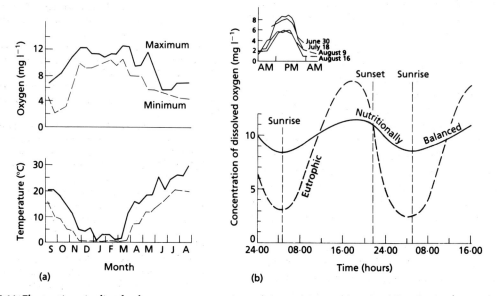

Fig. 5.11 Fluctuations in dissolved oxygen concentrations of rivers; (a) Annual cycle of dissolved oxygen concentration in a Michigan river; Red Cedar River, USA (after Ball & Bahr 1975). (b) Diurnal cycle of dissolved oxygen concentration in a nutritionally balanced and a eutrophic stream (after Gower 1980) (inset depicts diel variation of dissolved oxygen concentration in the Red Cedar River) (after Ball & Bahr 1975).

thetic activity causes high dissolved oxygen levels during the late afternoon, which is followed by oxygen depletion at night due to continued respiratory activity (Fig. 5.11(b)).

Results from the Red Cedar River (Ball & Bahr 1975) show that the magnitude of the diurnal variation in oxygen levels may vary from day to day according to flow rate and temperature (Fig. 5.11(b)). The diel cycle of oxygen levels may also differ between headwater and downstream environments (Macan 1974). Near the source, oxygen is lowest by day when the water is warmest and solubility is reduced, whereas further downstream the effects of decomposition of greater quantities of organic matter leads to minimum saturation levels at night.

Spatial variation in dissolved oxygen content between and within river systems will reflect the influence of physiographic and especially hydraulic factors. In particular, oxygen exchange between air and water is reduced at lower water velocities, and oxygen balance is affected by water depth and residence time (Gras *et al* 1983). Lower concentrations of dissolved oxygen could be anticipated in headwater streams, because of the influence of altitude on equilibrium concentration (see Fig. 5.10(a)), but this effect is often outweighed by the greater turbulence of upland streams compared with lowland rivers. At a micro-scale level, variations in dissolved oxygen concentration across the sediment−water interface and in relation to stones on the stream bed may be of particular importance to the habitats of many benthic invertebrates. Milne and Calow (1990), for example, reported a mean oxygen concentration of 5.4 mg l^{-1} under stones in a river during May 1987 when temperature ranged from 9.0 to 9.5°C. Oxygen concentrations in the water column were >11.6 mg l^{-1}, and it was found that the under-stone microhabitat of glossiphoniid leeches may become severely hypoxic during the breeding season.

The re-aeration properties of rivers are also strongly controlled by hydraulic conditions, since recovery from any oxygen deficit occurs more rapidly in shallow and turbulent reaches, where there is maximum contact between the water and the overlying air, than in deep and sluggish channel sections (Walling 1980). Natural and artificial changes in channel gradient, such as waterfalls and weirs, lead to an increase in dissolved oxygen concentrations (Gameson 1957). In many major lowland rivers, the spatial pattern of dissolved oxygen levels is also strongly affected by domestic and industrial point-source pollution.

REFERENCES

Ball RC, Bahr TG. (1975) Intensive survey: Red Cedar River, Michigan. In: Whitton BA (ed.) *River Ecology*, pp 431−60. Blackwell Scientific Publications, Oxford. [5.4]

Blakey JF. (1966) Temperature of surface waters in the conterminous United States. *United States Geological Survey Hydrological Investigations Atlas* HA-235, Washington, DC. [5.3]

Boon PJ, Shires SW. (1976) Temperature studies on a river system in north-east England. *Freshwater Biology* **6**: 23−32. [5.3]

Brett JR. (1960) Thermal requirements of fish — three decades of study, 1940−1970. In: *Biological Problems in Water Pollution*, Robert A. Taft Sanitary Engineering Centre Technical Report W60-3, pp 110−17. Cincinnati, Ohio. [5.3]

Brown GW. (1969) Predicting temperatures of small streams. *Water Resources Research* **5**: 68−75. [5.3]

Brown GW, Swank GW, Rothacher J. (1971) Water temperature in the Steamboat Drainage. *United States Department of Agriculture Forest Service Research Paper* PNW-119, pp 1−17. [5.3]

Burton TM, Turner RR, Harriss RC. (1977) Suspended and dissolved solids export from three North Florida watersheds in contrasting land use. In: Correll DL (ed.) *Watershed Research in Eastern North America*, pp 471−85. Chesapeake Bay Centre for Environmental Studies. [5.1]

Butcher RW, Pentelow FTK, Woodley JWA. (1930) Variations in composition of river waters. *International Reviews in Hydrobiology* **24**: 47−80. [5.4]

Butcher RW, Longwell J, Pentelow FTK. (1937) Survey of the River Tees. III−The non-tidal reaches — chemical and biological. *Technical Papers in Water Pollution Research, London* **6**: 189 pp. [5.4]

Calandro AJ. (1973) Analysis of stream-temperatures in Louisiana. *Louisiana Department of Public Works, Technical Report 6.* [5.3]

Calow P, Petts GE. (eds) (1994) *The Rivers Handbook*, vol. 2. Blackwell Scientific Publications, Oxford.

Castorina AR. (1980) Reservoir improvements as a key to source quality control. *Journal of the American Water Works Association* **72**: 28. [5.1]

Chang M, Roth FA, Hunt EV. (1982) Sediment production under various forest-site conditions. In:

Recent Developments in the Explanation and Prediction of Erosion and Sediment Yield. International Association of Hydrological Sciences Publication No. 137, pp 13–22. Wallingford, UK. [5.1]

Collings MR. (1969) Temperature analysis of a stream. *United States Geological Survey Professional Paper* 650-B, pp B174–B179. [5.3]

Collins WD. (1925) Temperature of water available for industrial use in the United States. *United States Geological Survey Water-Supply Paper* 520-F, pp 97–104. [5.3]

Comer LE, Greeney WJ, Dimhirn I. (1976) Stream temperature modeling. In: *Proceedings of the Symposium and Speciality Conference on Instream Flow Needs,* Rodeway Inn-Boise, Idaho, 3–6 May 1976, Volume II, pp 527–39. American Fisheries Society, Bethesda, Maryland. [5.3]

Crisp DT. (1988a) Prediction, from temperature, of eyeing, hatching and 'swim-up' times for salmonid embryos. *Freshwater Biology* **19**: 41–8. [5.3]

Crisp DT. (1988b) Water temperature data from streams and rivers in north east England. *Freshwater Biological Association Occasional Publication* **26**, pp 1–60. [5.3]

Crisp DT, Howson G. (1982) Effect of air temperature upon mean water temperature in streams in the north Pennines and English Lake District. *Freshwater Biology* **12**: 359–67. [5.3]

Crisp DT, Le Cren ED. (1970) The temperatures of three different small streams in North West England. *Hydrobiologia* **35**: 305–23. [5.3]

Crisp DT, Matthews AM, Westlake DF. (1982) The temperatures of nine flowing waters in southern England. *Hydrobiologia* **89**: 193–204. [5.3]

Department of the Environment (1981) *Examination of water and associated materials: colour and turbidity of waters.* HMSO, London. [5.2]

Dietrich WE, Dunne T. (1978) Sediment budget for a small catchment in mountainous terrain. *Zeitschrift fur Geomorphologie* **29**: 191–206. [5.1]

Dingman SL. (1972) Equilibrium temperatures of water surfaces as related to air temperature and solar radiation. *Water Resources Research* **8**: 42–9. [5.3]

Droppo IG, Ongley ED. (1989) Flocculation of suspended solids in southern Ontario rivers. In: *Sediment and the Environment,* International Association of Hydrological Sciences Publication No. 184, pp 95–103. [5.1]

Duijsings JJHM. (1985) *Streambank contribution to the sediment budget of a forest stream.* PhD thesis, University of Amsterdam. [5.1]

Duijsings JJHM. (1986) Seasonal variations in the sediment delivery ratio of a forested drainage basin in Luxembourg. In: *Drainage Basin Sediment Delivery,* International Association of Hydrological Sciences Publication No. 159, pp 153–64. [5.1]

Edinger JE, Duttweiler DW, Geyer JC. (1968) The response of water temperatures to meteorological conditions. *Water Resources Research* **4**: 1137–43. [5.3]

Edwards A, Martin D, Mitchell G. (eds) (1987) *Colour in upland waters. Proceedings of a workshop held at Yorkshire Water, Leeds, September 1987.* Yorkshire Water, Leeds; Water Research Centre, Medmenham. [5.2]

Everts CM. (1963) Temperature as a water quality parameter. In: *Water Temperature — Influences, Effects, and Control* (Proceedings of the 12th Pacific North West Symposium on Water Pollution Research, November 1963), pp 2–5. Corvallis, Oregon. [5.3]

Fowles CR. (1975) Temperature records from a small Canterbury stream. *Mauri Ora* **3**: 89–94. [5.3]

Fredriksen RL. (1970) Erosion and sedimentation following road construction and timber harvest on unstable soils in three small western Oregon watersheds. *US Forest Service Research Paper No. PNW 104.* [5.1]

Fry FEJ. (1964) Animals in aquatic environments: fishes. In: Dill DB, Adolph EF, Wilber CG (eds) *Adaptation to the Environment,* Handbook of Physiology Section 4, pp 715–28. American Physiology Society, Washington, DC. [5.3]

Gameson ALH. (1957) Weirs and the aeration of rivers. *Journal of the Institution of Water Engineers* **11**: 477–90. [5.4]

Geijskes DC. (1942) Observations on temperature in a tropical river. *Ecology* **23**: 106–10. [5.3]

Gippel CJ. (1989) *The Use of Turbidity Instruments to Measure Stream Water Suspended Sediment Concentration.* Monograph Series No. 4, Department of Geography and Oceanography, University College, Australian Defence Force Academy, Canberra. [5.1]

Gower AM. (1980) Ecological effects of changes in water quality. In: Gower AM (ed.) *Water Quality in Catchment Ecosystems,* pp 145–71. John Wiley & Sons, Chichester. [5.4]

Gras R, Albignat JP, Gosse Ph. (1983) The effects of hydraulic projects and their management on water quality. In: *Dissolved Loads of Rivers and Surface Quantity/Quality Relationships* (Proceedings of the Hamburg Symposium, August 1983), pp 313–32. IAHS Publication No. 141. Wallingford, UK. [5.4]

Hem JD. (1970) Study and interpretation of the chemical characteristics of natural water. Second edition. *United States Geological Survey Water-Supply Paper 1473.* [5.4]

Holtby LB. (1988) Effects of logging on stream temperatures in Carnation Creek, British Columbia, and associated impacts on the Coho salmon (*Oncorhynchus kisutch*). *Canadian Journal of Fisheries and Aquatic Sciences* **45**: 502–15. [5.3]

Hopkins CL. (1971) The annual temperature regime of a small stream in New Zealand. *Hydrobiologia* **37**:

397–408. [5.3]

Johnson FA. (1971) Stream temperatures in an alpine area. *Journal of Hydrology* **14**: 322–36. [5.3]

Lal R. (1982) Effects of slope length and terracing on runoff and erosion on a tropical soil. In: *Recent Developments in the Explanation and Prediction of Erosion and Sediment Yield*. International Association of Hydrological Sciences Publication No. 137, pp 23–31. [5.1]

Langford TE. (1972) A comparative assessment of thermal effects in some British and North American rivers. In: Oglesby RT, Carlson CA, McCann JA (eds) *River Ecology and Man*, pp 319–51. Academic Press, New York. [5.3]

Langford TE. (1983) *Electricity Generation and the Ecology of Natural Waters*. Liverpool University Press, Liverpool. [5.3]

Lehre AK. (1982) Sediment budget in a small coast range drainage basin in North-Central California. In: *Sediment Budgets and Routing in Forested Drainage Basins*, US Forest Service General Technical Report PNW-141. [5.1]

Macan TT. (1958) The temperature of a small stony stream. *Hydrobiologia* **12**: 89–106. [5.3]

Macan TT. (1974) *Freshwater Ecology* 2nd edn. Longman, London. [5.4]

McDonald AT, Edwards AMC, Naden PS, Martin D, Mitchell G. (1989) Discoloured runoff in the Yorkshire Pennines. In: *Proceedings of the Second National Hydrology Symposium*, pp 1.59–1.64, British Hydrological Society. [5.2]

MacKichan KA. (1967) Diurnal temperature fluctuations of three Nebraska streams. *United States Geological Survey Professional Paper 575-B*, B233–B234. [5.3]

Malcolm RLC. (1985) Geochemistry of stream fulvic and humic substances. In: Aiken GR, McKnight DM, Wershaw RL, Maccarthy P (eds) *Humic Substances in Soil and Water*, pp 181–209. John Wiley & Sons, New York. [5.2]

Meade RH, Parker RS. (1985) Sediment in rivers of the United States. In: *National Water Summary 1984*, US Geological Survey Water-Supply Paper 2275, pp 49–60. [5.1]

Milne IS, Calow P. (1990) Costs and benefits of brooding in glossiphoniid leeches with special reference to hypoxia as a selection pressure. *Journal of Animal Ecology* **59**: 41–56. [5.4]

Miyazawa T, Yamashita S, Kitagawa M, Sekine K. (1982) An evaluation and mapping of topographic factors affecting river water temperature in the upstream area of the Nagara River, Japan. *Beiträge zur Hydrologie Sonderheft* **3**: 83–95. [5.3]

Moore AM. (1967) Correlation and analysis of water-temperature data for Oregon streams. *United States Geological Survey Water-Supply Paper 1819 K*. [5.3]

Morrissy NM. (1971) Temperature relationships in small bodies of freshwater with special reference to trout streams in South Australia. *Bulletin of the Australian Society of Limnology* **4**: 8–20. [5.3]

Mosley MP. (1982) *New Zealand River Temperature Regimes*. Water and Soil Miscellaneous Publication No. 36, Water and Soil Division, Ministry of Works and Development for the National Water and Soil Conservation Organisation, Christchurch, New Zealand. [5.3]

Mosley MP. (1983) Variability of water temperatures in the braided Ashley and Rakaia rivers. *New Zealand Journal of Marine and Freshwater Research* **17**: 331–42. [5.3]

Muncy RJ, Atchison GJ, Bulkley RV, Menzel BW, Perry LG, Summerfelt RC. (1979) *Effects of Suspended Solids and Sediment on the Reproduction and Early Life of Freshwater Fishes: a Review*. US Environmental Protection Agency, Corvallis, Oregon. [5.1]

Naden PS, McDonald AT. (1987) Statistical modelling of water colour in the uplands: the Upper Nidd catchment 1979–1987. *Environmental Pollution* **60**: 141–63. [5.2]

O'Loughlin CL, Rowe LK, Pearce AJ. (1980) Sediment yield and water quality responses to clear felling of evergreen mixed forests in western New Zealand. In: *The Influence of Man on the Hydrological Regime with Special Reference to Representative and Experimental Basins*. International Association of Hydrological Sciences Publication No. 130, pp 285–92. [5.1]

Owens M, Edwards RW. (1961) The effects of plants on river conditions. II. Further studies and estimates of net productivity of macrophytes in a chalk stream. *Journal of Ecology* **49**: 119–26. [5.4]

Parker FL, Krenkel PA. (eds) (1969) *Engineering Aspects of Thermal Pollution*. Vanderbilt University Press, Portland, Oregon. [5.3]

Peart MR, Walling DE. (1986) Fingerprinting sediment source: the example of a drainage basin in Devon, UK. In: *Drainage Basin Sediment Delivery*, International Association of Hydrological Sciences Publication No. 159, pp 41–55. [5.1]

Petts GE. (1984) *Impounded Rivers: Perspectives for Ecological Management*. John Wiley & Sons, Chichester. [5.3]

Pluhowski EJ. (1970) Urbanization and its effect on the temperature of the streams on Long Island, New York. *United States Geological Survey Professional Paper 627-D*. [5.3]

Pluhowski EJ. (1972) Unusual temperature variations in two small streams in northern Virginia. In: *Geological Survey Research, 1972, United States Geological Survey Professional Paper 800-B*, pp B255–B258. [5.3]

Reid GK, Wood RD. (1976) *Ecology of Inland Waters and Estuaries*. Van Nostrand, New York. [5.3]

Reuter JH, Perdue EM. (1977) Importance of heavy

metal—organic matter interaction in natural waters. *Geochimica Cosmochimica Acta* **41**: 325–34. [5.2]

Ritchie JC. (1972) Sediment, fish and fish habitat. *Journal of Soil and Water Conservation* **27**: 125. [5.1]

Rose AH. (ed.) (1967) *Thermobiology.* Academic Press, London. [5.3]

Schmitz W. (1954) Grundlagen der Untersuchung der Temperaturverhältnisse in der Fliessgewässern. *Berliner Limnologie der Flusstation Freundenthal* **6**: 29–50. [5.3]

Shanley JB, Peters NE. (1988) Preliminary observations of streamflow generation during storms in a forested Piedmont watershed using temperature as a tracer. *Journal of Contaminant Hydrology* **3**: 349–65. [5.3]

Simonsen JF, Harremoës P. (1978) Oxygen and pH fluctuations in rivers. *Water Research* **12**: 477–89. [5.4]

Sioli H. (1964) General features of the limnology of Amazónia. *Verh. int. Ver. Limnol.* **15**: 1053–8. [5.3]

Skvortsov AF. (1959) River suspension and soils. *Soviet Soil Science* **4**: 409–16. [5.1]

Smith K. (1968) Some thermal characteristics of two rivers in the Pennine area of northern England. *Journal of Hydrology* **6**: 405–16. [5.3]

Smith K. (1972) River water temperatures: an environmental review. *Scottish Geographical Magazine* **88**: 211–20. [5.3]

Smith K. (1975) Water temperature variations within a major river system. *Nordic Hydrology* **6**: 155–69. [5.3]

Smith K. (1979) Temperature characteristics of British rivers and the effects of thermal pollution. In: Hollis GE (ed.) *Man's Impact on the Hydrological Cycle in the United Kingdom,* pp 229–42. GeoBooks, Norwich. [5.3]

Smith K. (1981) The prediction of river water temperatures. *Hydrological Sciences Bulletin* **26**: 19–32. [5.3]

Smith K, Lavis ME.(1975) Environmental influences on the temperature of a small upland stream. *Oikos* **26**: 228–36.

Steele TD. (1982) A characterization of stream temperatures in Pakistan using harmonic analysis. *Hydrological Sciences Journal* **4**: 451–67. [5.3]

Steele TD. (1983) A regional analysis of water quality in major streams of the United States. In: *Dissolved Loads of Rivers and Surface Water Quantity/Quality Relationships* (Proceedings of the Hamburg Symposium, August 1983), pp 131–43. International Association of Hydrological Sciences Publication No. 141. Wallingford, UK. [5.3]

Steele TD, Gilroy EJ, Hawkinson RO. (1974) An assessment of areal and temporal variations in streamflow quality using selected data from the National Stream Quality Accounting Network. *United States Geological Survey Open-File Report 74–217.* [5.3]

Stevens HH Jr, Ficke JF, Smoot GF. (1975) Water temperature—influential factors, field measurement, and data presentation. *United States Geological Survey Techniques of Water-Resources Investigations,* Book 1, Chapter D1. [5.3]

Swanson FJ, Janda RJ, Dunne T, Swanson DN. (eds) (1982) *Sediment Budgets and Routing in Forested Drainage Basins.* US Forest Service General Technical Report PNW-141. [5.1]

Trimble SW. (1976) Sedimentation in Coon Creek Valley, Wisconsin. In: *Proceedings of the Third Federal Interagency Sedimentation Conference,* pp 5-100–5-112. US Water Resources Council, Washington, DC. [5.1]

Trimble SW. (1981) Changes in sediment storage in Coon Creek Basin, Driftless Area, Wisconsin, 1853–1975. *Science* **214**: 181–3. [5.1]

Troxler RW, Thackston EL. (1977) Predicting the rate of warming of rivers below hydro-electric installations. *Journal of the Water Pollution Control Federation* August 1977, 1902–12. [5.3]

US Environmental Protection Agency (1979) *Impacts of Sediment and Nutrients on Biota in Surface Waters of the United States.* US Environmental Protection Agency, Athens, Georgia. [5.1]

Walker JH, Lawson JD. (1977) Natural stream temperature variation in a catchment. *Water Research* **11**: 373–7. [5.3]

Wall CJ, Wilding LP. (1976) Mineralogy and related parameters of fluvial suspended sediments in Northwestern Ohio. *Journal of Environmental Quality* **5**: 168–73. [5.1]

Walling DE. (1980) Water in the catchment ecosystem. In: Gower AM (ed.) *Water Quality in Catchment Ecosystems,* pp 1–47. John Wiley & Sons, Chichester. [5.4]

Walling DE. (1983) The sediment delivery problem. *Journal of Hydrology* **65**: 209–37. [5.1]

Walling DE. (1987) Rainfall, runoff and erosion of the land: a global view. In: Gregory KJ (ed.) *Energetics of the Physical Environment,* pp 89–117. John Wiley & Sons, Chichester. [5.1]

Walling DE. (1990) Linking the field to the river: sediment delivery from agricultural land. In: Boardman J, Foster IDL, Dearing JA (eds) *Soil Erosion on Agricultural Land,* pp 129–52. John Wiley & Sons, Chichester. [5.1]

Walling DE, Bradley SB. (1988) The use of caesium-137 measurements to investigate sediment delivery from cultivated areas in Devon, UK. In: *Sediment Budgets,* International Association of Hydrological Sciences Publication No. 174, pp 325–35. [5.1]

Walling DE, Carter R. (1980) River water temperatures. In: Doornkamp JC, Gregory KJ (eds) *Atlas of Drought in Britain 1975–76,* p 49. Institute of British Geographers, London. [5.3]

Walling DE, Kane P. (1982) Temporal variation of suspended sediment properties. In: *Recent Developments in the Explanation and Production of Erosion and Sediment Yield*, International Association of Hydrological Sciences Publication No. 137, pp 409–19. [5.1]

Walling DE, Moorehead PW. (1989) The particle size characteristics of fluvial suspended sediment: an overview. *Hydrobiologia* **176/177**: 125–49. [5.1]

Walling DE, Webb BW. (1983) Patterns of sediment yield. In: Gregory KJ (ed.) *Background to Palaeohydrology*, pp 69–100. John Wiley & Sons, Chichester. [5.1]

Ward JC. (1963) Annual variation of stream water temperature. *American Society of Civil Engineers Journal of the Sanitary Engineering Division* **89** (SA6): 1–16. [5.3]

Ward JV. (1985) Thermal characteristics of running waters. *Hydrobiologia* **125**: 31–46. [5.3]

Ward JV, Stanford JA. (1982) Thermal responses in the evolutionary ecology of aquatic insects. *Annual Reviews in Entomology* **27**: 97–117. [5.3]

Wark JW, Keller FJ. (1963) *Preliminary Study of Sediment Sources and Transport in the Potomac River Basin*. Interstate Commission on the Potomac River Basin. Edgewater, Maryland, USA. [5.1]

Wasson JG, Guyot JL, Dejoux C, Roche MA. (1989) *Regimen termico de los rios de Bolivia*. ORSTOM IHH-UMSA PHICAB IIQ-UMSA Senamhi Hidrobiologia-UMSA. La Paz, Bolivia. [5.3]

Weatherley NS, Ormerod SJ. (1990) Forests and the temperature of upland streams in Wales: a modelling exploration of the biological effects. *Freshwater Biology* **24**: 109–22. [5.3]

Webb BW. (1987) The relationship between air and water temperatures for a Devon river. *Reports and Transactions of the Devonshire Association for the Advancement of Science, Literature and Art* **119**: 197–222. [5.3]

Webb BW, Walling DE. (1982) The magnitude and frequency characteristics of fluvial transport in a Devon drainage basin and some geomorphological implications. *Catena* **9**: 9–23. [5.1]

Webb BW, Walling DE. (1985a) Temporal variation of river water temperatures in a Devon river system. *Hydrological Sciences Journal* **30**: 449–64. [5.3]

Webb BW, Walling DE. (1985b) Temperature characteristics of Devon rivers. *Proceedings of the Ussher Society* **6**: 237–45. [5.3]

Webb BW, Walling DE. (1986) Spatial variation of water temperature characteristics and behaviour in a Devon river system. *Freshwater Biology* **16**: 585–608. [5.3]

Williams GP. (1989) Sediment concentration versus water discharge during single hydrologic events in rivers. *Journal of Hydrology* **111**: 89–106. [5.1]

Wolman MG, Miller JC. (1960) Magnitude and frequency of forces in geomorphic processes. *Journal of Geology* **68**: 54–74. [5.1]

Wolman MG, Schick AP. (1967) Effects of construction on fluvial sediment, urban and suburban areas of Maryland. *Water Resources Research* **3**: 451–64. [5.1]

6: Water Quality
II. Chemical Characteristics

B.W.WEBB AND D.E.WALLING

A large number of different properties and parameters are available to describe the chemical characteristics of rivers. These range from general descriptors, such as measures of salinity and acidity, to composition in terms of major cation and anion content, and to the concentration of organic and inorganic micropollutants. The present chapter focuses on the chemical characteristics of rivers free from major point-source contamination by human activities, so that discussion of chemical parameters, which are most strongly influenced by river pollution and include trace metal concentrations, organic constituents and pesticide levels, is discussed in Calow and Petts (1994). Attention will be given to the behaviour of chemical properties in uncontaminated rivers in terms of both the concentrations and transport (flux) of materials. The emphasis is on a comparison *between* river systems from global to local scales. Chapter 7, on the other hand, deals with the dynamics of chemicals *within* river systems.

A fundamental distinction is drawn in the present chapter between chemical constituents that are present in dissolved form and those that are transported in sediment-associated form. Discussion of dissolved chemical species will include those elements that are vital to the health of plants and animals, are strongly involved in cycling processes between the inorganic and organic compartments of rivers and their drainage basins, and are often referred to by aquatic ecologists as nutrients. The definition of what constitutes a nutrient species is somewhat dependent on the perspective from which the river is being studied. For example, some geochemists (e.g. Meybeck 1982) view the major nutrients in rivers as comprising various species of nitrogen and phosphorus together with organic carbon, whereas investigators of forested and other stream ecosystems (e.g. Sanders 1972; Likens *et al* 1977) may define dissolved nutrient substances more widely. In the present chapter, the term solutes is employed to refer to dissolved chemical species, including nutrients.

The techniques available for determining the concentration and flux of chemical constituents in rivers have been comprehensively described elsewhere (e.g. Golterman & Clymo 1969; American Public Health Association 1971; Walling 1984). Considerable efforts have also been made to develop and refine laboratory procedures for reliably detecting simple and more complex chemical constituents present in low concentrations (e.g. Pereira *et al* 1987; Taylor 1987) and to standardize and harmonize sampling methods and analytical techniques between different laboratories and agencies involved in river quality investigations (e.g. Skougstad *et al* 1979; Simpson 1980).

6.1 SOLUTE BEHAVIOUR

Sources, processes and pathways

The transfer of chemical elements through the hydrological cycle involves a complex interaction of chemical, biological, and hydrological systems and processes (Neal & Hornung 1990). The nature of solute behaviour in river systems ultimately reflects the various sources and stores of dissolved material that are present in the drainage area and the different processes that mobilize and modify chemical constituents found in draining waters.

Although in-stream transformations of solutes through biological, chemical and physical processes can exert a considerable influence on water chemistry, solute behaviour is determined largely by the interaction of hydrological and biogeochemical processes at a basin-wide rather than a channel scale (Walling 1980). This interaction is schematically represented in Fig. 6.1(a), and of particular influence on stream chemistry are the different pathways, volumes and flux rates of water which may be involved in the transformation of precipitation to river runoff.

Atmospheric inputs

Rivers may receive a substantial part of their solute load from the atmosphere. This can be in solid, liquid or gaseous form, and is derived from a variety of different sources. A distinction is often made between dry and wet fallout of material from the atmosphere (Walling 1980; Cryer 1986). The former involves deposition of relatively large particles (>20 μm in diameter) under the influence of gravity, whereas the latter comprises the chemical constituents of wet precipitation, which occur through the solution of particles that have acted as condensation nuclei for raindrops (rain-out) or through solution of atmospheric particles below cloud level which have been impacted by falling raindrops (wash-out). An important source of atmospheric aerosols and gases is marine-derived material, which is transferred from the sea surface via spray and gas exchange. This is the predominant source of Na^+

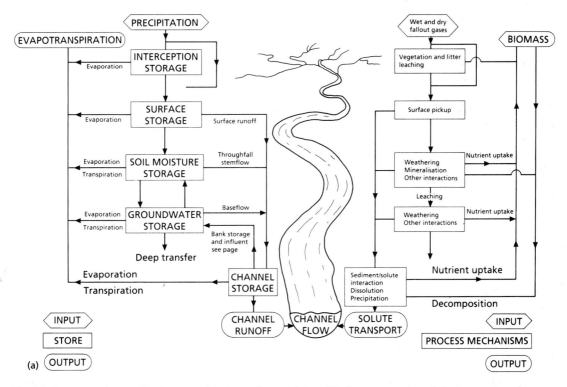

Fig. 6.1 Sources and controls of stream and river solutes. (a) Simplified representation of the interaction of hydrological and biogeochemical processes operating in a drainage basin. (b) (*overpage*) Evolution of the ionic content of atmospheric precipitation in France, south of the Cherbourg–Basel line (from Meybeck 1986). (c) Streamwater quality for a Virginian stream, Mill Run, USA superimposed on the $K_2O-Al_2O_3-SiO_2$ solubility diagram (from Afifi & Bricker 1983). (d) Annual calcium budget for an aggrading (northern hardwood) forested ecosystem at Hubbard Brook; standing crop values (boxed) and calcium fluxes expressed as $kg\,ha^{-1}$ and $kg\,ha^{-1}$ $year^{-1}$, respectively; values in parentheses represent annual accretion rates (after Likens *et al* 1977).

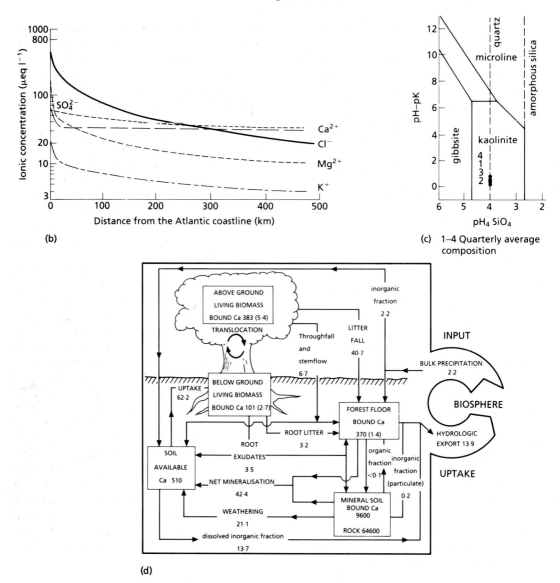

(b)

(c)   1–4 Quarterly average
      composition

(d)

Fig. 6.1 (*contd*)

and Cl⁻, is a major contributor to Mg^{2+} and to K^+, and is a significant source of sulphur in the atmosphere. Material derived from the terrestrial environment in the form of blown soil and dust, and through biological emissions from living vegetation and the burning of organic matter, is also an important source of aerosols and gases in the atmosphere, especially in the case of Ca^{2+}, NH_4^+, NO_3^-, NO_2^-, HCO_3^- and SO_4^{2-}. Emissions from

anthropogenic sources and from volcanoes also may significantly contribute to atmospheric fallout of material.

There is a general tendency for the atmospheric input of terrestrially derived material to increase and for the fallout of cyclic salts (marine-derived material) to decrease with increasing distance from the coastal zone (e.g. Eriksson 1960; Stallard & Edmond 1981), as data from a number of stations in France (Meybeck 1986) have clearly demonstrated (Fig. 6.1(b)). Complex physico-chemical processes, which are involved in the transport and deposition of atmospheric materials and include fractionation and relative enrichment mechanisms (e.g. Bloch *et al* 1966; Belot *et al* 1982), can significantly modify the composition of atmospheric inputs received at a given location. Spatial patterns of solute input to rivers (e.g. Junge & Werby 1958; Stevenson 1968; Munger & Eisenreich 1983) may be further diversified by the particular storm directions and air-mass trajectories that affect a river basin. Atmospheric inputs may also vary over time in response to seasonal changes in meteorological conditions and to the type and duration of precipitation events (Walling 1980). Interception of wet and dry atmospheric fallout by a vegetation canopy can, through solution and cation exchange processes (e.g. Best & Monk 1975; White 1981; Foster & Grieve 1984; Reynolds & Pomeroy 1988; Ferrier *et al* 1990; Rodà *et al* 1990), further modify the content and composition of solute input to river systems.

The significance of atmospheric inputs to stream and river solute content varies depending on the ionic constituent involved and the environmental setting of the river basin. At a global scale, Meybeck (1983) calculated that natural wet and dry fallout contributes only 4% and 6% of river transport to the oceans for Si and Ca^{2+} respectively but 53% and 72% in the case of Na^+ and Cl^- respectively. Although information on worldwide variations in the nutrient content of unpolluted precipitation is relatively limited, it has been estimated that in the case of some nutrient elements the atmospheric input far exceeds river output. Fallout of dissolved inorganic and organic nitrogen, for example, is 170% of total dissolved nitrogen transport in world rivers (Meybeck 1983). The impact of atmospheric inputs at more local scales will depend on the importance of other solute sources together with hydrological conditions, so that the atmospheric influence will be greatest in drainage basins underlain by rock types resistant to chemical breakdown and where precipitation is routed quickly to the river channel.

The contribution of atmospherically derived solutes to rivers under natural conditions may be obscured by the occurrence of atmospheric pollution. Considerable attention has been given in recent years to increasing industrial emissions of SO_2 and NO_x which have resulted from combustion of fossil fuels, smelting of non-ferrous metals, fertilizer and sulphuric acid manufacture, and other industrial processes, and have given rise to problems of acid precipitation and the acidification of freshwater environments (e.g. Seip & Tollan 1985; Buijsman *et al* 1987; Young *et al* 1988; Meybeck *et al* 1989). Anthropogenic emissions of oxidized sulphur, for example, are estimated to have risen for the globe as a whole from $75-80 \times 10^6$ t year^{-1} in the late 1970s to about 90×10^6 t year^{-1} in 1985 (Möller 1984; Varheyli 1985).

A distinction can be drawn between those constituents of atmospheric fallout, both natural and anthropogenic, that affect river solutes in a passive manner and those that have an impact in an active way (Cryer 1986). In the former case, soluble material from the atmosphere is simply added in an unchanged form to the water-course, although its concentration may be increased by the effects of evapotranspiration losses. In the latter case, the components of atmospheric input may react chemically with the vegetation, soil and rock materials of a catchment area and mobilize new solute species in weathering and other processes. The role of complex biogeochemical ion exchanges and weathering reactions between precipitation water and permeable material in catchments has been highlighted, for example, in recent studies of stream acidification (Edwards *et al* 1990; Neal *et al* 1990b; Rosenqvist 1990).

Chemical reactions

In many cases, a large proportion of the solute load carried by streams and rivers is derived from chemical reactions between water and rock and soil minerals. At the global scale, it is estimated that 60% of the major dissolved constituents carried by rivers to the oceans is derived from chemical weathering (Walling & Webb 1986a). Hem (1970) identified four groups of chemical reactions which are important in establishing and maintaining the composition of natural water:

1 Reversible solution and deposition reactions,

including ion-exchange processes, in which water is not chemically altered. Cation exchange and, to a lesser extent, anion exchange, which involve reversible transfer of ionic constituents between soil particles and draining water, represent important mechanisms whereby the chemistry of precipitation is modified in its translation to river runoff.

2 Reversible solution and deposition reactions in which water molecules break down into H^+ and OH^- ions. These include simple hydrolysis reactions, such as the solution of carbonates.

3 Reversible solution and deposition reactions and ion reactions which involve changes in oxidation state, such as the reduction of ferric to ferrous iron.

4 Less readily reversible reactions which include complex processes of hydration and hydrolysis such as those involved in the breakdown of aluminosilicate rock-forming minerals. Absorption of ions by clay minerals and sesqui-oxides/hydroxides and the complexing of metal cations by organic matter ligands are processes that may also be placed in this category.

The release of solutes in weathering processes is fundamentally governed by the thermodynamics and kinetics of the reactions involved as well as by the residence time of water in the soil or rock body (Walling 1980). The latter factor influences whether weathering reactions can proceed to an equilibrium stage. Given sufficient data on soil and rock mineralogy, precipitation and streamflow volume, and water chemistry, it is possible to quantify the contribution of individual weathering reactions to river solutes (e.g. Garrels & Christ 1965; Waylen 1979). A study of Mill Run (Afifi & Bricker 1983), a small forested drainage basin in Virginia, USA, demonstrated, for example, that stream-water chemistry is controlled by the dissolution of orthoclase, plagioclase, chlorite, amphibole and pyrite minerals, by the precipitation of microcrystalline gibbsite and FeOOH, and by the formation of an SiO_2-rich residue. Mineral equilibria diagrams (Fig. 6.1(c)) suggest that cation concentrations in the stream water of this catchment are controlled by the kinetics of dissolution–precipitation reactions, rather than by thermodynamic equilibria with primary rock-forming minerals.

Biological processes

Water chemistry may also be strongly affected by biotic processes associated with the soil and vegetation cover of a drainage basin ecosystem. Plant respiration and bacterial degradation, for example, may influence CO_2 concentrations in soil pores, and terrestrial insect populations may regulate solute transfer processes through an enhancement of litter decomposition (Swank 1986). A variety of microbial transformations may strongly influence processes of mobilization and immobilization of nutrients and other elements in catchment ecosystems. The dynamics of nitrogen in the soil, for example, provides a very good example of the influence that bacterial mineralization has on the balance of nitrogen species in soil water which, in turn, supplies the river channel. Organic nitrogen contained in organic matter can be converted to salts of NH_4^+ through decomposition by heterotrophic bacteria. NH_4^+ ions, in turn, may be nitrified by autotrophic bacteria; this involves oxidation initially to NO_2^- by *Nitrosomonas* and finally to highly soluble NO_3^- by *Nitrobacter*. Other biologically mediated transformations may also affect the cycling of nitrogen in catchments under forest or grassland (Roberts *et al* 1983; Royal Society 1983), and these include the take-up and immobilization by microflora and fauna of NH_4^+ ions released in the process of mineralization; reduction of nitrate into nitrogen oxides and dinitrogen by a wide variety of microorganisms in the process of denitrification (Swank 1986), and nitrogen fixation which involves reduction of dinitrogen in the atmosphere to NH_3 through the action of certain prokaryotic microorganisms.

Solute uptake and incorporation in biomass above and below ground is a particularly important mechanism in forested drainage basins (Stevens *et al* 1988). A study in a small granitic basin in Colorado, USA (Lewis & Grant 1979) showed that most chemical elements (Ca, Na, K, N, P, S) are stored within the soils and vegetation of the catchment, and an investigation of chemical mass balances in a small stream in central Java, Indonesia (Bruijnzeel 1983) indicated that a vigorously growing plantation forest of *Agathis dammara* acts as a long-term sink for chemical elements, with annual uptake ranging from 71.9 to 1.1 kg ha^{-1} year^{-1} in the case of SiO_2 and Mn, respectively. In the Hubbard Brook Experimental Forest, the importance of biomass accumulation in the vegetation and forest floor has been shown to be greatest for nitrogen and least for sodium, with storage accounting for 80% and <3% respectively of the inputs from weathering and the atmosphere (Likens *et al* 1977). In forested drainage basins underlain by resistant lithologies, storage of solutes and processes of internal cycling within the ecosystem, involving litter fall, canopy leaching, root mortality and resorption mechanisms, may be more important in regulating stream chemistry than inputs from the atmosphere or from weathering. In such circumstances, construction of detailed chemical budgets (Fig. 6.1(d)) of the kind formulated from long-term studies of watersheds in the Hubbard Brook Experimental Forest (Likens *et al* 1977) and the Coweeta Hydrologic Laboratory (Swank & Douglass 1977) may be required to interpret fully the biological, physical and chemical factors controlling stream solute behaviour.

Biological processes operating within the river channel itself may also modify water chemistry (Swank 1986). These include uptake of nutrients and silica by aquatic organisms, nitrogen fixation, nitrification of ammonia and denitrification by bacteria living in stream sediments, and the processing of detrital material by macroinvertebrates, which may add dissolved ions, such as K^+ and Ca^{2+}, to the solute load of the stream (Webster & Patten 1979). Recent studies of the Afon Hafren in mid-Wales (Fiebig *et al* 1990) have also highlighted the role that the stream bed biota have in immobilizing dissolved organic carbon which enters the water-course in soil waters.

Spatial variation and controls

The solute content and composition of rivers varies greatly in space depending on the particular sources, processes and pathways that are dominant in a given drainage basin. In general, Meybeck and Helmer (1989) have suggested that six environmental factors, comprising the occurrence of highly soluble or easily weatherable minerals, distance to the marine environment, aridity, terrestrial primary productivity, ambient temperature and rates of tectonic uplift, are the most important environmental controls on river chemistry. In detail, the extent of variation and the controlling factors responsible will vary depending on the scale of the area under consideration.

World and continental variations

The load-weighted average total dissolved solids content of world river water has been calculated as 120 mg l^{-1} from a sample of 496 rivers in which pollution is of minor significance (Walling & Webb 1986a). World averages of individual ionic constituents in river water have also been computed (Meybeck & Helmer 1989), both as a global discharge-weighted natural content and as a global most-common natural content (Table 6.1). The latter is defined a the median of a cumulative distribution curve of concentrations assembled from information available for major world rivers and their tributaries. The dissolved content of rivers is dominated by HCO_3^-, SO_4^{2-}, Ca^{2+} and SiO_2, and 97.3% of global runoff has been classified as being of the calcium bicarbonate type (Meybeck 1981).

Despite the dominance of a few chemical constituents in world rivers, enormous spatial variability can be encountered in the chemistry of river water. Total dissolved solids concentrations, for example, are less than 10 mg l^{-1} in some Amazonian tributaries (Meybeck 1976) but reach 60 000 mg l^{-1} in the Leben River, Tunisia (Colombani 1983). Furthermore, more than 20 water types, defined by the relative importance of the major cation and anion constituents, have been recognized across the globe (Meybeck *et al* 1989). Worldwide variations in water chemistry have been accounted for by Gibbs (1970) in terms of dominance by rock weathering, by atmospheric precipitation and by evaporation–crystallization processes, and climatic and geological factors are often cited as the most important environmental controls on stream solutes at a global scale. The influence of climate acting through water availability is seen in an inverse relationship between discharge-weighted total dissolved solids concentration and mean annual runoff (Fig. 6.2(b)) which

Table 6.1 Average chemical composition of unpolluted world rivers and variations in chemistry of pristine streams according to rock type (from Meybeck & Helmer 1989)

Parameter (units)	Global averages		Common rock types*						Rarer rock types†			
	DWNC	MCNC	1	2	3	4	5	6	A	B	C	D
Tz$^+$ (μeq l^{-1})	1200	800	166	207	435	223	770	3247	2700	40 700	312 000	4130
pH	–	–	6.6	6.6	7.2	6.8	–	7.9	–	–	8.0	7.4
SiO$_2$ (μmol l^{-1})	170	180	150	130	200	150	150	100	332	116	20	1660
Ca^{2+} (μeq l^{-1})	670	400	39	60	154	88	404	2560	150	4350	30 350	245
Mg^{2+} (μeq l^{-1})	275	200	31	57	161	63	240	640	2490	10 000	5 640	40
Na$^+$ (μeq l^{-1})	225	160	88	80	105	51	105	34	52	26 100	276 000	3830
K$^+$ (μeq l^{-1})	33	27	8	10	14	21	20	13	5	260	197	184
Cl$^-$ (μeq l^{-1})	162	110	0	0	0	0	20	0	93	420	266 000	1653
SO$_4^{2-}$ (μeq l^{-1})	172	100	31	56	10	95	143	85	472	29 000	27 700	290
HCO$_3^-$ (μeq l^{-1})	850	500	128	136	425	125	580	3195	2020	10 700	3 000	2230
DOC (mg l^{-1})	5.75	4.2										
N-NH$_4^+$ (mg l^{-1})	–	0.015										
N-NO$_3^-$ (mg l^{-1})	–	0.10										
N organic (mg l^{-1})	–	0.26										
P-PO$_4^{3-}$ (mg l^{-1})	–	0.010										

Tz$^+$, cation sum; DOC, dissolved organic carbon; DWNC, global discharge-weighted natural concentration; * MCNC, most common natural concentration; * corrected for oceanic cyclic salts: 1 granite, 2 gneiss, 3 volcanics, 4 sandstone, 5 shale, 6 carbonates; † uncorrected for oceanic salts: A batholith, B coal shale, C salt rock, D hydrothermal.

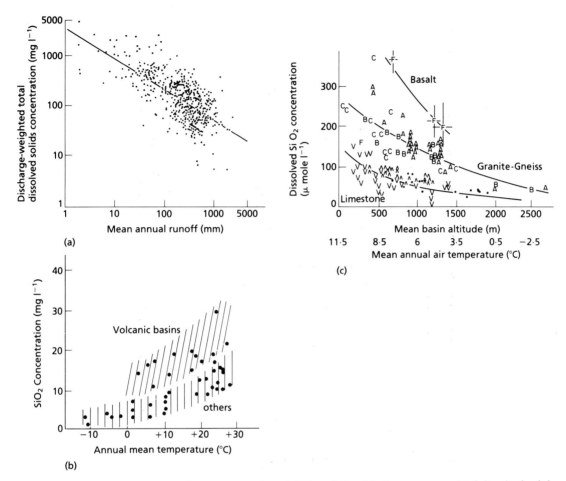

Fig. 6.2 Spatial variations in river solute concentrations. (a) The relationship between mean total dissolved solids concentration and mean annual runoff for a sample of 496 world rivers. (b) The relationship between SiO_2 content of river water and mean annual temperature (after Meybeck 1981). (c) Variation of dissolved silica concentration as a function of mean drainage basin altitude and corresponding mean annual air temperature in France (from Meybeck 1986). A granite; B gneiss; C micaschist; F basalt; v limestone (chalk excluded); ∧ Jura limestone; ● limestone watersheds.

indicates a general dilution effect as runoff volumes increase. Mean annual temperature is also an important control on the SiO_2 content of rivers (Fig. 6.2(b)), and the impact of climate on chemical composition of river water at a global level is demonstrated by the distinctiveness of the ionic balance in rivers of the arid zone compared with other morphoclimatic regions (Walling & Webb 1986b). A general climatic control is also evident in the worldwide variation of dissolved organic as well as dissolved inorganic constituents of river

runoff. Median dissolved organic carbon concentrations, for example, have been found to be 10 mg l^{-1} in taiga rivers, 6 mg l^{-1} in humid tropical environments and 3 mg l^{-1} in temperate and semiarid areas (Meybeck 1982), and this probably reflects variations in the organic content of soils in different climatic zones.

Contrasts in rock type may often moderate the effects of climate on river solutes as indicated by the contrast in SiO_2 concentration between volcanic and other rock types over a range of annual mean temperatures (Fig. 6.2(b)). The variability of dissolved major elements in pristine streams draining common and rarer rock types is presented in Table 6.1, although worldwide contrasts in solute content associated with lithological differences are less marked than those associated with contrasting climates (Walling & Webb 1986a).

Spatial variability of stream chemistry has also been investigated across large geographical areas at a subglobal scale. For example, nutrient and total alkalinity levels have been mapped for the surface waters of the conterminous USA and general correlations established with land-use type (Omernik 1977; Omernik & Powers 1983); Hu and colleagues (1982) investigated the major ion composition of some large Chinese rivers and found their chemistry to be generally dominated by the weathering of carbonates and evaporites rather than aluminosilicates, and Stallard and Edmond (1983) demonstrated that substrate lithology and the nature of the erosional regime exert the strongest control on the solute content of rivers within the Amazon Basin.

National, regional and local variations

A narrowing of the focus to consider spatial variability of river solutes at countrywide and regional scales often reveals the occurrence of more complex patterns of variation and the influence of additional controls compared with global and continental scales. Marked geographical patterns of total and individual solute concentrations have been mapped for mainland Britain, where specific electrical conductance has been found to vary between 35 and 1200 μS cm^{-1}, nitrate-nitrogen to vary between 0.1 and 15.0 mg l^{-1} and chloride to vary between 5 and 200 mg l^{-1} in small and unpolluted tributaries (Walling & Webb 1981). The pattern of spatial variation in the case of conductance levels was found to be strongly related to geological controls, in the case of nitrate-nitrogen levels to the intensity of agricultural activity, and in the case of chloride concentrations to the distance from the coast and the balance between precipitation and runoff (Walling & Webb 1981). In France, the effect of increasing basin altitude and, in turn, decreasing mean annual air temperature on the dissolved SiO_2

content of streams is apparent (Fig. 6.2(c)), and this trend is independent of catchment geology (Meybeck 1986).

The likely occurrence of acid waters, which has been mapped for Wales (United Kingdom Acid Waters Review Group 1988), provides a good example of how soils, geology and land use interact to control water chemistry at a regional scale. In this area, a transition from acidic soils overlying rocks with little or no buffering capacity to acidic soils on rocks with low, moderate and infinite buffering capacities, and to non-acidic soils over any lithology, sees a transition from the occurrence of acid waters at all flow levels to the absence of acidic conditions at any discharge value.

Stream chemistry may also vary considerably between the tributaries of a single river system. Mapping of total and individual solute concentrations in the Exe Basin, Devon, UK under stable low-flow conditions, for example, has revealed marked and distinctive geographical patterns (Webb 1983; Webb & Walling 1983). At this scale of investigation, the influence of topographic factors, such as basin size, slope and altitude, may operate in addition to geological and land-use controls (Miller 1961; Foggin & Forcier 1977).

The influence of flow conditions

Solute concentrations in river channels are strongly influenced by flow conditions and may respond dramatically to temporal changes in river discharge levels. River flow exerts a strong control on water chemistry by determining the volume of water available to dilute dissolved concentrations (Hem 1970), although the precise relationship between solute behaviour and streamflow discharge will also reflect the origins and routes by which runoff is produced within the drainage basin system (Burt 1986).

Rating relationships

The response of total or individual solute levels to changing river flow is often represented by the construction of a rating relationship. Most commonly, this consists of a simple power function:

$$C = aQ^b \tag{1}$$

where C is solute concentration, Q is discharge, and a and b are the constant and exponent of the function. Most rivers exhibit decreasing total dissolved solids (TDS) concentrations with increasing flow (Fig. 6.3(a)) and therefore are characterized by a negative value for the b exponent of the rating relationship (Gregory & Walling 1973). For example, more than 97% of relationships between TDS concentration and discharge reported from a global sample of 370 rivers (Walling & Webb 1983) evidenced a dilution effect, and this widespread phenomenon may be explained in general terms by reference to the sources and routes by which water and solutes are supplied to the river channel under different flow conditions. At low flows, solute concentrations are high because of evapotranspirational losses from the channel and because runoff is supplied from the lower soil profile and groundwater reservoir where water has a long residence time and solute release is promoted. During higher discharges, runoff is generally translated much more rapidly to the river channel, has less opportunity for solute pickup, and therefore has a lower dissolved solids content.

The slope of solute-rating relationships can vary considerably between catchments and between individual chemical constituents. Several factors, including the nature of groundwater circulation (Talsma & Hallam 1982; Andrews 1983), the availability and location of readily soluble material in vegetation, soil and rock (Edwards 1973; Carling 1983), and soil buffering processes (Johnson *et al* 1969) may influence the slope of the rating relationship. Positive exponent values, which indicate the occurrence of a 'concentration' rather than a 'dilution' effect at higher discharges, are occasionally found for TDS concentrations, especially in rivers where solute concentrations are low and atmospheric fallout is the dominant source (Cryer 1976; Foster 1979a) or in drainage basins where unusual geological or hydrological conditions cause surface runoff to be more highly mineralized than subsurface flows (Iorns *et al* 1965; Cryer 1980). A positive relationship between concentration and flow is more frequently encountered for some individual ionic species, especially those such as NO_3^-, PO_4^{3-} and

K^+ which are actively involved in the nutrient cycles of forested and moorland ecosystems and are mobilized in surface runoff through the action of vegetation leaching (Bond 1979; Waylen 1979). The extension of the drainage network to areas not drained under dry conditions, where a store of readily soluble material produced by the breakdown of surface material by biological and microbiological processes can be tapped, may also contribute to increasing concentrations of nutrients at higher discharges. Concentrations of dissolved organic matter and dissolved organic carbon commonly exhibit a positive correlation with discharge level (Baker *et al* 1974; Grieve 1984), which is explained by the greater dissolved organic content of flows from the upper and more organic horizons of the soil that supply runoff at times of high discharge (Grieve 1990). Investigation of the variation in rating curves on a micro-scale in subcatchments of the Hubbard Brook Experimental Forest, New Hampshire, USA (Lawrence *et al* 1988; Lawrence & Driscoll 1990) has shown that the form of dissolved concentration–discharge relationships for parameters such as H^+, Al and dissolved organic carbon may vary markedly over short distances from the source of small forested streams, and this behaviour is explained by vertical variations in the subsurface flow paths involved and areal variations in soil solution chemistry and flow response.

Although the relationship between solute concentration and river discharge is reasonably well defined for many rivers, scattering of data points around the rating line does occur. Scatter may be reduced by subdividing data values on the basis of season (Foster 1978; Oborne *et al* 1980; Al-Jabbari *et al* 1983), stage conditions (Oxley 1974; Loughran & Malone 1976) and flow components (Walling 1974). In some catchments, however, the simple power function relationship is an oversimplified model of the general response of solute concentrations to changing flow conditions, and more complex segmented and polynomial functions (e.g. Steele 1968; Finlayson 1977; Foster 1980), and looped and trapezoidal rating plots (O'Connor 1976; Carbonnel & Meybeck 1975; Collins 1979) have to be adopted.

Storm-period responses

Complexity and variability of solute response to changing flow conditions is often apparent when fluctuations in water chemistry are examined during storm events (e.g. Miller & Drever 1977; Cornish 1982; Muraoka & Hirata 1988; Jenkins 1989). Intra- and intercatchment contrasts in storm-period response may be evident for particular dissolved constituents (e.g. Foster 1979b, Webb & Walling 1983; Reynolds *et al* 1989), and strong differences are often apparent between the chemographs of individual ions during flood events. The latter type of variability reflects the different origins and storage locations of different chemical species in a drainage basin and the extent to which they are accessed by runoff from different sources. Reid and colleagues (1981), for example, have shown in the Glendye catchment of northeast Scotland that dissolved species derived from chemical weathering (e.g. Ca^{2+}, HCO_3^-) are strongly diluted during storm events by waters originating in the upper layers of the soil, but that runoff from these surface horizons increases the concentration of Fe and related species in the river channel. The different flow pathways, their chemistry, and the residence time of water along them have been invoked to explain the episodic pattern of pH depression, which is associated with storm events in many acidic headwater streams. Monitoring of fluctuations of the major chemical constituents in the Svartberget catchment of northern Sweden (Fig. 6.3(b)), together with measurements of hillslope hydrology and chemistry, have suggested that the 'acid episodes' in the stream arise from the passage of runoff through organic-rich forest mor and streambank vegetation (Bishop *et al* 1990). Solute response may also vary between storm events for an individual chemical constituent at a particular river site. Such variation is often typical of streams dominated by atmospheric inputs of soluble material (Fig. 6.3(c)), where interstorm differences in precipitation chemistry strongly influence stream solute concentration (Dupraz *et al* 1982). Marked seasonal variation in storm-period response of nitrate-nitrogen concentrations have been recorded in the River Dart, Devon, UK, where

marked dilution occurs in winter months but concentration effects are typical of the summer period. This contrast in behaviour reflects the fact that in winter months storm runoff is low but baseflows are relatively high in nitrate-nitrogen, whereas in the summer period, the nitrate-nitrogen content of baseflow is low but that of storm runoff, which originates as throughflow in the soil, is relatively high (Webb & Walling 1985).

Complexity in solute behaviour during storm events is often associated with the occurrence of hysteretic effects, whereby chemical concentrations differ markedly at the same value of discharge on the rising and falling limbs of the hydrograph, and a looped trend in the relationship between concentration and flow results for an individual storm event (e.g. Hendrickson & Krieger 1964; Webb *et al* 1987; Meybeck *et al* 1989). Hysteresis reflects contrasts in solute and discharge responses during storm events with respect to both their timing and form (Walling & Webb 1986a), and this may be promoted by first and subsequent flushes of soluble material accumulated in a prestorm period (Walling & Foster 1975; Klein 1981; Muscutt *et al* 1990), especially after periods of prolonged drought (Slack 1977; Walling & Foster 1978), by the exhaustion of solute stores within a drainage basin during a sequence of flood hydrographs (Walling 1978), and by variations throughout a storm in the contribution of flows from vertically and areally differentiated catchment sources, which differ in their chemical characteristics (Burt 1979; Anderson & Burt 1982; Johnson & East 1982; Burt & Arkell 1986; Lawrence & Driscoll 1990). Additional complexity may arise through the modification of soil-water chemistry in its passage through the riparian zone to the stream channel (Stanford & Ward 1988; Fiebig *et al* 1990). Increasingly, stream-water chemistry is being viewed as the result of a variable mixture of soil-water end-members (Lawrence *et al* 1988; Christophersen *et al* 1990a), and Hooper and colleagues (1990), for example, have explained fluctuations in stream chemistry at the Panola Mountain catchment, Georgia, USA as a varying mixture of groundwater, water draining from the

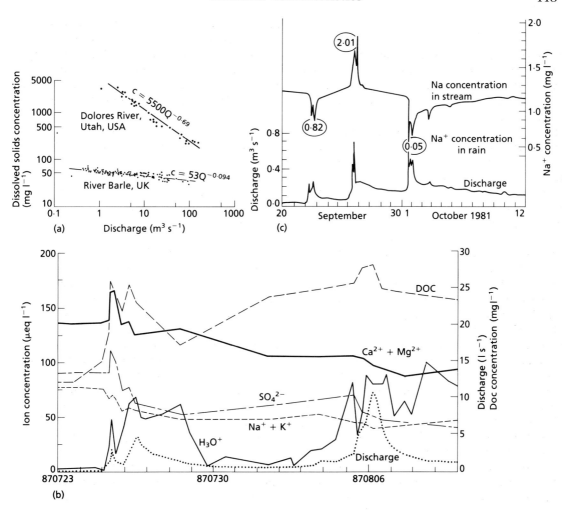

Fig. 6.3 Flow-related variations in river solute concentrations. (a) Solute concentration in discharge rating relationships (data for Dolores River; from Iorns *et al* 1965). (b) Storm-period changes in water chemistry in the Swedish catchment, Svartberget (after Bishop *et al* 1990). (c) Variation in storm-period solute response of Na^+ according to precipitation chemistry in a French drainage basin; Valet des Cloutasses (after Dupraz *et al* 1982).

hillslope and flows from the organic horizon.

Storm-period solute responses may show additional complexity when large, rather than small, drainage basins are considered. Studies in the Exe Basin, Devon, UK (Walling & Webb 1980; Webb & Walling 1982a), have demonstrated that fluctuations in water chemistry at downstream sites during storm events are strongly affected by the spatial origins of storm runoff in the upstream catchment area and by the routing of flows through the channel network which results in a

kinematic differential between floodwave and floodwater velocities (Glover & Johnson 1974).

Time trends

Solute levels also exhibit variations through time which are not related to the occurrence of storm events. Marked diurnal oscillations have been recorded for total and individual solute concentrations (Walling 1975) and have been accounted for by changes in discharge, which are caused by increasing evaporation and a lowering of the water-table during the day and in turn promote a daily cycle of accumulation and redissolution of soluble material (Hem 1948). An investigation of diurnal variations of electrical conductivity in the Kassjoan basin in Central Sweden suggests that the effects of evapotranspiration may be supplemented by the influence of earth tides, which are assumed to affect fissures within the bedrock and in turn change the composition of groundwater contributing to river runoff (Calles 1982).

A clear annual cycle of solute behaviour is apparent for many rivers (e.g. Sutcliffe & Carrick 1973; Foster & Walling 1978; Houston & Brooker 1981; Neal *et al* 1990a) and can be readily identified when chemical concentrations of river samples collected throughout the year are plotted up against discharge in the form of 'elliptical doughnut' and 'Q-c-t' (Fig. 6.4(a)) diagrams (Gunnerson 1967; Davis & Keller 1983). In some catchments, the annual march of stream solute levels is a simple inverse reflection of the discharge regime, but in other rivers an annual hysteresis in concentrations will be present, reflecting such factors as autumn flushing of soluble material accumulated over summer months, spring release of meltwater from a winter snowpack, and seasonal variations in the chemistry of incoming precipitation (Feller & Kimmins 1979; Williams *et al* 1983). The annual cycle of biological activity may also strongly influence water chemistry through the impact of autumnal vegetation dieback and leaf fall (Slack & Feltz 1968), the seasonal uptake by plants and animals (Edwards 1974; Casey & Ladle 1976; Casey *et al* 1981) and the effects of microbial populations on the reactions that build up and decompose organic matter in catchment soils (e.g. Blackie & Newson 1986; Stevens *et al* 1989). Dissolved organic matter concentrations may exhibit seasonal changes which are much more pronounced than those occurring during storm events. Results from the Loch Fleet catchment in south-west Scotland (Grieve 1990), for example, indicate storm-period variations in dissolved organic carbon of about 2 mg l^{-1} but a seasonal fluctuation with an amplitude of $8-9$ mg l^{-1}. Dissolved organic carbon and dissolved organic matter often reach maximum levels in the summer and autumn period, which can be explained by a concentration effect due to summer drying (Moore 1987) but also more strongly by increased decomposition of soil organic materials at higher temperatures during the summer period (Grieve 1990). The nature of the annual cycle of water chemistry may also vary in character from year to year. Stream concentrations of nitrate-nitrogen may show large contrasts between the autumn periods of different years, depending on the amount of accumulation in the catchment over the summer months and on the sequence by which soils are wetted up and nitrogen is mobilized to water-courses with the onset of the winter season (Webb & Walling 1985). Data on dissolved reactive phosphate concentrations, collected on a weekly basis over a 16-year period in the River Frome, Dorset, UK (Casey & Clarke 1986), although indicating a spring depression in levels also reveal large year-to-year variations in the annual regime of this nutrient (Fig. 6.4(b)).

Significant changes in solute concentrations may also be evident over longer time periods where records of sufficient quality and length are available to establish a trend. Smith and colleagues (1987) examined water quality trends in major US rivers over the period 1974−81, and found in the case of alkalinity (Fig. 6.4(c)), for example, that significant decreases were more common than significant increases. In general, it was observed that the reductions in alkalinity did not result from the introduction of strong acids associated with acid deposition from the atmos-

Fig. 6.4 (*Opposite*) Temporal trends in river solute concentrations. (a) Annual hysteresis in total hardness concentrations over period from October 1961 to September 1962 in the Snake River, Wawawai, USA (after Gunnerson 1967) (numbers 1−12 indicate calendar month during which sample was taken). (Inset depicts a 'Q-c-t' diagram which shows the annual cycle of specific conductance in Basin 3, Alptal, Switzerland; after Davis & Keller 1983.) (b) Variation in the annual regime of dissolved reactive phosphate concentration in a Dorset river; River Frome, UK, 1965−80 (after Casey & Clarke 1986). (c) Trends in flow-adjusted concentrations of alkalinity (1974−81) at NASQAN stations in the USA (after Smith *et al* 1987). (d) Long-term trends in nitrate-nitrogen concentrations of selected European rivers (after Dykzeul 1982; Hagebro *et al* 1983; Roberts & Marsh 1987). (e) Fertilizer usage in western Europe (after Roberts & Marsh 1987).

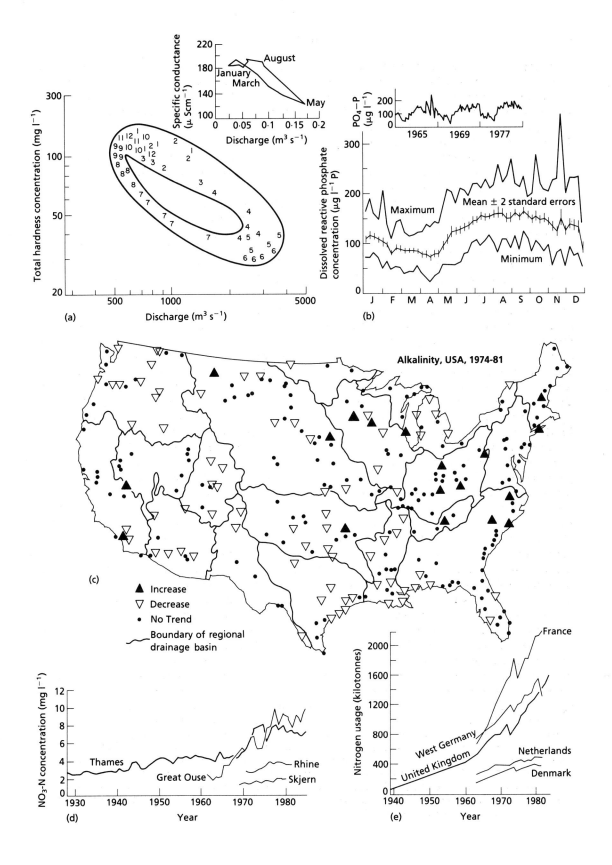

(a)

(b)

(c) Alkalinity, USA, 1974-81

▲ Increase
▽ Decrease
• No Trend
— Boundary of regional drainage basin

(d)

(e)

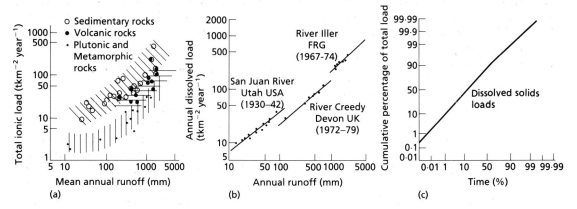

Fig. 6.5 Variations in river solute loads. (a) World rivers: the relationship between total ionic load and mean annual runoff for different rock types (from Meybeck 1981). (b) Selected catchments: the relationship between dissolved load and annual runoff for three rivers in different climatic zones. (c) River Creedy, UK: a dissolved load duration curve for a Devon river.

phere or with acid mine drainage, but rather reflected other processes occurring in the soils and vegetation of drainage basin, including the reduction of soil carbon dioxide and uptake of base cations during tree growth. A steady decline in calcium $(1.1–1.4\ \mu eq\ l^{-1}\ year^{-1})$ and magnesium $(0.6–0.8\ \mu eq\ l^{-1}\ year^{-1})$ have been recorded for the period 1972–87 in the Birkenes catchment of southern Norway and ascribed to soil acidification in the drainage area (Christophersen *et al* 1990b). Significant increases in nitrate-nitrogen concentrations have been monitored in the surface waters of many western European countries over the past 30 years (Fig. 6.4(d)) and an intensification of agricultural practice, involving land-use changes and the increased use of nitrogenous fertilizers (Fig. 6.4(e)), is identified as a major contributor to this trend (Dykzeul 1982; Hagebro *et al* 1983; Roberts & Marsh 1987; Burt *et al* 1988; Meybeck *et al* 1989). Analysis of changes in nitrate-nitrogen concentrations at 383 rivers across the USA over a shorter period (1974–81) also indicated widespread increases, but suggested that increased emissions of NO_x gases may have played a large role in elevating stream nitrate-nitrogen levels in the midwestern and mid-Atlantic regions (Smith *et al* 1987).

Solute fluxes

Considerable quantities of dissolved material can be transported over time by a river system. On a global scale, it is estimated that total dissolved load transport to the oceans is 3.7×10^9 t year^{-1} (Walling & Webb 1987), although for individual rivers solute loads may range from <1.0 t km^{-2} year^{-1} in catchments underlain by rocks resistant to chemical attack to values of 6000 t km^{-2} year^{-1} in basins underlain by halite deposits (Meybeck 1984). The mean annual solute load tends to increase with increasing mean annual runoff (Walling & Webb 1983), although there is considerable scatter in this relationship (Fig. 6.5(a)) which can be largely attributed to variation in basin geology (Meybeck 1981). Spatial patterns of solute transport become more complex and are explicable in terms of additional environmental factors, such as land use and topographic characteristics, when variations over more restricted geographical areas are considered (e.g. Walling & Webb 1981).

Solute flux also fluctuates over time as well as in space. A close relationship is often evident between annual dissolved load and annual runoff as results from three contrasted rivers indicate (Fig. 6.5(b)). Transport of solutes in river systems is generally less biased towards episodic and high flows than is the case for particulate material. A long-term record from the River Creedy, Devon, UK (Fig. 6.5(c)) has indicated that 50% and 90% of solute flux occurred in 12% and 56%, respectively, of the 7.75-year study period (Webb & Walling 1982b). Storm events may influence the transport of dissolved organic material more strongly than that of dissolved inorganic constituents because flows from surface or near-surface horizons of the soil are most effective in mobilizing organic material in solution. Results from a small catchment in the Ochil Hills in eastern Scotland suggested that half of the annual load of dissolved organic matter is carried by flows occurring less than 6% of the time (Grieve 1984).

6.2 SEDIMENT-ASSOCIATED TRANSPORT

The preceding discussion of water chemistry and material transport by rivers relates to substances carried in solution. This is conventionally defined as material passing through a 0.45-μm filter and may in reality include colloidal material as well as true solutes. It is, however, important to emphasize that the suspended sediment carried by a river, which was discussed in section 3.1 and which is traditionally studied in isolation as a physical water-quality characteristic, also has important implications for water chemistry and material transport. Suspended sediment may exhibit a complex chemical composition both as a mineral or organic substrate and in terms of the substances 'attached' to or 'incorporated' into the sediment particles. Interactions between material in solution and the sediment may exert an important control on water chemistry. Thus, for example, sediment-associated substances may be released into solution and, conversely, substances in solution may be sorbed by sediment. In some circumstances, increases and decreases in concentration associated with the dissolved phase could therefore be buffered by uptake or release associated with the particulate phase. Any assessment of material transport by rivers must also consider both the sediment-associated and the dissolved components, because the former may dominate the transport of many substances. Förstner (1977) has, for example,

estimated that 98% of the Fe and Al transport by rivers in the USA and Germany occurs in association with the particulate load.

Sediment geochemistry and bonding mechanisms

The natural suspended sediment load of a river will comprise both inorganic and organic material, and a distinction may be made between the allochthonous and autochthonous components. The former component is commonly dominant and represents material washed into the river from the surrounding drainage basin, whereas the latter component comprises material formed in the water-body itself. The inorganic material will be largely allochthonous in origin and comprises clay minerals, mineral and rock fragments and precipitates (e.g. SiO_2 and $CaCO_3$). The organic matter consists of micro-organisms (phytoplankton, zooplankton and bacteria), the remains of macrophytes and other large-sized organisms, detritus derived from decaying material, and organic matter associated with eroded soil. A large proportion of the organic material may therefore be autochthonous in origin. The composition of individual particles may be complex, involving both organic and inorganic coatings of the particle substrate.

Bonding of substances to sediment occurs as a result of two major groups of processes, namely physical and chemical sorption (*cf.* Golterman *et al* 1983). The former primarily involves electrostatic attraction and ion exchange, whereas the latter includes exchange interactions with the OH groups of clay minerals and metal hydroxides and with the COOH and OH groups of organic substances, chelation, interactions with Fe/Mn oxyhydrates and coprecipitation.

Sediment – water quality interactions

With some sorption processes operating in natural waters, adsorption equilibria may be attained between suspended particulates and fine-bed material and certain dissolved components. These equilibria are commonly regulated by temperature, pH and the nature of the clay minerals and they can control the trace element concentration in the water. Thus, for example, Golterman *et al* (1983) indicated that if large quantities of zinc were to be released into a river to raise concentrations to in excess of 1 mg l^{-1}, these would return to the normal level of 10–100 µg l^{-1}, within a downstream distance of 10 km. Similar equilibria have also been widely reported for phosphorus, such that a reduction in dissolved phosphate concentrations can trigger desorption and vice versa (e.g. Stumm & Leckie 1971). This behaviour can be quantified experimentally using adsorption isotherms (*cf.* Olsen & Watanabe 1957). In the case of phosphorus, Froelich (1988) suggested that the process of sorption and desorption by sediment involves two components that operate at different speeds. With sorption, the first 'fast' reaction occurs rapidly over timescales of minutes to hours and represents adsorption on to an exchangeable ion surface, whilst the second 'slow' reaction operates on timescales of days to months or perhaps years, and represents a slow diffusion into the particle. Because of this 'slow' second step, reversal of the process and therefore desorption may occur less readily.

As indicated above, sorption of substances on to sediment frequently operates as a reversible process. In some circumstances, however, sorbed or incorporated substances are more firmly fixed. Nevertheless, biological activity and changes in the river environment may cause release from the particulate to the dissolved phase and in this way the particulate phase may again influence the dissolved phase (*cf.* Table 6.2).

When considering both sediment bonding and interaction between the dissolved and particulate phase, the nature and composition of the sediment particles will exert an important control on their precise significance. In general, the 'chemical activity' of sediment increases with decreasing grain size due to the increased specific surface area (m^2 g^{-1}) of fine clay particles (*cf.* Fig. 6.6(a)). Such fine particles will, for example, therefore exhibit greatly increased levels of cation exchange capacity (*cf.* Fig. 6.6(b)). Analyses of the copper content of several size fractions of suspended sediment from the River Amazon reported by Gibbs (1977) and presented in Fig. 6.6(c) demonstrate the preferential association of this element with the finest fractions of the sediment trans-

ported by this river. Similar results for the total concentration of a variety of major and trace elements in the bed sediments of the Niagara River in Canada have been reported by Mudroch and Duncan (1986), although in this case the cutoff for the finest fraction was much larger (13 µm). Variations in particle composition and the presence or absence of surface coatings may also be important. Thus in the case of phosphorus sorption, Froelich (1988) suggested that the initial 'fast' step is largely dependent on the surface area and charge balance of the sediment particles, whereas the second 'slow' diffusion step is very dependent on the composition of the solid phase. Sediments containing iron and aluminium oxides, or with surfaces coated with these substances, display a much higher 'slow' sorption capacity, probably due to the reaction of phosphate with the oxides. Thus 'pure' clays (e.g. kaolinite) have only a limited ability to sorb phosphorus beyond the initial step, while gibbsite or clays containing natural oxide coatings have a much higher capacity for phosphorus sorption. As noted in section 5.1, the particle-size distribution and other properties of suspended sediment can be expected to vary temporally, in response to season and flow magnitude (*cf.* Fig. 6.6(d)), and these may in turn have important implications for sediment–water interactions.

Sediment-associated loads

When investigating catchment nutrient budgets or other aspects of material export from drainage basins, it is important to recognize the need to consider sediment-associated loads as well as the transport of material in solution. In some cases the former may represent a major proportion of the total load. For example, Duffy *et al* (1978) reported that the sediment phase accounts for 64–76% of the total phosphorus yield from five small forested watersheds in north-central Mississippi, USA, and Schuman and colleagues (1976) indicated that sediment losses from small contour-cropped watersheds at Treynor, Iowa, USA account for 85% of the total discharge of phosphorus. Meybeck (1984) has also emphasized the important role of suspended sediment in geochemical budgets and in the overall flux of el-

Table 6.2 Fate of particulate phases when submitted to various environmental changes

Environmental changes	Electrostatically adsorbed	Specifically adsorbed	Bound to particulate organic matter	Bound to carbonates	Occluded in Fe, Mg oxides and hydroxide	Bound to sulphides	Silicates and other residuals
Bacterial degradation	0	0	-	0	0	+ anaerobic - aerobic	0
Establishment of oxidizing conditions	0	0	0	0	+	-	0
Establishment of reducing conditions	0	0	0	0	-	+	0
Small pH variations	+/-	+/-	0	+/-	0	+/-	0, -(1)
Transfer to brackish or saline water	-	0	0	0	0	0	0
Intestinal tract of organisms after ingestion	-	-	-	-	-	-	0

0, no marked change; -, release from particulate matter to dissolved phase; +, gain from dissolved phase to particulate phase; +/-, release or gain; (1), in the case of diatoms. Based on Unesco (1978).

Chapter 6

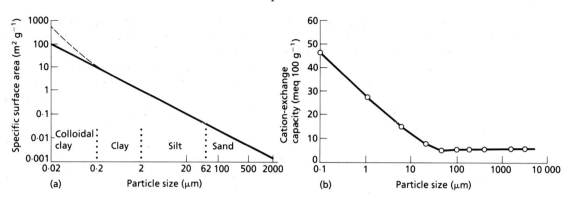

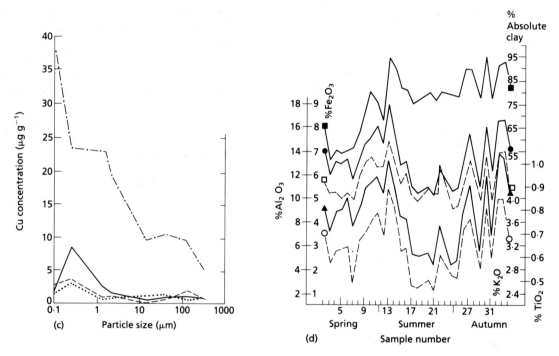

Fig. 6.6 The significance of the particle-size composition of suspended sediment in controlling (a) specific surface area. (b) Cation exchange capacity, Mattole River, California, USA. (c) Trace metal content in the River Amazon. —·—· crystal; ——— coat; ---- organic; ····· exchange. (d) Seasonal variation of the chemistry and clay content of suspended sediment transported by Wilton Creek, Ontario, Canada. ■ Clay; ○ Al_2O_3; ▲ TiO_2; ● Fe_2O_3; □ K_2O. (After (b) Malcolm & Kennedy 1970; (c) Gibbs 1977; and (d) Ongley *et al* 1981).

ements from the land surface of the globe to the oceans. He estimated the typical elemental composition of suspended sediment transported by world rivers and used this to calculate the likely contribution of sediment-associated transport to the total land–ocean flux of individual elements. Some of his results presented in Table 6.3 provide information on the typical composition of suspended sediment and indicate, for example, that suspended sediment probably accounts for more than 90% of the transport of Al, Fe, Ti, Mn, Si and P, and a major proportion of the flux of several other elements. Suspended sediment also accounts for a major proportion of the transport of organic substances, such as organic nitrogen and organic phosphorus. Such estimates can clearly be only approximate at the present time, since detailed data on sediment composition are still lacking for many rivers of the world, but their general order of magnitude is likely to be correct.

Because of the episodic nature of high suspended sediment loads, sediment-associated transport also tends to be episodic and dominated by storm events. For this reason it is important that sampling programmes are designed carefully, to document the main periods of transport. Programmes of infrequent regular sampling are unlikely to provide meaningful load data (*cf.* Walling & Webb 1985). The need to collect substantial quantities of sediment for chemical analysis may also necessitate the use of specialized equipment such as a continuous flow centrifuge, which permits the recovery of sediment from bulk water samples (e.g. Horowitz *et al* 1989).

When evaluating sediment-associated loads of specific substances, it is frequently useful to assemble information on the types of sediment bonding involved. This could, for example, be particularly important in studies of sediment–water quality interactions, when it is necesary to predict the potential interchange between the particulate and dissolved phases. Equally, when

Table 6.3 Role of suspended sediment in the global transport of major elements from the land to the oceans by rivers (after Meybeck 1984).

Element	Sediment-associated transport / Total transport (%)	Typical sediment content (mg g^{-1})
Al	99.87	90.0
Fe	99.8	51.5
Ti	99.6	5.8
Mn	98	1.0
Si	96	281
P	95	1.15
K	86.5	21.0
F	80	0.97
B	60	0.07
N	57	1.2
Mg	56	11.0
Ca	43	24.5
C	34	20.0
Sr	32	0.15
Na	28	7.1
S	13	2.15
Cl	1	0.5
Inorganic C	28	10.0
Organic C	45	10.0
Organic N	68	1.2
Organic P	94	0.45
Inorganic P	97	0.70

investigating the fate of sediment deposited in a lake or other depositional sinks it could be important to determine whether a substance may be released from the sediment or whether it is permanently bound and therefore unlikely to participate in future biogeochemical cycling. Bioavailability assessments assume particular importance in the case of toxic contaminants, but such pollutants lie beyond the scope of this discussion.

To take the example of sediment-associated phosphorus, fractionation procedures exist to separate the apatite phosphorus (A-P), the non-apatite inorganic phosphorus (NAI-P) and the organic phosphorus (O-P) (*cf.* Allan 1979). The A-P may be viewed as mineral phosphorus associated with rock particles and this is essentially inert and unavailable for biological uptake or dissolved-particulate phase interactions. The O-P includes all phosphorus in organic forms and may have both autochthonous and allochthonous origins. This is not immediately available for biological uptake, but is a pool of potentially available phosphorus. Bacterial degradation of algae will, for example, release orthophosphorus. NAI-P consists of orthophosphorus adsorbed by or occluded within sediment particles and is the fraction which is readily bioavailable. Other workers have developed procedures for further fractionating the NAI-P, for example into loosely sorbed phosphorus and Fe- and Al-bound phosphorus (*cf.* Pettersson 1986). Similar fractionation procedures exist for determining the speciation of sediment-associated metals (e.g. Tessier *et al* 1979) and Table 6.4 demonstrates their potential utility. Some of these procedures have been used by

Gibbs (1977) to fractionate the copper content of suspended sediment transported by the River Amazon, as shown in Fig. 6.6(c). Where interest focuses primarily on the potential bioavailability of substances such as phosphorus, rather than their precise chemical speciation, bioassay techniques may also be employed (*cf.* Lee *et al* 1980; de Pinto *et al* 1981). In this case the crystalline fraction may be viewed as essentially inert, whereas the other fractions are more readily available, although to different degrees.

Temporal variations in sediment sources and erosional processes can be expected to cause variations in both the total content of specific sediment-associated substances and the relative importance of the various forms in which they are held within the sediment. To use again the example of phosphorus, Fig. 6.7 demonstrates seasonal variation in both the total-P content of suspended sediment and the contribution of O-P, A-P and NAI-P to the total-P content, as reported by Ongley (1978) for Wilton Creek, Ontario, Canada. A-P does not demonstrate a marked seasonal trend but reaches its highest concentrations in spring when the erosion associated with spring melt contributes large quantities of detrital material. The O-P content of suspended sediment evidences a marked seasonal cycle and rises rapidly in the spring and maintains high values throughout the summer in response to increased algal productivity. NAI-P similarly increases through the summer, probably reflecting the increased residence time of sediment within the channel during periods of low flow.

Table 6.4 Sequential extraction procedures used for the fractionation of sediment-associated trace metals

Extraction method	Extracted phase
H_2O	Easily soluble fraction
$BaCl_2$/TEA pH 8.1	Exchangeable cations
NaOH	Humates, fulvates
Acidic cation exchange	Carbonate fraction
$NH_2OH/HCl/HNO_3$	Mn-oxides, amorphous Fe-oxides
H_2O_2/NH_4OAc	Organic residues and sulphides
NH_2OH/HCl/acetic acid	Hydrous Fe-oxides
$HF/HClO_4$ digestion	Inorganic residues

After Allan (1979).

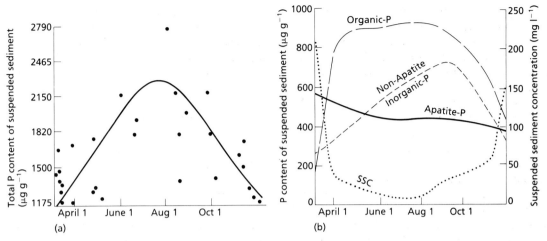

Fig. 6.7 Seasonal variation of (a) the total-P content of suspended sediment, and (b) sediment-associated P forms in Wilton Creek, Ontario, Canada (based on data from Ongley 1978, reported in Allan 1979).

REFERENCES

Afifi AA, Bricker OP. (1983) Weathering reactions, water chemistry and denudation rates in drainage basins of different bedrock types: 1 — Sandstone and shale. In: *Dissolved Loads of Rivers and Surface Water Quantity/Quality Relationships* (Proceedings of the Hamburg Symposium, August 1983), pp 193–203. International Association of Hydrological Sciences Publication No. 141. Wallingford, UK. [6.1]

Al-Jabbari MH, Al-Ansari NA, McManus J. (1983) Variation in solute concentration within the River Almond and its effect on the estimated dissolved load. In: *Dissolved Loads of Rivers and Surface Water Quantity/Quality Relationships* (Proceedings of the Hamburg Symposium, August 1983), pp 21–9. International Association of Hydrological Sciences Publication No. 141. Wallingford, UK. [6.1]

Allan RJ. (1979) Sediment-related fluvial transmission of contaminants: some advances by 1979. *Inland Waters Directorate, Environment Canada, Scientific Series No. 107.* Ottawa, Canada. [6.2]

American Public Health Association (1971) *Standard methods for the examination of water and waste water* 13th edn. APHA, Washington, DC. [6.1]

Anderson MG, Burt TP. (1982) The contribution of throughflow to storm runoff: an evaluation of a chemical mixing model. *Earth Surface Processes and Landforms* 7: 565–74. [6.1]

Andrews ED. (1983) Denudation of the Piceance Creek Basin, Colorado. In: *Dissolved Loads of Rivers and Surface Water Quantity/Quality Relationships* (Proceedings of the Hamburg Symposium, August 1983), pp 205–15. International Association of Hydro-logical Sciences Publication No. 141. Wallingford, UK. [6.1]

Baker CD, Bartlett PD, Farr IS, Williams GI. (1974) Improved methods for the measurement of dissolved and particulate organic carbon in fresh water and their application to chalk streams. *Freshwater Biology* 4: 467–81. [6.1]

Belot Y, Caput C, Gauthier C. (1982) Transfer of americium from sea water to atmosphere by bubble bursting. *Atmosphere and Environment* 16: 1463–6. [6.1]

Best GR, Monk CD. (1975) Cation flux in hardwood and white pine watersheds. In: Howell FG, Gentry JB, Smith MH (eds) *Mineral Cycling in South Eastern Ecosystems*, pp 847–61. United States Energy Research and Development Administration, Washington, DC. [6.1]

Bishop KH, Grip H, O'Neill A. (1990) The origins of acid runoff in a hillslope during storm events. In: Neal C, Hornung M (eds) *Special Issue: Transfer of Elements through the Hydrological Cycle. Journal of Hydrology* 116: 35–61. [6.1]

Blackie JR, Newson MD. (1986) The effects of forestry on the quantity and quality of runoff in upland Britain. In: Solbe JF de LG (ed.) *Effects of Land Use on Fresh Waters*, pp 398–412. Ellis Horwood Series in Water and Wastewater Technology, Water Research Centre Publication, Chichester. [6.1]

Bloch MRD, Kaplan V, Kertes V, Schnerb J. (1966) Ion separation in bursting air bubbles: an explanation for the irregular ion ratios in atmospheric precipitations. *Nature* 209: 802–3. [6.1]

Bond HW. (1979) Nutrient concentration patterns in a stream draining a montane ecosystem in Utah.

Ecology **60**: 1184–96. [6.1]

Bruijnzeel LA. (1983) The chemical mass balance of a small basin in a wet monsoonal environment and the effect of fast-growing plantation forest. In: *Dissolved Loads of Rivers and Surface Water Quantity/Quality Relationships* (Proceedings of the Hamburg Symposium, August 1983), pp 229–39. International Association of Hydrological Sciences Publication No. 141. Wallingford, UK. [6.1]

Buijsman E, Mass HFM, Asman WAH. (1987) Anthropogenic NH₃ emissions in Europe. *Atmosphere and Environment* **21**: 1009–22. [6.1]

Burt TP. (1979) The relationship between throughflow generation and the solute concentration of soil and stream water. *Earth Surface Processes* **4**: 257–66. [6.1]

Burt TP (1986) Runoff processes and solutional denudation rates on humid temperate hillslopes. In: Trudgill ST (ed.) *Solute Processes*, pp 193–249. John Wiley and Sons, Chichester, UK. [6.1]

Burt TP, Arkell BP. (1986) Variable source areas of stream discharge and their relationship to point and non-point sources of nitrate pollution. In: *Monitoring to Detect Changes in Water Quality Series* (Proceedings of the Budapest Symposium, July 1986), pp 155–64. International Association of Hydrological Sciences Publication No. 157. Wallingford, UK. [6.1]

Burt TP, Arkell BP, Trudgill ST, Walling DE. (1988) Stream nitrate levels in a small catchment in southwest England over a period of 15 years (1970–1985). *Hydrological Processes* **2**: 267–84. [6.1]

Calles UM. (1982) Diurnal variations in electrical conductivity of water in a small stream. *Nordic Hydrology* **13**: 157–64. [6.1]

Calow P, Petts GE. (eds) (1994) *The Rivers Handbook*, vol. 1. Blackwell Scientific Publications, Oxford. [6.1]

Carbonnel JP, Meybeck M. (1975) Quality variations of the Mekong River at Phnom Penh, Cambodia, and chemical transport in the Mekong Basin. *Journal of Hydrology* **27**: 249–65. [6.1]

Carling PA. (1983) Particulate dynamics, dissolved and total load, in two small basins, northern Pennines, UK. *Hydrological Sciences Journal* **28**: 355–75. [6.1]

Casey H, Clarke RT. (1986) The seasonal variation of dissolved reactive phosphate concentrations in the River Frome, Dorset, England. In: *Monitoring to Detect Changes in Water Quality Series* (Proceedings of the Budapest Symposium, July 1986), pp 257–65. International Association of Hydrological Sciences Publication No. 157. Wallingford, UK. [6.1]

Casey H, Ladle M. (1976) Chemistry and biology of the South Winterbourne, Dorset, England. *Freshwater Biology* **6**: 1–12. [6.1]

Casey H, Clarke RT, Marker AFH. (1981) The seasonal variation in silicon concentration in chalk-streams in relation to diatom growth. *Freshwater Biology* **11**: 335–44. [6.1]

Christophersen N, Neal C, Hooper RP, Vogt RD, Andersen S. (1990a) Modelling streamwater chemistry as a mixture of soilwater end-members—a step towards second-generation acidification models. In: Neal C, Hornung M (eds) *Special Issue: Transfer of Elements through the Hydrological Cycle. Journal of Hydrology* **116**: 307–20. [6.1]

Christophersen N, Robson A, Neal C, Whitehead PG, Vigerust B, Henriksen A. (1990b) Evidence for long-term deterioration of streamwater chemistry and soil acidification at the Birkenes catchment, southern Norway. In: Neal C, Hornung M (eds) *Special Issue: Transfer of Elements through the Hydrological Cycle. Journal of Hydrology* **116**: 63–76. [6.1]

Collins DN. (1979) Hydrochemistry of meltwaters draining from an alpine glacier. *Arctic and Alpine Research* **11**: 307–24. [6.1]

Colombani J. (1983) Evolution de la concentration en matières dissolutes en Afrique. Deux exemples opposés: les fleuves du Togo et la Medjerdah en Tunisie. In: *Dissolved Loads of Rivers and Surface Water Quantity/Quality Relationships* (Proceedings of the Hamburg Symposium, August 1983), pp 51–69. International Association of Hydrological Sciences Publication No. 141. [6.1]

Cornish PM. (1982) The variations of dissolved ion concentration with discharge in some New South Wales streams. In: O'Loughlin EM, Bren LJ (eds) *The First National Symposium on Forest Hydrology*, Melbourne, 11–13 May 1982, pp 67–71.

Cryer R. (1976) The significance and variation of atmospheric nutrient inputs in a small catchment system. *Journal of Hydrology* **29**: 121–37. [6.1]

Cryer R. (1980) The chemical quality of some pipeflow waters in upland mid-Wales and its implications. *Cambria* **6**: 28–46. [6.1]

Cryer R. (1986) Atmospheric solute inputs. In: Trudgill ST (ed.) *Solute Processes*, pp 15–84. John Wiley and Sons, Chichester, UK. [6.1]

Davis JS, Keller HM. (1983) Dissolved loads in streams and rivers—discharge and seasonally related variations. In: *Dissolved Loads of Rivers and Surface Water Quantity/Quality Relationships* (Proceedings of the Hamburg Symposium, August 1983), pp 79–89. International Association of Hydrological Sciences Publication No. 141. Wallingford, UK. [6.1]

De Pinto JV, Young TC, Martin SC. (1981) Algal-available phosphorus in suspended sediments from lower Great Lakes tributaries. *Journal of Great Lakes Research* **7**: 311–25. [6.2]

Duffy PD, Schreiber JD, McClurkin DC, McDowell LL. (1978) Aqueous and sediment phase phosphorus yield from five southern pine watersheds. *Journal of Environmental Quality* **7**: 45–50. [6.2]

Dupraz C, Lelong F, Trop JP, Dumazet B. (1982) Comparative study of the effects of vegetation on the hydrological and hydrochemical flows in three minor catchments of Mount Lozère (France) — methodological aspects and first results. In: *Hydrological Research Basins and their Use in Water Resources Planning*, pp 671–82. Landeshydrologies, Berne. [6.1]

Dykzeul A. (1982) *The water quality of the River Rhine in the Netherlands over the period 1970–81*. Government Institute for Waste Water Treatment, Report No. 82–061, pp 43–6. [6.1]

Edwards AMC. (1973) The variation of dissolved constituents with discharge in some Norfolk rivers. *Journal of Hydrology* **18**: 219–42. [6.1]

Edwards AMC. (1974) Silicon depletions in some Norfolk rivers. *Freshwater Biology* **4**: 267–74. [6.1]

Edwards RW, Gee AS, Stoner JH. (1990) *Acid Waters in Wales*. Kluwer Academic Publishers, Dordrecht, Netherlands. [6.1]

Eriksson E. (1960) The yearly circulation of chloride and sulfur in nature; meteorological, geochemical and pedological implications, part II. *Tellus* **12**: 63–109. [6.1]

Feller MC, Kimmins JP. (1979) Chemical characteristics of small streams near Haney in Southwestern British Columbia. *Water Resources Research* **15**: 247–58. [6.1]

Ferrier RC, Walker TAB, Harriman R, Miller JD, Anderson HA. (1990) Hydrological and hydrochemical fluxes through vegetation and soil in the Allt a'Mharcaidh, Western Cairngorms, Scotland: their effect on streamwater quality. In: Neal C, Hornung M (eds) *Special Issue: Transfer of Elements through the Hydrological Cycle. Journal of Hydrology* **116**: 251–66. [6.1]

Fiebig DM, Lock MA, Neal C. (1990) Soil water in the riparian zone as a source of carbon for a headwater stream. In: Neal C, Hornung M (eds) *Special Issue: Transfer of Elements through the Hydrological Cycle. Journal of Hydrology* **116**: 217–37. [6.1]

Finlayson B. (1977) *Runoff contributing areas and erosion*. Research Papers, School of Geography, University of Oxford, No. 18. [6.1]

Foggin GT, Forcier LK. (1977) *Using topographic characteristics to predict total solute concentrations in streams draining small forested watersheds in western Montana*. University of Montana Joint Water Resources Research Center Report No. 89. [6.1]

Förstner U. (1977) Trace metals. In: Shear H, Watson AEP (eds) *The Fluvial Transport of Sediment-Associated Nutrients and Contaminants*, pp 219–33. International Joint Commission on the Great Lakes, Windsor, Ontario. [6.2]

Foster IDL. (1978) Seasonal solute behaviour of stormflow in a small agricultural catchment. *Catena* **5**: 151–63. [6.1]

Foster IDL. (1979a) Chemistry of bulk precipitation throughfall, soil water and stream water in a small catchment in Devon, England. *Catena* **6**: 145–55. [6.1]

Foster IDL. (1979b) Intra-catchment variability in solute response, an East Devon example. *Earth Surface Processes* **4**: 381–94. [6.1]

Foster IDL. (1980) Chemical yields in runoff, and denudation in a small arable catchment, East Devon, England. *Journal of Hydrology* **47**: 349–68. [6.1]

Foster IDL, Grieve IC. (1984) Some implications of small catchment solute studies for geomorphological research. In: Burt TP, Walling DE (eds) *Catchment Experiments in Fluvial Geomorphology*, pp 359–98. GeoBooks, Norwich. [6.1]

Foster IDL, Walling DE. (1978) The effects of the 1976 drought and autumn rainfall on stream solute levels. *Earth Surface Processes* **3**: 393–406. [6.1]

Froelich PN. (1988) Kinetic control of dissolved phosphate in natural rivers and estuaries: a primer on the phosphate buffer mechanism. *Limnology and Oceanography* **33**: 649–68. [6.2]

Garrels RM, Christ CL. (1965) *Solutions, Minerals and Equilibria*. Harper and Row, New York. [6.1]

Gibbs R. (1970) Mechanisms controlling world water chemistry. *Science* **170**: 1088–90. [6.1]

Gibbs RJ. (1977) Transport phases of transition metals in the Amazon and Yukon Rivers. *Geological Society of America Bulletin* **88**: 829–43. [6.2]

Glover BJ, Johnson P. (1974) Variations in the natural chemical concentrations of river water during flood flows and the lag effect. *Journal of Hydrology* **22**: 303–16. [6.1]

Golterman HL, Clymo RS (eds) (1969) Methods for chemical analysis of fresh waters. *International Biological Programme Handbook 8.* [6]

Golterman HL, Sly PG, Thomas RL. (1983) *Study of the Relationship Between Water Quality and Sediment Transport*. Unesco Technical Papers in Hydrology No. 26, Unesco, Paris. [6.2]

Gregory KJ, Walling DE. (1973) *Drainage Basin Form and Process: A Geomorphological Approach*. Edward Arnold, London. [6.1]

Grieve IC. (1984) Concentrations and annual loading of dissolved organic matter in a small moorland stream. *Freshwater Biology* **14**: 533–7. [6.1]

Grieve IC. (1990) Seasonal, hydrological, and land management factors controlling dissolved organic carbon concentrations in the Loch Fleet catchments, southwest Scotland. *Hydrological Processes* **4**: 231–9. [6.1]

Gunnerson CG. (1967) Streamflow and quality in the Columbia River Basin. *Proceedings American Society of Civil Engineers, Journal of the Sanitary Engineering Division* **93** (SA6): 1–16. [6.1]

Hagebro C, Bang S, Somer E. (1983) Nitrate load/discharge relationships and nitrate load trends in Danish rivers. In: *Dissolved Loads of Rivers and Surface Water Quantity/Quality Relationships* (Proceedings of the Hamburg Symposium, August 1983), pp 377–86. International Association of Hydrological Sciences

Publication No. 141. Wallingford, UK. [6.1]

Hem JD. (1948) Fluctuations in concentrations of dissolved solids in some southwestern streams. *Transactions of the American Geophysical Union* **29**: 80–4. [6.1]

Hem JD. (1970) Study and interpretation of the chemical characteristics of natural water. *United States Geological Survey Water-Supply Paper 1473*. [6.1]

Hendrickson GE, Krieger RA. (1964) Geochemistry of natural waters of the Blue Grass Region, Kentucky. *United States Geological Survey Water-Supply Paper 1700*. [6.1]

Hooper RP, Christophersen N, Peters NE. (1990) Modelling streamwater chemistry as a mixture of soilwater end-members—an application to the Panola Mountain catchment, Georgia, USA. In: Neal C, Hornung M (eds) *Special Issue: Transfer of Elements through the Hydrological Cycle. Journal of Hydrology* **116**: 321–43. [6.1]

Horowitz AJ, Elrick KA, Hooper RC. (1989) A comparison of instrumental dewatering methods for the separation on concentration of suspended sediment for subsequent trace element analysis. *Hydrological Processes* **3**: 163–84. [6.2]

Houston JA, Brooker MP. (1981) A comparison of nutrient sources and behaviour in two lowland subcatchments of the River Wye. *Water Research* **15**: 49–57. [6.1]

Hu M, Stallard RF, Edmond JM. (1982) Major ion chemistry of some large Chinese rivers. *Nature* **298**: 550–3. [6.1]

Iorns WV, Hembree CH, Oakland GL. (1965) Water resources of the Upper Colorado basin. *United States Geological Survey Professional Paper 441*. [6.1]

Jenkins A. (1989) Storm period hydrochemical response in an unforested Scottish catchment. *Hydrological Sciences Journal* **34**: 393–404. [6.1]

Johnson FA, East JW. (1982) Cyclical relationships between river discharge and chemical concentration during flood events. *Journal of Hydrology* **57**: 93–106. [6.1]

Johnson NM, Likens GE, Bormann FH, Fisher DW, Pierce RS. (1969) A working model for the variation in stream water chemistry at the Hubbard Brook Experimental Forest, New Hampshire. *Water Resources Research* **5**: 1353–63. [6.1]

Junge CE, Werby RT. (1958) The concentration of chloride, sodium, potassium, calcium and sulfate in rain water over the United States. *Journal of Meteorology* **15**: 417–25. [6.1]

Klein M. (1981) Dissolved material transport—the flushing effect in surface and subsurface flow. *Earth Surface Processes and Landforms* **6**: 173–8. [6.1]

Lawrence GB, Driscoll CT. (1990) Longitudinal patterns of concentration—discharge relationships in stream water draining the Hubbard Brook Experimental Forest, New Hampshire. In: Neal C, Hornung M (eds) *Special Issue: Transfer of Elements through the Hydrological Cycle. Journal of Hydrology* **116**: 147–65. [6.1]

Lawrence GB, Driscoll CT, Fuller RD. (1988) Hydrologic control of aluminium chemistry in an acidic headwater stream. *Water Resources Research* **24**: 659–69. [6.1]

Lee GF, Jones RA, Rast W. (1980) Availability of phosphorus to phytoplankton and its implications for phosphorus management strategies. In: Loehr RC, Martin CS, Rast W (eds) *Phosphorus Management Strategies for Lakes*. Ann Arbor Science Publishers. [6.2]

Lewis WM, Grant MC. (1979) Changes in the output of ions from a watershed as a result of the acidification of precipitation. *Ecology* **60**: 1093–7. [6.1]

Likens GE, Bormann FH, Pierce RS, Eaton JS, Johnson NM. (1977) *Biogeochemistry of a Forested Ecosystem*. Springer-Verlag, New York. [6.1]

Loughran RJ, Malone KJ. (1976) *Variations in some stream solutes in a Hunter Valley catchment*. Research Papers in Geography, University of Newcastle, New South Wales, No. 8. [6.1]

Malcolm RL, Kennedy VC. (1970) Variation of cation exchange capacity and rate with particle size in stream sediment. *Journal of the Water Pollution Control Federation* **2**: 153–60. [6.2]

Meybeck M. (1976) Total dissolved transport by world major rivers. *Hydrological Sciences Bulletin* **21**: 265–84. [6.1]

Meybeck M. (1981) Pathways of major elements from land to ocean through rivers. In: *River Inputs to Ocean Systems*, pp 18–30. UNEP/Unesco Report. [6.1]

Meybeck M. (1982) Carbon, nitrogen and phosphorus transport by world rivers. *American Journal of Science* **282**: 401–50. [6, 6.1]

Meybeck M. (1983) Atmospheric inputs and river transport of dissolved substances. In: *Dissolved Loads of Rivers and Surface Water Quantity/Quality Relationships* (Proceedings of the Hamburg Symposium, August 1983), pp 173–92. International Association of Hydrological Sciences Publication No. 141. Wallingford, UK. [6.1]

Meybeck M. (1984) *Les fleuves et le cycle géochimique des éléments*. Thèse de Doctorat d'Etat Université Pierre et Marie Curie, Paris. [6.1, 6.2]

Meybeck M. (1986) Composition chimique des ruisseaux non pollués de France. *Sci. Géol. Bull.* **39**: 3–77. [6.1]

Meybeck M, Helmer R. (1989) The quality of rivers: from pristine stage to global pollution. *Palaeogeography, Palaeoclimatology, Palaeoecology (Global and Planetary Change Section)* **75**: 283–309. [6.1]

Meybeck M, Chapman D, Helmer R. (eds) (1989) *Global Freshwater Quality: A First Assessment*. Blackwell Reference, Oxford. [6.1]

Miller JP. (1961) Solutes in small streams draining single rock types, Sangre de Cristo Range, New Mexico. *United States Geological Survey Water-Supply Paper 1535-F.* [6.1]

Miller WR, Drever JI. (1977) Water chemistry of a stream following a storm, Absaroka Mountains, Wyoming. *Geological Society of America Bulletin* **88**: 286–90. [6.1]

Möller D. (1984) Estimation of the global man-made sulphur emission. *Atmosphere and Environment* **18**: 19–27. [6.1]

Moore TR. (1987) Patterns of dissolved organic matter in sub-Arctic peatlands. *Earth Surface Processes and Landforms* **12**: 387–97. [6.1]

Mudroch A, Duncan GA. (1986) Distribution of metals in different size fractions of sediment from the Niagara River. *Journal of Great Lakes Research* **12**: 117–26. [6.2]

Munger JW, Eisenreich SJ. (1983) Continental-scale variations in precipitation chemistry. *Environmental Science and Technology* **17**: 32A–42A. [6.1]

Muraoka K, Hirata T. (1988) Streamwater chemistry during rainfall events in a forested basin. *Journal of Hydrology* **102**: 235–53. [6.1]

Muscutt AD, Wheater HS, Reynolds B. (1990) Stormflow hydrochemistry of a small Welsh upland catchment. In: Neal C, Hornung M (eds) *Special Issue: Transfer of Elements through the Hydrological Cycle. Journal of Hydrology* **116**: 239–49. [6.1]

Neal C, Hornung, M. (1990) Special Issue: Transfer of Elements through the Hydrological Cycle. *Journal of Hydrology* **116**. [6.1]

Neal C, Smith CJ, Walls J, Billingham P, Hill S, Neal M. (1990a) Hydrogeochemical variations in Hafren forest stream waters, mid-Wales. In: Neal C, Hornung M (eds) *Special Issue: Transfer of Elements through the Hydrological Cycle. Journal of Hydrology* **116**: 185–200. [6.1]

Neal C, Mulder J, Christophersen N *et al* (1990b) Limitations to the understanding of ion-exchange and solubility controls for acidic Welsh, Scottish and Norwegian sites. In: Neal C, Hornung M (eds) *Special Issue: Transfer of Elements through the Hydrological Cycle. Journal of Hydrology* **116**: 11–23. [6.1]

Oborne AC, Brooker MP, Edwards RW. (1980) The chemistry of the River Wye. *Journal of Hydrology* **52**: 59–70. [6.1]

O'Connor DJ. (1976) The concentration of dissolved solids and river flow. *Water Resources Research* **12**: 279–94. [6.1]

Olsen SR, Watanabe FS. (1957) A method to determine a phosphorus adsorption isotherm of soils as measured by the Langmuir adsorption isotherm. *Soil Science Society of America Proceedings* **21**: 144–9. [6.2]

Omernik JM. (1977) *Nonpoint source — stream nutrient level relationships: a nationwide study.* EPA-600/3–77–105. Corvallis Environmental Research Laboratory, United States Environmental Protection Agency, Corvallis, Oregon. [6.1]

Omernik JM, Powers CF. (1983) Total alkalinity of surface waters — a national map. *Annals of the Association of American Geographers* **73**: 133–6. [6.1]

Ongley ED. (1978) *Sediment-related phosphorus, trace metals and anion flux in two rural southeastern Ontario catchments.* Inland Waters Directorate Research Subvention Program Progress Report (unpublished). [6.2]

Ongley ED, Bynoe MC, Percival JB. (1981) Physical and geochemical characteristics of suspended solids, Wilton Creek, Ontario. *Canadian Journal of Earth Science* **18**: 1365–79. [6.2]

Oxley NC. (1974) Suspended sediment delivery rates and the solute concentration of stream discharge in two Welsh catchments. In: Gregory KJ, Walling DE (eds) *Fluvial Processes in Instrumented Watersheds*, pp 141–54. Institute of British Geographers Special Publication No. 6. [6.1]

Pereira WE, Rostad CE, Updegraff DM, Bennett JL. (1987) Anaerobic microbial transformations of azaarenes in groundwater at hazardous-waste sites. In: Averett RC, McKnight DM (eds) *Chemical Quality of Water and the Hydrologic Cycle*, pp 111–23. Lewis Publishers, Chelsea, Michigan. [6]

Pettersson K. (1986) The fractional composition of phosphorus in lake sediments of different characteristics. In: Sly PG (ed.) *Sediments and Water Interactions*, pp 149–55. Springer-Verlag, New York. [6.2]

Reid JM, MacLeod DA, Cresser MS. (1981) Factors affecting the chemistry of precipitation and river water in an upland catchment. *Journal of Hydrology* **50**: 129–45. [6.1]

Reynolds B, Pomeroy AB. (1988) Hydrogeochemistry of chlorine in an upland catchment in mid-Wales. *Journal of Hydrology* **99**: 19–32. [6.1]

Reynolds B, Hornung M, Hughes S. (1989) Chemistry of streams draining grassland and forest catchments at Plynlimon, mid-Wales. *Hydrological Sciences Journal* **34**: 667–86. [6.1]

Roberts G, Marsh T. (1987) The effects of agricultural practices on the nitrate concentrations in the surface water domestic supply sources of Western Europe. In: *Water for the Future: Hydrology in Perspective* (Proceedings of the Rome Symposium, April 1987), pp 365–80. International Association of Hydrological Sciences Publication No. 164. Wallingford, UK. [6.1]

Roberts G, Hudson JA, Blackie JR. (1983) Nutrient cycling in the Wye and Severn at Plynlimon. *Institute of Hydrology Report* No. 86. Wallingford, UK. [6.1]

Rodà F, Avila A, Bonilla D. (1990) Precipitation, throughfall, soil solution and streamwater chemistry in a holm-oak (*Quercus ilex*) forest. In: Neal C, Hornung M (eds) *Special Issue: Transfer of Elements through the Hydrological Cycle. Journal of Hydrology*

116: 167–83. [6.1]

Rosenqvist ITh. (1990) From rain to lake: water pathways and chemical changes. In: Neal C, Hornung M (eds) *Special Issue: Transfer of Elements through the Hydrological Cycle. Journal of Hydrology* **116**: 3–10. [6.1]

Royal Society (1983) *The Nitrogen Cycle of the United Kingdom.* Report of a Royal Society Study Group, London. [6.1]

Sanders WM III. (1972) Nutrients. In: Oglesby RT, Carlson CA, McCann JA (eds) *River Ecology and Man*, pp 389–415. Academic Press. New York. [6]

Schuman GE, Piest RF, Spomer RG. (1976) Physical and chemical characteristics of sediment originating from Missouri valley loess. In: *Proceedings of the Third Federal Interagency Sedimentation Conference*, 3-28–3-40. US Water Resources Council, Washington, DC. [6.2]

Seip HM, Tollan A. (1985) Acid Deposition. In: Rodda JC (ed.) *Facets of Hydrology Volume II*, pp 69–98. John Wiley & Sons, Chichester, UK. [6.1]

Simpson EA. (1980) The harmonization of the monitoring of the quality of rivers in the United Kingdom. *Hydrological Sciences Bulletin* **25**: 13–23. [6]

Skougstad MW, Fishman MJ, Friedman LC, Erdmann OE, Duncan SS. (1979) Methods for the determination of inorganic substances in water and fluvial sediments. *United States Geological Survey Techniques of Water-Resources Investigations 5*, Chapter A1. [6]

Slack JG. (1977) River water quality in Essex during and after the 1976 drought. *Effluent and Water Treatment Journal* **17**: 575–8. [6.1]

Slack KV, Feltz HR. (1968) Tree leaf control on lowflow water quality in a small Virginia stream. *Environmental Science and Technology* **2**: 126–31. [6.1]

Smith RA, Alexander RB, Wolman MG. (1987) Analysis and interpretation of water-quality trends in major US rivers, 1974–81. *United States Geological Survey Water-Supply Paper 2307*. [6.1]

Stallard RF, Edmond JM. (1981) Geochemistry of the Amazon 1. Precipitation chemistry and the marine contribution to the dissolved load at the time of peak discharge. *Journal of Geophysical Research* **86** (C10): 9844–58. [6.1]

Stallard RF, Edmond JM. (1983) Geochemistry of the Amazon 2. The influence of geology and weathering environment on the dissolved load. *Journal of Geophysical Research* **88** (C14): 9671–88. [6.1]

Stanford JA, Ward JV. (1988) The hyporheic habitat of river ecosystems. *Nature* **355**: 64–5. [6.1]

Steele TD. (1968) Digital-computer applications in chemical quality studies of surface water in a small watershed. *Publications de l'Association Internationale d'Hydrologie Scientifique* **80**: 203–14. [6.1]

Stevens PA, Adamson JK, Anderson MA, Hornung M. (1988) Effects of clearfelling on surface water quality and site nutrient status. In: Usher MB, Thomson DBA (eds) *Ecological Changes in the Uplands*, pp 289–93. Special Publication M. British Ecological Society, Blackwell, Oxford. [6.1]

Stevens PA, Hornung M, Hughes S. (1989) Solute concentrations in a mature sitka spruce plantation in Beddgelert forest, North Wales. *For. Ecol. Management* **27**: 1–20. [6.1]

Stevenson CM. (1968) An analysis of the chemical composition of rain-water and air over the British Isles and Eire for the years 1959–1964. *Royal Meteorological Society Quarterly Journal* **94**: 56–70. [6.1]

Stumm W, Leckie JO. (1971) Phosphate exchange with sediment: its role in the productivity of surface waters. *Advances in Water Pollution Research* **2** (3): 26/1–26/16. [6.2]

Sutcliffe DW, Carrick TR. (1973) Studies on mountain streams in the English Lake District II. Aspects of water chemistry in the River Duddon. *Freshwater Biology* **3**: 543–60. [6.1]

Swank WT. (1986) Biological control of solute losses from forest ecosystems. In: Trudgill ST (ed.) *Solute Processes*, pp 85–139. John Wiley & Sons, Chichester, UK. [6.1]

Swank WT, Douglass JE. (1977) Nutrient budgets for undisturbed and manipulated hardwood forest ecosystems in the mountains of North Carolina. In: Correll DL (ed.) *Watershed Research in Eastern North America: A Workshop to Compare Results*, pp 343–63. Smithsonian Institution, Edgewater, Maryland. [6.1]

Talsma T, Hallam PM. (1982) Stream water quality of forest catchments in the Cotter Valley, ACT. In: O'Loughlin EM, Bren LJ (eds) *The First National Symposium on Forest Hydrology*, Melbourne, 11–13 May 1982, pp 50–9. [6.1]

Taylor HE. (1987) Analytical methodology for the measurement of the chemical composition of snow cores from the Cascade/Sierra Nevada mountain ranges. In: Averett RC, McKnight DM (eds) *Chemical Quality of Water and the Hydrologic Cycle*, pp 55–69. Lewis Publishers, Chelsea, Michigan. [6]

Tessier A, Campbell P, Bisson M. (1979) Sequential extraction procedure for the speciation of particulate trace metals. *Analytical Chemistry* **51**: 844–51. [6.2]

Unesco (1978) *Monitoring of particulate matter quality in rivers and lakes. Recommendations of a workshop on the assessment of particulate matter contamination in rivers and lakes.* Unesco (GEMS/WATER; Med-IX) BUD/Report 1. [6.2]

United Kingdom Acid Waters Review Group (1988) *Acidity in United Kingdom Fresh Waters.* Second report of the UK Acid Waters Review Group. Her Majesty's Stationery Office, London. [6.1]

Varheyli G. (1985) Continental and global sulphur

budgets 1: anthropogenic SO_2 emissions. *Atmosphere and Environment* **19**: 1029–40. [6.1]

Walling D.E. (1974) Suspended sediment and solute yields from a small catchment prior to urbanization. In: Gregory KJ, Walling DE (eds) *Fluvial Processes in Instrumented Watersheds*, pp 169–92. Institute of British Geographers Special Publication No. 6. [6.1]

Walling DE. (1975) Solute variations in small catchment streams: some comments. *Transactions of the Institute of British Geographers* **64**: 141–7. [6.1]

Walling DE. (1978) Suspended sediment and solute response characteristics of the River Exe, Devon, England. In: Davison-Arnott R, Nickling W (eds) *Research in Fluvial Geomorphology*, pp 169–97. GeoBooks, Norwich. [6.1]

Walling DE. (1980) Water in the catchment ecosystem. In: Gower AM (ed.) *Water Quality in Catchment Ecosystems*, pp 2–47. John Wiley and Sons, Chichester, UK. [6.1]

Walling DE (1984) Dissolved loads and their measurement. In: Hadley RF, Walling DE (eds) *Erosion and Sediment Yield: Some Methods of Measurement and Modelling*, pp 111–77. GeoBooks, Norwich. [6]

Walling DE, Foster IDL. (1975) Variations in the natural chemical concentration of river water during flood flows, and the lag effect: some further comments. *Journal of Hydrology* **26**: 237–44. [6.1]

Walling DE, Foster IDL. (1978) The 1976 drought and nitrate levels in the River Exe Basin. *Journal of the Institution of Water Engineers and Scientists* **32**: 341–52. [6.1]

Walling DE, Webb BW. (1980) The spatial dimension in the interpretation of stream solute behaviour. *Journal of Hydrology* **47**: 129–49. [6.1]

Walling DE, Webb BW. (1981) Water quality. In: Lewin J (ed.) *British Rivers*, pp 126–69. George Allen & Unwin, London. [6.1]

Walling DE, Webb BW. (1983) The dissolved loads of rivers: a global overview. In: *Dissolved Loads of Rivers and Surface Water Quantity/Quality Relationships* (Proceedings of the Hamburg Symposium, August 1983), pp 3–20. International Association of Hydrological Sciences Publication No. 141. Wallingford, UK. [6.1]

Walling DE, Webb BW. (1985) Estimating the discharge of contaminants to coastal waters by rivers. *Marine Pollution Bulletin* **16**: 488–92. [6.2]

Walling DE, Webb BW. (1986a) Solutes in river systems. In: Trudgill ST (ed.) *Solute Processes*, pp 251–327. John Wiley & Sons, Chichester, UK. [6.1]

Walling DE, Webb BW. (1986b) Solute transport by rivers in arid environments: an overview. *Journal of Water Resources* **5**: 800–22. [6.1]

Walling DE, Webb BW. (1987) Material transport by the world's rivers: evolving perspectives. In: *Water for the Future: Hydrology in Perspective* (Proceedings of the Rome Symposium, April 1987), pp 313–29. International Association of Hydrological Sciences Publication No. 164. Wallingford, UK. [6.1]

Waylen MJ. (1979) Chemical weathering in a drainage basin underlain by Old Red Sandstone. *Earth Surface Processes* **4**: 167–78. [6.1]

Webb BW. (1983) Factors influencing spatial variation of background solute levels in a Devon river system. *Reports and Transactions of the Devonshire Association for the Advancement of Science Literature and Arts* **115**: 51–69. [6.1]

Webb BW, Walling DE. (1982a) Catchment scale and the interpretation of water quality behaviour. In: *Hydrological Research Basins and their use in Water Resources Planning*, pp 759–70. Landeshydrologie, Berne. [6.1]

Webb BW, Walling DE. (1982b) The magnitude and frequency characteristics of fluvial transport in a Devon drainage basin and some geomorphological implications. *Catena* **9**: 9–23. [6.1]

Webb BW, Walling DE. (1983) Stream solute behaviour in the River Exe basin, Devon, UK. In: *Dissolved Loads of Rivers and Surface Water Quantity/Quality Relationships* (Proceedings of the Hamburg Symposium, August 1983), pp 153–69. International Association of Hydrological Sciences Publication No. 141. Wallingford, UK. [6.1]

Webb BW, Walling DE. (1985) Nitrate behaviour in streamflow from a grassland catchment in Devon, UK. *Water Research* **19**: 1005–16. [6.1]

Webb BW, Davis JS, Keller HM. (1987) Hysteresis in stream solute behaviour. In: Gardiner V (ed.) *International Geomorphology 1986 Part I*, pp 767–82. John Wiley & Sons, Chichester, UK. [6.1]

Webster JR, Patten BC. (1979) Effects of watershed perturbation on stream potassium and calcium dynamics. *Ecological Monographs* **49**: 51–72. [6.1]

White EM. (1981) Nutrient contents of precipitation and canopy throughfall under corn, soybeans and oats. *Water Resources Bulletin* **17**: 708–12. [6.1]

Williams AG, Ternan JL, Kent M. (1983) Stream solute sources and variations in a temperate granite drainage basin. In: *Dissolved Loads of Rivers and Surface Water Quantity/Quality Relationships* (Proceedings of the Hamburg Symposium, August 1983), pp 299–310. International Association of Hydrological Sciences Publication No. 141. Wallingford, UK. [6.1]

Young JR, Ellis EC, Hidy GM. (1988) Deposition of airborne acidifiers in the western environment. *Journal of Environmental Quality* **17**: 1–26. [6.1]

7: Cycles and Spirals of Nutrients

J.D. NEWBOLD

7.1 INTRODUCTION

Rivers play a major role in global biogeochemical cycling by transporting elements from terrestrial environments to the sea (Walling & Webb 1994). Many of these elements are essential nutrients and are utilized by river biota. A number of the major ions found in river water, such as Ca, Mg, K, Na, Si and Cl, are often present well in excess of any biological demands within the river, and may pass through the river system virtually unaffected. Other elements — notably, carbon, phosphorus and nitrogen, or particular chemical forms of these elements — may be in relatively short supply and undergo considerable utilization as they pass downstream.

Biota remove nutrients from river water, but they also regenerate nutrients to the water. This cycling of nutrients within the river may proceed intensively and yet produce small or negligible net effects on nutrient concentrations. In fact, for nutrients such as phosphorus, which does not exchange with the atmosphere, biota cannot alter the long-run total transport substantially. On the other hand, biota do influence the chemical and physical forms of nutrients, and the timing of nutrient transport. These effects may interact, in turn, with physical transport processes. For example, the biota might speed downstream transport by converting particulate-bound nutrients to dissolved forms or reducing detrital particles to smaller, more easily transported, sizes. Carbon, nitrogen and, to a limited extent, sulphur exchange with the atmosphere. For these elements, biota may strongly influence long-run total transport. This chapter considers the dynamics of selected chemicals (nutrients) *within* river systems. For related reviews see Meyer *et al* (1988) and Stream Solute Workshop (1990).

Cycling is of interest not only because the biota may affect nutrient concentrations but because nutrient concentrations may affect the biota. These interactive influences, however, occur on a template of continual downstream transport, so that biotic processes in upstream reaches may influence those in downstream reaches. As a nutrient atom undergoes a series of transformations, completing a 'cycle' by returning to a previous state, it also traverses some distance downstream. This open, or longitudinally displaced, cycling has been termed 'spiralling' (Webster 1975; Wallace *et al* 1977; Webster & Patten 1979). Within the framework of the spiralling concept we can view cycling as involving exchanges and transformations that can be quantified on an areal or volumetric basis; it does not involve downstream transport (Fig. 7.1(a)). In this sense, cycling occurs at any point in the river, regardless of the fact that the atoms involved in the cycle represent an ever-changing population, and no individual atoms remain in place long enough to complete a cycle. Spiralling, on the other hand, involves measures of both cycling and downstream transport and refers explicitly to the longitudinal scale over which cycles occur (Fig. 7.1(b)).

7.2 OVERVIEW OF CYCLING AND SPIRALLING

Cycles

Figure 7.2 provides a simplified overview of the cycling of the three major elements we shall

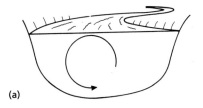

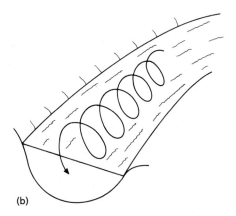

Fig. 7.1 (a) The nutrient cycle as viewed from a perspective that does not 'see' downstream transport. (b) The nutrient cycle, in conjunction with downstream transport, describes a spiral.

consider here: carbon, phosphorus and nitrogen. In this view of the river ecosystem, the metabolic activity within the river depends upon two major sources of organic carbon: primary production that occurs within the river (autochthonous carbon) and organic carbon supplied from the terrestrial environment (allochthonous carbon). Organic carbon passes through the food web (within which are significant subcycles not represented here) to an ultimate fate of being respired to dissolved inorganic carbon (DIC). Because DIC exchanges relatively freely with atmospheric carbon dioxide, we represent here only the organic-carbon 'half-cycle', which represents the primary pathway of energy flow in the ecosystem. The base of the carbon food web consists of both primary producers (algae, cyanobacteria, macrophytes, bryophytes; see Reynolds (1994), Fox (1994) and Wetzel & Ward (1994)), and microbial consumers of the allochthonous carbon (bacteria and fungi; see Maltby, 1994). Both groups use inorganic phosphorus and nitrogen from the river water, and these elements flow through the food web in rough stoichiometric proportion to the flow of carbon and energy. As the carbon is respired to DIC, the phosphorus and nitrogen are regenerated as inorganic forms to be recycled to the algae and microbes.

In many ecosystems, a large fraction of the metabolism is supported by recycled nutrient (Pomeroy 1970), i.e. total nutrient utilization

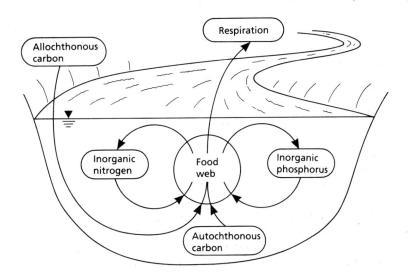

Fig. 7.2 Cycles of nitrogen and phosphorus are coupled to the 'half-cycle' of organic carbon.

within the ecosystem exceeds the supply of new nutrient from outside the ecosystem (Dugdale & Goering 1967). By delineating ecosystem boundaries and comparing imports and exports of nutrients to internal storage and utilization, much can be inferred about the significance of nutrient cycling to the productivity of the ecosystem. How does this logic apply to a stream or river? If the 'ecosystem' is a very short river reach, upstream imports and downstream exports may overwhelm internal utilization, suggesting that cycling is unimportant. Yet if the reach is very long, the opposite might be the case. To address these issues, we turn to the spiralling concept.

Spirals

A given nutrient atom, as it passes downstream, may be used again and again, the amount of utilization depending on the 'tightness' of the spirals (Webster & Patten 1979) or the downstream displacement from one cycle to the next. This in turn depends not only on how quickly cycling occurs, but also on the retentiveness of the ecosystem, or the degree to which the downstream transport of nutrient is retarded relative to that of water. That is, if it takes an average time, t_c, for the average nutrient atom to complete a cycle, while it moves downstream at an average velocity, v_T, then the cycle is completed over a downstream distance of:

$$S = v_T t_C \qquad (1)$$

where S is the spiralling length (Elwood *et al* 1983). The velocity, v_T, may be near that of water, v_W, in large rivers, but very much lower in streams and rivers where nutrients reside in the sediments for a high proportion of the time.

The cycle of a nutrient such as phosphorus or nitrogen can be thought of as consisting of two components: (1) the biological assimilation ('uptake') of dissolved inorganic nutrients from the water column; and (2) the subsequent biological processing and movement through the food web leading eventually to regeneration to the inorganic form (Fig. 7.3). Spiralling length consists of the average downstream distance travelled by a dissolved nutrient atom until uptake (the 'uptake length') plus the downstream distance travelled within the biota until regeneration (the 'turnover length'). To illustrate the relationship between spiralling length, nutrient fluxes and nutrient retentiveness, we shall simplify the river ecosystem to two compartments: water (W), by which we mean the inorganic nutrient dissolved in the water, and biota (B), by which we mean nutrient in living tissue.

Suppose that the biotic utilization of nutrients from the water compartment is U, expressed as mass per unit area per unit time ($M\ L^{-2}\ T^{-1}$), and the downstream flux of dissolved nutrient is F_W, as mass per unit width of river per unit time ($M\ L^{-1}\ T^{-1}$), which can also be expressed as $F_W = C_W\ v_W\ d$, where C_W is the dissolved nutrient concentration, v_W is the water velocity, and d is river depth. Thus, in each unit distance of river, a

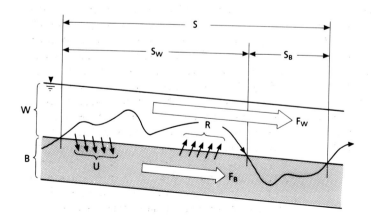

Fig. 7.3 Spiralling in a simplified river ecosystem consisting of two compartments: water (W) and biota (B). The spiralling length, S, is the average distance a nutrient atom travels downstream during one cycle through the water and biotic compartments. It is the sum of the uptake length (S_w) and the turnover length (S_B), and can be calculated from the downstream nutrient fluxes (F_W and F_B) and the exchange fluxes (U and R), as described in the text.

proportion, k_L $(L^{-1}) = U/F_w$, of the downstream flux is taken up by the biota. Depending on the level of nutrient regeneration and on external sources of nutrients, F_w may decline, remain uniform, or increase in the downstream direction. Provided, however, that the ratio U/F_w (i.e. k_L) remains uniform, then the particular atoms, F_w^*, passing any point in the river will disappear from the water column exponentially with distance, x:

$$F_w^* = F_{w0}^* \exp{(-k_L x)} \qquad (2)$$

where $F_{w0}^* = F_w^*$ at $x = 0$. For example, if a tracer (such as $^{32}PO_4$ for phosphorus) is injected into a river, equation (2) describes the longitudinal disappearance of the tracer and can be used to estimate k_L (Ball & Hooper 1963; Newbold *et al* 1981). The average travel distance, or uptake length, is given by $S_w = 1/k_L$, which can be verified by a simple integration of equation (2). Thus, since $k_L = U/F_w$:

$$S_w = F_w/U \qquad (3)$$

If followed downstream at the water velocity, v_w, the nutrient atoms disappear exponentially with time (as well as distance), at the rate $k_w = k_L v_w$. Substituting for k_L, we can write $S_w = 1/k_L = v_w/k_w$, or, given that the residence time in the water is $t_w = 1/k_w$, $S = v_w T_w$. The stock of inorganic nutrient, on a unit-area basis, is $X_w = C_w d$, so that the uptake can be written as $U = k_w X_w$.

The turnover length, or distance of travel in the biotic compartment, is analogous to the uptake length. Suppose that the standing stock of nutrient in the biota is X_B $[M L^{-2}]$, and that a fraction, k_B, is regenerated per unit time giving a residence time for nutrient atom of $1/k_B$, and a regeneration flux of R $[M L^{-2} T^{-1}] = k_B X_B$. The biotic compartment is distributed between that in the water column X_{sus} (moving downstream at v_w) and that in the sediments, X_{sed} (with zero velocity). That is, $X_B = X_{sus} + X_{sed}$. The weighted average velocity at which the compartment as a whole moves downstream, then is $v_B = (X_{sus}/X_B)v_w$. Therefore, during its residence time of $1/k_B$ in the biotic compartment, a nutrient atom will travel an average distance:

$$S_B = v_B/k_B \qquad (4)$$

which is the turnover length.

The downstream flux (per unit width) of the biotic compartment is $F_B[M L^{-1} T^{-1}] = v_w X_{sus} = v_B X_B$. From this, and the definition of R above, it can be seen that equation (4) is equivalent to:

$$S_B = F_B/R \qquad (5)$$

The total downstream displacement during one cycle, or spiralling length, then is:

$$S = S_w + S_B = v_w/k_w + v_B/k_B = F_w/U + F_B/R \qquad (6)$$

In the idealized case in which the river is at steady state and is longitudinally uniform on the scale of the spiralling length, the regeneration flux equals the uptake flux $(R = U)$ so that:

$$S = F_T/U \qquad (7)$$

where F_T is the total downstream nutrient flux, or $F_w + F_B$. Thus, the spiralling length represents the distance of river over which utilization is equal to the downstream flux of nutrients or, alternatively, over which the downstream flux is cycled, on average, only once.

Equation (7) quantifies the intuitive idea that more intensive nutrient utilization involves cycling over shorter distances, or 'tighter' spiralling. The total time for completing a cycle is $t_c = 1/k_w + 1/k_B$, while the average downstream velocity for the entire nutrient stock is, $v_T = F_T/X_T$, where $X_T = X_w + X_B$. By combining these definitions with equation (6) and assuming $R = U$, one can obtain equation (1), i.e. $S = v_T t_c$.

An actual analysis might involve several ecosystem compartments (see section 7.6). Each compartment has a turnover rate and a downstream velocity, and hence a turnover length. The total spiralling length can be determined from the individual turnover lengths by weighting each compartmental turnover length by the proportion of the total uptake flux that passes through it, and summing these weighted values (Newbold *et al* 1983a). The biotic compartment presented here, and its turnover length, may be thought of as an aggregation of the entire portion of the cycle in which the nutrient is not in the dissolved inorganic form, and hence is temporarily unavailable for plant and microbial uptake. This may include nutrient in dead tissue and nutrient in less available organic molecules. However, if the biotic compartment actually consists of several compartments with differing turnover rates and transport velocities, the equation $S_B = F_B/R$ is only an approximation unless the ecosystem is at steady state and without longitudinal gradients in standing stock.

7.3 NUTRIENT LIMITATION AND NUTRIENT RETENTION

Primary productivity in most of the world's lakes is limited by phosphorus, nitrogen, or both, even when these nutrients are supplied at relatively high levels of anthropogenic enrichment (Dillon & Rigler 1974; Vollenweider 1976; Elser *et al* 1990). Nutrient limitation of primary productivity and microbial activity in streams and smaller rivers, however, appears to be far less widespread and largely restricted to near-pristine conditions. Phosphorus limitation of algal growth or production has been reported in several streams (Stockner & Shortreed 1978; Elwood *et al* 1981b; Horner & Welch 1981; Pringle & Bowers 1984; Bothwell 1985, 1988; Peterson *et al* 1985; Freeman 1986; Knorr & Fairchild 1987; Pringle 1987; Keithan *et al* 1988; Biggs & Close 1989; Hart & Robinson 1990; Horner *et al* 1990; Lock *et al* 1990), and there have been a few reports of phosphorus limitation of heterotrophic microbial activity (Elwood *et al* 1981b; Klotz 1985; Lock *et al* 1990). In all cases concentrations of soluble reactive phosphorus (SRP; see section 7.6) were $\leq 15\,\mu g\,l^{-1}$ and frequently $<5\,\mu g\,l^{-1}$. Mixed results have been reported in the range of $3-50\,\mu g\,l^{-1}$ (Wurhmann & Eichenberger 1975; Wong & Clark 1976; Horner & Welch 1981; Horner *et al* 1983; Bothwell 1985, 1989; Klotz 1985; Hullar & Vestal 1989), while higher concentrations have not revealed limitation (Pringle *et al* 1986; Knorr & Fairchild 1987; Kiethan *et al* 1988; Munn *et al* 1989).

Periphyton growth has responded to nitrogen enrichment in several streams where inorganic nitrogen concentrations were less than $60\,\mu g\,l^{-1}$ (Grimm & Fisher 1986; Hill & Knight 1988; Triska *et al* 1989a), and has failed to respond in streams where inorganic nitrogen exceeded $50\,\mu g\,l^{-1}$ (Wurhmann & Eichenberger 1975; Pringle *et al* 1986; Knorr & Fairchild 1987; Kiethan *et al* 1988; Munn *et al* 1989). The ap-

parently consistent transition at $50-60\,\mu g\,l^{-1}$ should be viewed with caution because of the small number of studies involved, and because inorganic nitrogen includes both ammonium and nitrate, which are assimilated with differing efficiencies (see section 7.7). Meyer and Johnson (1983) inferred that microbial activity in decomposing leaves was nitrogen limited at $8\,\mu g\,l^{-1}$, whereas Newbold *et al* (1983b) saw no effect at $30\,\mu g\,l^{-1}$.

One explanation for the lower incidence of nutrient limitation in streams than in lakes is that water velocity and associated turbulence have an enriching effect (Ruttner 1963). Uptake of nutrients by periphyton is enhanced at higher velocities (Whitford & Schumacher 1961, 1964), presumably because diffusion barriers to nutrient transport are reduced. Yet this explanation is not sufficient because: (1) planktonic algae regularly deplete nutrients to levels far below those at which nutrient limitation has been observed in streams; and (2) algal mats or biofilms present a diffusion barrier that at least partially offsets the enriching effect of velocity (Bothwell 1989).

An alternative explanation for the lower incidence of nutrient limitation in streams is that they are open systems with a large capacity to retain nutrients. Lakes are vulnerable to nutrient limitation in large part because they act, in the short run, as closed systems. The total nutrient in the water column remains relatively constant but, as plankton populations increase, dissolved available nutrient is incorporated into biomass. Net incorporation of nutrient necessarily ceases when nearly all of the nutrient is sequestered. It is this very low residual concentration, rather than the potentially high initial concentration, that actually limits algal growth.

In a stream with a continual supply of nutrient from the watershed, benthic organisms may deplete the available nutrient supply in three ways: (1) by temporarily reducing downstream flux (F_T) during nutrient accumulation (Grimm 1987); this depletion ceases at steady state; (2) diverting nutrients to the atmosphere or terrestrial environment; this is rarely significant except for denitrification (section 7.7); and (3) detaching and carrying the nutrient downstream in unavailable form, i.e. as the flux, F_B. In the latter case, the flux of dissolved inorganic nutrient at steady state is $F_W = F_T - F_B$. Recalling from the previous section that the total biomass X_B migrates downstream at the velocity v_B, and that $F_B = v_B X_B$, we see that nutrient depletion increases as the retentiveness of the stream or river decreases. A retentive stream is characterized by a short turnover length relative to the uptake length since, if $R = U$, equation (6) yields:

$$S_W / S_B = F_W / F_B \qquad (8)$$

Many streams and smaller rivers are highly retentive ($v_B \ll v_W$), supporting large standing stocks with little nutrient depletion (Newbold *et al* 1982b; Newbold 1987; section 7.6). Thus, we would expect nutrient limitation to occur only as a transient phenomenon during rapid growth, or when nutrient is supplied from the watershed at concentrations already low enough to be limiting.

Growth rates of many algae and bacteria are saturated by nutrient concentrations substantially below the limiting concentrations observed in the field studies reviewed above. To explain this discrepancy, Bothwell (1989) suggested and experimentally supported the hypothesis that nutrient concentrations can be locally depleted to limiting levels within algal biofilms. A second possible explanation for the discrepancy — applicable to phosphorus limitation — is that soluble reactive phosphorus measurements may overestimate PO_4-P concentrations (section 18.6).

In larger, less retentive, rivers we might expect nutrient depletion and nutrient limitation to be more prevalent. In an entirely planktonic river, where $v_B = v_W$, nutrient in biomass depletes the dissolved inorganic nutrient on a one-to-one basis, as in a closed algal culture or the epilimnion of a lake. River phytoplankton has been reported to deplete phosphorus from municipal discharges to limiting levels (Décamps *et al* 1984). But in several rivers, phytoplankton has been observed not to be nutrient limited (Burkholder-Crecco & Bachmann 1979; Megard 1981; Elser & Kimmel 1985; Krogstad & Løvstad 1989; Wiley *et al* 1990). Three factors — non-algal turbidity, extremely high nutrient concentrations and water residence time — seem to account for the absence of nutrient limitation. Søballe and Kimmel (1987), in a comparison of 126 rivers with 149 natural lakes in the USA, found that in rivers with residence times (i.e. time-of-travel from the river source) greater than about 50 days, the relationship between total phosphorus and phytoplankton populations approximated that of lakes (strongly implying nutrient limitation), whereas in rivers with shorter residence times, phytoplankton populations were lower than in lakes with equivalent phosphorus concentrations.

7.4 MODELLING SPATIAL AND TEMPORAL VARIATION

The spiralling concept as presented so far does not account for spatial and temporal variation in nutrient fluxes and stocks. Mathematical simulations are widely used to describe such variations, as well as other aspects of ecosystem dynamics in rivers. Many river models express the behaviour of the concentration, C, of a solute or suspensoid in the water column as a one-dimensional advection–dispersion equation, one form of which is:

$$\frac{\partial C}{\partial t} = \underbrace{\frac{-Q}{A} \frac{\partial C}{\partial x}}_{\text{Advection}} + \underbrace{\frac{1}{A} \frac{\partial}{\partial x} \left[AD \frac{\partial C}{\partial x} \right]}_{\text{Dispersion}}$$

$$+ \underbrace{\frac{\partial Q}{\partial x} (C_L - C)}_{\text{Inflow and dilution}} + \underbrace{N}_{\text{Net of sources and sinks}} \qquad (9)$$

where x is downstream distance, Q is stream flow, A is cross-sectional area, D is the coefficient of longitudinal dispersion $[L^2 T^{-1}]$, C_L is the concentration of influent water, and $N [M L^{-3} T^{-1}]$ is the net of gains from all losses to other ecosystem compartments, the atmosphere, or the terrestrial environment.

Differences among models include the forms of the source and sink terms, whether the channel and flow characteristics can vary with space and time, the number of solutes and suspensoids that are modelled simultaneously, and the number of additional, non-transporting compartments that are coupled to the source and sink terms. Only the simplest forms have analytical solutions; most require numerical solution by computer.

Figure 7.4 presents a model for the simple

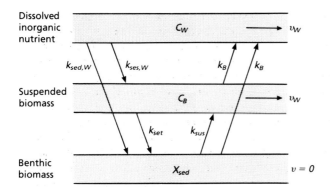

Dissolved inorganic nutrient

C_W → v_W

$k_{sed,W}$ $k_{ses,W}$ k_B k_B

Suspended biomass

C_B → v_W

k_{set} k_{sus}

Benthic biomass

X_{sed} $v = 0$

$$\frac{\partial C_W}{\partial t} = -v_W \frac{\partial C_W}{\partial x} + \frac{\partial}{\partial x} D \frac{\partial C_W}{\partial x} - k_W C_W + \frac{1}{d} k_B X_B$$

$$\frac{\partial C_B}{\partial t} = -v_W \frac{\partial C_B}{\partial x} + \frac{\partial}{\partial x} D \frac{\partial C_B}{\partial x} - (k_B + k_{set}) C_B + \frac{1}{d} k_{sus} X_{sed} + k_{ses,W} C_W$$

$$\frac{\partial X_{sed}}{\partial t} = -(k_B + k_{sus}) X_{sed} + d k_{set} C_B + d k_{sed,W} C_W$$

where

$X_B = X_{sed} + dC_B$

$k_W = k_{ses,W} + k_{sed,W}$

Uptake length, $S_W = v_W/k_W$

Biomass transport velocity, $v_B = k_{sus}/(k_{set} + k_{sus})$

Turnover length, $S_B = v_B/k_B$

Fig. 7.4 Linear, donor-controlled (or tracer) model for a two-compartment, longitudinally uniform, stream ecosystem.

two-compartment ecosystem used above to introduce spiralling. The biotic compartment is subdivided into a suspended portion and a sedimentary portion. The advection–diffusion equations are simplified, relative to equation (9), by assuming uniform flow and cross-section. The equation for the sediment compartment contains no transport terms and is expressed in terms of mass per unit area. As illustrated in Fig. 7.4, spiralling length can be computed from the model parameters.

An important aspect of this model is that all transfers among compartments vary linearly with the nutrient content of the donor compartment. A linear, donor-controlled model cannot incorporate many mechanisms known to control exchanges among compartments. It does not, for example, include an influence of biomass on nutrient uptake, or any means for saturating uptake at high nutrient concentrations. A model of this type is, however, appropriate for simulating the

dynamics of a tracer (such as a radioisotope) added to a stream or river in which other aspects of the ecosystem, including the cycling of the naturally occurring nutrient, are relatively constant (O'Neill 1979). This is because a tracer, by definition, is added at a level too low to affect the dynamics of naturally occurring nutrient or any other aspect of the ecosystem. The model, therefore, simulates only the simple mixing of the tracer through compartments at rates determined by the steady fluxes of the naturally occurring nutrient. Model coefficients estimated by simulating a tracer injection, therefore, quantify the cycling of the ecosystem under the conditions of the experiment, but may poorly predict responses to altered conditions, such as addition of polluting levels of a nutrient.

Where the objective is to describe or investigate the dynamics of nutrient concentration over a range of concentrations, such as might be produced by an experimental nutrient addition, non-linear functions may be used. For example, to model uptake by attached algae of nitrate added to experimental flumes, Kim *et al* (1990) used an additional water compartment with no downstream velocity representing an 'exchange zone' near the streambed. Nutrient transfer into the exchange zone from the water column was governed by first-order kinetics, while nutrient uptake occurred from the exchange zone according to Michaelis–Menten kinetics, i.e.:

$$U = X_{\text{B}} \, V_{\text{max}} \left(\frac{C_{\text{cx}}}{K_{\text{s}} + C_{\text{cx}}} \right) \qquad (10)$$

where C_{cx} is the NO_3-N concentration in the exchange zone, K_{s} is the half-saturation constant at which uptake is half its maximum rate, V_{max} [$M \, T^{-1}$ per mass of X_{B}]. Note that equation (10) predicts that uptake length will increase with increasing concentrations, as has been observed experimentally by Mulholland *et al* (1990). In non-biological applications, other types of exchanges, such as sorption isotherms (Bencala 1984), and chemical submodels (Chapman 1982), have been coupled to equation (9). Functions used in modelling trace contaminant behaviour have been reviewed by Thomann (1984), Kuwabara and Helliker (1988), and O'Connor (1988a, 1988b, 1988c).

The dynamics of an added nutrient, unlike those of a tracer, may interact with other aspects of the ecosystem. For example, using equation (10) to describe an increase in uptake due to added nutrients raises the question of whether the biomass, X_{B}, increases in response. This, in turn, may depend on whether there is a secondary effect on consumer populations that might keep X_{B} in check (section 7.8). Considerations such as these lead to increasingly complex ecosystem models in which the dynamics of a single nutrient are only a small part.

Such complexity appears, to varying degrees, in water quality simulation models, designed to predict effects of potential waste discharges, or of reducing existing discharges (Chapra & Reckhow 1983; Orlob 1983; Bowie *et al* 1985; Thomann & Mueller 1987; McCutcheon 1989). Several water quality models simulate phosphorus and nitrogen dynamics, including release of these nutrients from discharged organic matter, uptake by and influence on the growth of algae, releases from decaying algae, and exchanges with sediments. Often, however, the focus is on the net effects on water-column concentrations, rather than on cycling *per se*, and in some models the sediments are regarded as potentially infinite sources or sinks without mass balance constraints. Water quality simulation models are typically constructed by using functional relationships and parameter values available in the scientific literature (see Bowie *et al* 1985) and then adjusted to bring model simulations into agreement with data from a particular river. These data are often limited, either in quantity or in dynamic range, so that unequivocal parameter estimations and rigorous testing of the model are rarely feasible. None the less, the models are undeniably effective in simulating major water quality parameters, and they serve to identify areas in which scientific understanding of ecosystem processes is limited.

7.5 ORGANIC CARBON SPIRALLING

The carbon cycle in rivers differs from that of phosphorus and nitrogen in that: (1) the inorganic phase of the cycle (CO_2, HCO_3^-, and CO_3^{2-}) exchanges freely with atmospheric carbon dioxide; and (2) a variable, but sometimes very large, proportion of the organic phase begins not with uptake of inorganic carbon (i.e. photosynthesis), but with lateral accrual of organic carbon from the terrestrial environment. In describing spiralling of carbon, therefore, we focus on the portion of the cycle, or half-cycle, involving organic carbon, beginning with its entry into the river from terrestrial sources or formation by photosynthesis and ending in respiration to carbon dioxide. The organic matter turnover length, S_{C}, is the expected downstream travel distance associated with this half-cycle (Newbold *et al* 1982a). As we shall see below, both computation and meaningful interpretation of S_{C} are made problematic by the very great diversity of organic carbon forms in a river.

For the moment, however, we assume that all organic carbon in the river is of similar quality, degrading at the rate k_{C}, and migrating downstream at the velocity, v_c. The turnover length, then (as derived above), is $S_c = v_c/k_c$. If X_c is the areal standing stock of organic carbon (hereafter 'carbon'), then the areal rate of carbon loss to respiration is $R = k_c X_c$, and the downstream flux of carbon per unit width is $F_T = v_c X_c$. Simple substitutions yield:

$$S_c = F_T/R_w \qquad (11)$$

This allows turnover length to be estimated from measurements of downstream transport and areal respiration in the river. S_c, like spiralling length, measures the combined effects of retentiveness (v_c) and rate of biological processing (k_c) in determining how effectively a unit of river bottom processes nutrient supplied from upstream. Moreover, under our simplifying assumption of uniform carbon quality, the turnover length describes how carbon entering the river at a particular location affects downstream metabolism. For example, if I^* is a steady input (per unit area) to a unit length of river at location $x = 0$, then the respiration $R^*(x)$ at downstream distance x attributable to I^* is given by $R^* = (I_0^*/S_c)$ exp $(-x/S_c)$. This is essentially the same model that Streeter and Phelps (1925) used to describe the downstream effects of organic matter in sewage effluent. Streeter and Phelps assumed that all organic matter remained in suspension ($v_c = v_w$), expressed R in terms of oxygen utilization, and coupled this model with one describing oxygen depletion and re-aeration. The Streeter–Phelps approach, although sometimes highly elaborate, remains the basis for many currently used water quality models (Gromiec *et al* 1983; *cf.* section 7.4).

Estimates of organic carbon turnover length, based on equation (11), have been made for several streams and rivers (Newbold *et al* 1982a; Minshall *et al* 1983, 1992; Edwards & Meyer 1987; Naiman *et al* 1987; Meyer & Edwards 1990; Richey *et al* 1990). Values ranged from 2.9 km for a small wooded stream in New Hampshire, USA (data from Fisher & Likens 1973), to 4000 km for the Amazon River (Richey *et al* 1990). Minshall *et al* (1983) estimated the turnover lengths of only particulate carbon (i.e. excluding dissolved organic matter). Their estimates ranged from 1.0 to 10 km in headwater North American streams

to 250 km in the McKenzie River, Oregon, with an average flow of 55 m^3 s^{-1} (Minshall *et al* 1983). The major influence on turnover length is stream size. Areal respiration often, but not always, increases in the downstream direction, and the variation usually remains within one order of magnitude (Minshall *et al* 1983, 1992; Bott *et al* 1985; Naiman *et al* 1987; Meyer & Edwards 1990). Water-column organic matter concentration may increase somewhat with stream size but on the whole remains remarkably constant (Schlessinger & Melack 1981; Mulholland & Watts 1982). It is, therefore, the simple variation in depth and velocity that accounts for most of the size effect on turnover length (since $F_T = v_w d C_c$, where C_c is total organic carbon concentration).

Carbon turnover lengths are generally longer than the stream or river in which they were measured. That is, most of the carbon entering a river is transported, either to a larger downstream river or to an estuary or sea. Much of the carbon reaching the sea is highly refractory (Ittekot 1988), and probably undergoes very little degradation within the river system. Clearly, much of the metabolism in a river is of more labile forms, such as algae and fresh leaf litter, which contribute little to transport. This diversity of forms affects the estimation of turnover length. Equation (11), in effect, yields an average of the individual turnover of the various forms of carbon, weighted by their relative contribution to the measured respiration. Respiration of forms whose turnover length is short relative to river length can be expected to be about equal to the rate of lateral input (on an annual basis), while respiration of forms with very long turnover lengths must be far less than lateral inputs. Thus, a turnover length weighted by carbon inputs rather than by respiration would be much longer than estimated by equation (11).

The promise of analysing organic carbon dynamics from the perspective of spiralling is that it might provide a measure of the upstream–downstream interdependence of the river ecosystem. Such a measure would be useful from a practical standpoint, for example in evaluating the importance of protecting headwater stream reaches to maintain a downstream fishery. It

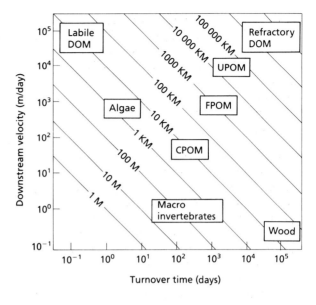

Fig. 7.5 Postulated turnover lengths (diagonal isoclines) for carbon of various forms in a medium-size river (e.g. with a flow of 3–30 m³ s⁻¹). Values represent rough averages from various sources and some speculation. Data sources include: Waters 1961, 1965; Fisher 1977; Marker & Gunn 1977; Minshall *et al* 1983, 1992; Bott *et al* 1985; Webster & Benfield 1986. CPOM, course particulate organic matter (>1 mm); FPOM, fine particulate organic matter (50 μm to 1 mm); UPOM, ultrafine particulate organic matter (<50 μm); DOM, dissolved organic matter.

would also provide an approach to analysing longitudinal variations in ecosystem structure, such as those postulated by the river continuum concept (Vannote *et al* 1980). If natural carbon were uniform in character, the turnover length, in conjunction with the simple exponential decay model presented above, would provide such a measure as it does for the case of organic pollution. But the diversity of organic carbon prevents this interpretation. If, for example, a turnover length of 100 km is an average of two kinds of carbon with turnover lengths of 1 km and 199 km, respectively, the turnover length tells us little about the longitudinal interdependence of the river ecosystem. What is necessary, is to describe carbon spiralling in terms of specific carbon forms, accounting not only for the individual turnover lengths of these forms, but for the transformation of carbon among them.

Such an analysis has not been conducted, but sufficient information exists to illustrate its potential. Figure 7.5 presents rough turnover times $(1/k_c)$ and downstream transport velocities (v_c) for a small river (e.g. baseflow 3–30 m³ s⁻¹), representing a composite of several studies. Turnover lengths for the individual forms range over approximately eight orders of magnitude from ~100 m for macroinvertebrates to ~10⁶ km or more for refractory dissolved organic matter. Note

that labile DOM and woody debris, which occupy opposite extremes on the turnover time and velocity axes, both have intermediate turnover lengths in the range of 10^2–10^4 m. The downstream influence of carbon of a given form depends not only on the turnover length shown here, but also on its conversion to other forms. For example, carbon initially in algae and coarse particulate organic matter (CPOM) may remain local if assimilated by macroinvertebrates, but pass downstream if it is leached as labile DOM or converted to fine particles (FPOM). In rivers of larger or smaller size, we would expect turnover times to remain roughly as in Fig. 7.5, but downstream velocities would differ dramatically. Transport velocities of CPOM, and FPOM in first-order streams, for example, are of the order of 0.1 and 1–10 m day⁻¹, respectively.

Figure 7.5 suggests that longitudinal linkages — or influences of upstream ecosystems on downstream ecosystems — may be transmitted primarily through labile DOM, algae, CPOM, FPOM, and wood; that is, the forms whose turnover lengths are on the scale of the river length. In contrast, most of the downstream transport of carbon in unpolluted streams and rivers occurs as refractory DOM and particles smaller than 50 μm (UPOM) (Wallace *et al* 1982; Minshall *et al* 1983), which are forms with extremely long turnover lengths. A complete analysis will require coupling the turnover-length concept with budgetary studies of individual reaches (e.g. Fisher & Likens 1973; Fisher 1977; Mulholland 1981; Cummins *et al* 1983; Richey *et al* 1990), and with geomorphic approaches which consider the role of all channels, of all sizes, within a river network (e.g. Cummins *et al* 1983; Naiman *et al* 1987; Meyer & Edwards 1990).

7.6 PHOSPHORUS

Forms and concentrations

Phosphorus is an essential nutrient centrally involved in energy transformations within organisms, making up roughly 0.1–1% of organic matter. Figure 7.6 illustrates the major aspects of the phosphorus cycle in rivers. Phosphorus in water is normally categorized as being either dissolved or particulate, depending on whether it passes 0.45-μm membrane filter. 'Dissolved' phosphorus, therefore, may include a substantial colloidal component. Within the dissolved fraction, inorganic P (DIP) occurs as orthophosphate (PO_4), which is usually estimated by variations of the molybdenum blue method (Strickland & Parsons 1972). Because this method may overestimate the orthophosphate by hydrolysing organic and colloidally bound forms (Rigler 1966; Stainton 1980; Tarapchak 1983), measurements based on the molybdenum blue method are often referred to as 'soluble reactive phosphorus' (SRP). 'Total dissolved phosphorus' (TDP) represents a molybdenum blue assay for phosphorus following an acid digestion that releases the dissolved phosphorus in organic forms. The organic fraction (DOP) is not well characterized, but in lake water some consists of inositol hexaphosphate (Herbes *et al* 1975; Eisenreich & Armstrong 1977) and DNA fragments (Minear 1972).

In unpolluted rivers, SRP averages about 10 μg l^{-1} (Meybeck 1982) on a worldwide basis, while total dissolved phosphorus averages about 25 μg l^{-1}. Dissolved phosphorus concentrations may increase with discharge, but rarely by a factor of more than 2–4 during peak flows (Kunishi *et al* 1972; Leonard *et al* 1979; Meyer & Likens 1979; Saunders & Lewis 1988). Agricultural activities may increase dissolved phosphorus to the range of 50–100 μg l^{-1} (Omernik 1977; Smart *et al* 1985; Mason *et al* 1990), and to >500 μg l^{-1} during snowmelt (Rekolainen 1989). Municipal effluents, however, may increase concentrations to the range of 1000 μg l^{-1} (Meybeck 1982).

Particulate phosphorus includes phosphorus

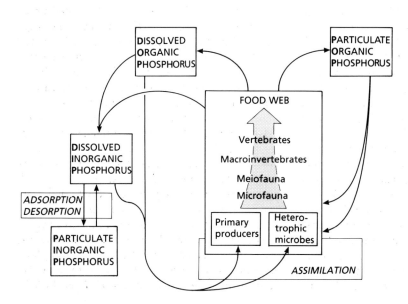

Fig. 7.6 Schematic of the phosphorus cycle in streams and rivers.

incorporated into mineral structures, adsorbed on to surfaces, primarily of clays, and incorporated into organic matter, and averages worldwide about 500 µg l^{-1} (Meybeck 1982). Concentrations of suspended particulates vary greatly with land use and erodibility of the watershed (Cosser 1989; Karlsson & Löwgren 1990) and increase dramatically with stormflows (Kunishi *et al* 1972; Verhoff & Melfi 1978; Meyer & Likens 1979; Munn & Prepas 1986; Prairie & Kalff 1988; Krogstad & Løvstad 1989).

Dissolved-to-particulate transfers

Movement from the dissolved inorganic (DIP) compartment to other ecosystem compartments is perhaps best observed by adding a radio-isotope tracer, ^{33}P-PO$_4$, directly to a stream river. Unfortunately, this is not generally feasible, and such experiments have been conducted in only a few streams and one relatively small river. The uptake length can be estimated from equation (2) and the longitudinal disappearance of ^{32}P relative to a hydrological tracer such as ^{3}H-H$_2$O or chloride. In Walker Branch, a first-order woodland stream in Tennessee, USA (baseflow 3–8 l s^{-1}), ^{32}P-estimated uptake lengths ranged from 21 to 165 m, varying roughly inversely with the quantity of leaf detritus on the stream bottom (Mulholland *et al* 1985b). Corresponding rates of phosphorus uptake were 1.8–22 mg^{-2} day^{-1}. In the Sturgeon River, Michigan, USA, with a flow increasing downstream from 1.1 to 1.4 m^3 s^{-1}, Ball and Hooper (1963) estimated uptake lengths between 1100 and 1700 m in successive reaches (with no downstream trend). From their data, areal uptake of phosphorus averaged about 10 mg^{-2} day, or near the middle of the range of estimates from Walker Branch.

Chambers, or microcosms, offer a promising but underexploited alternative method for using ^{32}P to estimate phosphorus uptake (Corning *et al* 1989; Paul *et al* 1989; Paul & Duthie 1989; Stream Solute Workshop 1990). Corning *et al* (1989) incubated rocks from Canadian streams and rivers under *in situ* conditions of temperature and flow in closed 6-litre recirculating chambers. Initial loss of ^{32}P from solution is first order until regeneration of ^{32}P becomes significant. Corning

et al reported uptake rates by the epilithon ranging from 0.5 to 4.3 mg m^{-2} day^{-1}.

Estimates of phosphorus removal based on tracer data may be erroneously high if the true levels of PO$_4$ are below the value estimated as SRP. This problem has proved particularly vexing in lakes, where SRP may overestimate PO$_4$ by more than an order of magnitude (Rigler 1966; White *et al* 1981). Peters (1981), using a radiological bioassay, found that PO$_4$ constituted between 6% and 44% of SRP in water from various river waters, but the radiobiological method (based on Rigler 1966) has been criticized by Tarapchak and Herche (1988). Jordan and Dinsmore (1985), also using radiobioassay, concluded that in the River Main in Northern Ireland the readily available phosphorus (presumably PO$_4$) accounted for 76% of the SRP. Newbold *et al* (1983a) inferred from exchange rates with periphyton of known pool size, that true PO$_4$ concentrations in Walker Branch were at least 70% of the average SRP.

Several studies have observed the uptake of stable phosphorus (^{31}P) experimentally injected into small streams (McColl 1974; Meyer 1979; Aumen *et al* 1990; Mulholland *et al* 1990; Munn & Meyer 1990; D'Angelo *et al* 1991). In a few cases, the observed net uptake of phosphorus has been negligible, indicating that whatever uptake occurs in the stream is saturated at ambient phosphorus concentrations. In most cases, however, considerable retention of the added phosphorus occurred, with uptake lengths in the range of 5–200 m. An uptake length measured by adding stable phosphorus is equivalent to one measured using ^{32}P only if the uptake flux of phosphorus increases in direct proportion to the increase in concentration (i.e. F/U remains constant). Mulholland *et al* (1990) found that an addition which increased PO$_4$ concentration from a background of 2.9 µg l^{-1} to 7.4 µg l^{-1} yielded an uptake length 55% longer than that measured by ^{33}P at the ambient concentration (2.9 µg l^{-1}). Estimates made using this approach, therefore, provide only an upper bound for uptake length.

A number of mass balance studies have shown phosphorus removal from solution downstream from municipal and agricultural sources of phosphorus (Keup 1968; Taylor & Kunishi 1971; Johnson *et al* 1976; Harms *et al* 1978; Hill 1982).

In these studies, DIP concentrations declined longitudinally from peaks sometimes exceeding 1000 µg l^{-1} (Harms *et al* 1978), with net uptake rates in the range of 10 to >100 mg m^{-2} day^{-2} (in most cases, these uptake rates are inferred because few studies reported net uptake directly). Phosphorus apparently accumulates in the sediments by physical adsorption (as discussed below) during periods of steady flow, and is then transported downstream as particulate load during storms (Cahill *et al* 1974; Harms *et al* 1978; Verhoff & Melfi 1978; Verhoff *et al* 1979).

As phosphorus leaves the dissolved form in the water column, it may be transferred either to the sediments or to suspended solids (seston) within the water column. Uptake by seston may be negligible in very small streams (e.g. 1.3% of total uptake; Newbold *et al* 1983a), but increases in importance with stream size (Ball & Hooper 1963; Paul *et al* 1989) and may account for nearly all of the uptake in large rivers with true phytoplankton populations (Décamps *et al* 1984). As in marine and lacustrine environments, most of the sestonic uptake appears to occur within the smallest size fractions (e.g. <1 µm; Corning *et al* 1989). Phosphorus transferred to seston may subsequently settle to the sediments. Simmons and Cheng (1985) concluded that this was the major pathway for removal of phosphorus from the water column in a river receiving sewage effluent.

Biological uptake

Phosphorus is removed from the water by algae (including cyanophyte bacteria), heterotrophic microbes, macrophytes, bryophytes and riparian plants. The kinetics of phosphorus uptake by planktonic algae have received extensive attention (see Cembella *et al* 1984a, 1984b), and the fundamentals undoubtedly apply to attached algae. Uptake generally follows Michaelis–Menten kinetics (see equation (10)). Bothwell (1985) found that K_s for periphytic river diatoms ranged from 0.5 to 7.2 µg l^{-1}, or below that for most phytoplankton populations. Algae (and bacteria) can take up phosphorus at a much greater rate than they can utilize it for growth so that V_{max} can vary greatly, declining as cells accumulate phosphorus (Rhee 1974). As a result, steady-

state growth is saturated by concentrations very much lower than K_s (Droop 1973; Rhee 1974). Uptake of phosphorus by attached algae is also influenced by transport of phosphorus to and into the algal biofilms (section 7.3).

Few data exist to evaluate the relative importance of ecosystem components in taking up phosphorus from the water column. In Walker Branch, which is well shaded, 60% of the phosphorus uptake in the summer was accounted for by large detritus (>1 mm), 35% by fine particles (<1 mm) and 5% by the epilithon, consisting primarily of diatoms (Newbold *et al* 1983a). The phosphorus content of leaf detritus increases during decomposition (Kaushik & Hynes 1968; Meyer 1980), and the rate of phosphorus uptake correlates with measures of metabolic activity (Gregory 1978; Mulholland *et al* 1984; Elwood *et al* 1988). In less shaded streams and rivers, the epilithon and associated algae are probably the dominant fate of phosphorus (Ball & Hooper 1963; Peterson *et al* 1985). Ball and Hooper (1963) did not make a mass accounting for the phosphorus taken up, but they did find that weight-specific uptake by macrophytes (*Chara*, *Fontinalis* and *Potomogeton*) was far slower than by periphyton.

Adsorption

Adsorption and desorption of phosphorus on to organic and inorganic surfaces are continual kinetic processes. Thus, some portion of the phosphorus cycle (e.g. as observed via ^{32}P injection) involves entirely abiotic processes. Unfortunately, few studies have addressed phosphorus adsorption from a kinetic standpoint (except see Pomeroy *et al* 1965; Li *et al* 1972; Furumai *et al* 1989), so that it is difficult to evaluate the magnitude of this cycling. Studies of ^{32}P uptake by epilithon and by particulate organic matter have found that uptake is small or negligible under sterile conditions (Gregory 1978; Elwood *et al* 1981a; Paul *et al* 1989). These studies, however, involved very low phosphorus concentrations and did not include large quantities of inorganic fine particles.

Most studies of phosphorus adsorption in rivers have focused not on the kinetic exchange of phosphorus, but on the role of sediments in controlling or 'buffering' dissolved phosphorus concentrations. Adsorption of phosphorus on to natural river sediments and suspensoids generally corresponds to a Langmuir isotherm (Green *et al* 1978; McCallister & Logan 1978; Stabel & Geiger 1985; Furumai *et al* 1989), which has the form:

$$X = bkC/(1 + kC) \qquad (12)$$

in which X is the quantity of adsorbed phosphorus per mass of sediment, C is the dissolved phosphorus concentration, b is the 'adsorption maximum' in units of X, and k is the 'adsorption energy' in units of $1/C$. At phosphorus concentrations below $0.1–1.0$ mg l^{-1}, the isotherms are approximately linear (i.e. $kC \ll 1$). For fine particles (<0.1 mm), which account for nearly all of the sorption capacity (Meyer 1979), the slope (bk) of the isotherms ranges from about 0.1 to 1.0 µg P g^{-1} sediment per µg P l^{-1} water.

Sediment particles are often characterized in terms of their 'equilibrium phosphorus concentration' (EPC), which is the water concentration at which phosphorus is neither adsorbed nor desorbed (White & Beckett 1964; Klotz 1988). The EPC reflects the quantity, X, of adsorbed phosphorus and, in terms of the Langmuir isotherm, represents the value of C that satisfies equation (12). Sediments that enter rivers may either desorb phosphorus to the water (Mayer & Gloss 1980; Gloss *et al* 1981; Grobbelaar 1983; Viner 1988), or adsorb phosphorus from the water (Kunishi *et al* 1972) depending on whether the initial EPC exceeds the water phosphorus concentration. The EPC of stream sediments is often near the stream water phosphorus concentration (Taylor & Kunishi 1971; Meyer 1979; Mayer & Gloss 1980; Hill 1982; Klotz 1985, 1988; Munn & Meyer 1990), indicating rapid equilibration. This equilibration has frequently been interpreted as evidence that the sediments control or 'buffer' stream water concentrations. Such buffering, however, implies a net uptake or release of phosphorus by the sediments which, under steady conditions, will diminish as the sediment EPC adjusts to the concentration of the source water. Thus, it would seem premature to conclude that the sediment EPC controls the stream water concentration (rather than the converse) without additional information about source water concentrations, the actual quantity of sediments in the streambed, or both.

In general, we would expect sediments to influence phosphorus concentrations when: (1) there is an abundance of fine inorganic particles (e.g. several hundred kg m^{-2}); and (2) a large discrepancy exists between the EPC of sediments entering the river and the concentration of phosphorus entering the river. These conditions would explain the prolonged periods of high net phosphorus removal downstream from pollution sources discussed above. In streams and rivers with few fine inorganic sediments and low dissolved phosphorus inputs, however, the role of adsorption in controlling phosphorus concentrations may be minimal, and perhaps exceeded by the influence of biological processes. Mulholland *et al* (1990), for example, observed an apparent transition from biotically dominated uptake to sorption-dominated uptake at about 5 µg l^{-1}, a concentration that might have saturated biological uptake capacity.

Regeneration

Where field injections of ^{32}P have been made, the loss of ^{32}P from the water has not been accompanied by commensurate depletion of stable phosphorus (Ball & Hooper 1963; Newbold *et al* 1983a; Mulholland *et al* 1985b), implying that regeneration of phosphorus approximately balances gross uptake. Other evidence for regeneration comes from direct observation of ^{32}P in stream water for several weeks following a ^{32}P release (Elwood & Nelson 1972), and from ^{32}P accumulation in detritus and periphyton placed into a stream after the initial labelling (Ball & Hooper 1963; Elwood & Nelson 1972). Similar results have been obtained in laboratory streams (Short & Maslin 1977; Paul & Duthie 1989). Studies with ^{32}P have shown that phosphorus associated with epilithon and detritus returns to the water with a turnover time of a week or less, whereas turnover in fine sediments may be an order of magnitude slower (Ball & Hooper 1963; Elwood & Nelson 1972; Newbold *et al* 1983a).

Regeneration of phosphorus from the biota may occur via: (1) excretion of phosphorus (either as DIP or DOP) from living algae and bacteria;

(2) release of phosphorus upon death and lysis of cells; and (3) ingestion followed by egestion, excretion, and death of animal consumers. Although the first two pathways may in some cases be significant (Barsdate *et al* 1974; Lean & Nalewajko 1976; DePinto 1979), there is a growing consensus from marine and lacustrine studies that animal consumers, particularly Protozoa and other microscopic animals, represent the major pathway. Regeneration by consumers is discussed in section 7.8.

A portion of the phosphorus lost from organisms may be in the form of dissolved organic phosphorus (DOP). Much of the DOP released may be hydrolysed rapidly to DIP by alkaline phosphatase and ultraviolet degradation (Francko 1986). Alkaline phosphatase, which is capable of hydrolysing a range of organic phosphorus compounds, has been detected in the sediments of some streams (Sayler *et al* 1979; Klotz 1985). However, the bulk of the DOP that is found in natural waters is not hydrolysed by alkaline phosphatase (Herbes *et al* 1975; Jordan & Dinsmore 1985; Hino 1989). Mulholland *et al* (1988) found that about 12% of ^{32}P released from decomposing leaf detritus was organic with molecular size >5000 daltons. This organic fraction appeared to be far less available for utilization than PO_4, with only about 10% utilized within 24 hours of incubation. High molecular weight organic phosphorus has been identified in other river and lake waters (Lean 1973; Downes & Paerl 1978; Peters 1978, 1981; White *et al* 1981) where it has similarly been shown to be biologically available, but at a slower rate than PO_4 (Pearl & Downes 1978; Peters 1981). It remains unclear whether utilization of DOP can occur directly (Smith *et al* 1985), possibly involving enzymatic activity on cell surfaces (Ammerman & Azam 1985), or whether the DOP is first hydrolysed within the water. Whatever the ecological significance of this distinction, it is clearly important to the interpretation of ^{32}P kinetics. The potential role of DOP in cycling has special significance for rivers because DOP is highly transportable yet unavailable for rapid uptake and thus could contribute substantially to spiralling length.

Phosphorus spiralling in a woodland stream

An extensive quantification of phosphorus spiralling has been attempted for only one stream, Walker Branch, Tennessee (Newbold *et al* 1983a). This description was obtained by fitting a multicompartmental model to the ^{32}P data to estimate transfer coefficients among all compartments (Fig. 7.7). From the model, spiralling within each compartment is described with three parameters: the residence time in the compartment, the fraction of the cycling phosphorus flux passing through the compartment, and the downstream velocity of the compartment (Table 7.1). The spiralling length was 190 m, consisting of 165 m of travel in the water (as described above), and 25 m of travel in the particulate compartments (CPOM, FPOM and epilithon). Phosphorus taken up by consumers had a downstream travel distance of 2 m, but since only 2.8% of the flux passed through this compartment, the contribution to total spiralling length was only 0.06 m. Although dissolved phosphorus accounted for most of the spiralling length, its turnover time of 75 min was a small fraction of the 18 days required to complete a cycle.

Walker Branch is highly retentive of phosphorus. In effect, the biotic phase 'slows' the average movement of phosphorus downstream by a factor of about 300 relative to the water velocity. Stated another way, the standing stock of phosphorus is 300 times greater than in the absence of retention. The snail *Elimia* (=*Goniobasis*) *claeviformis* plays a large role in this retention, accounting for 23% of the standing stock exchangeable phosphorus, and drifting downstream at a velocity of <1 cm day^{-1}.

7.7 NITROGEN

Nitrogen cycle

Nitrogen is a fundamental constituent of protein and, like phosphorus, its availability can frequently limit algal and microbial productivity. The major pathways of the nitrogen cycle in aquatic systems are shown in Fig. 7.8 (see also Kaushik *et al* 1981; Sprent 1987), although not all of these pathways have been studied extensively

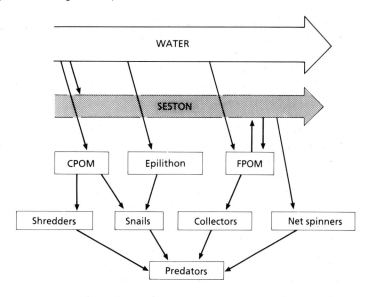

Fig. 7.7 Ecosystem compartments and flows used to analyse spiralling of phosphorus in a small woodland stream (from Newbold *et al* 1983a). In addition to the phosphorus fluxes indicated by arrows on the diagram, every compartment releases phosphorus directly back to the water compartment. These arrows are omitted for simplicity. Dynamics of the water and seston compartments were described by partial differential equations similar to equation (9) and coupled to other compartments as illustrated in Fig. 7.3.

in rivers. The nitrogen cycle in rivers is more complex than the phosphorus cycle in several important respects. First, the processes of nitrogen fixation and denitrification involve exchanges with atmosphere, so that cycling of nitrogen cannot be considered closed with respect to the air–water interface. (Atmospheric exchanges of ammonia may also be significant under some

Table 7.1 Calculation of spiralling indices for phosphorus in Walker Branch (from Newbold *et al* 1983a)

Compartment	t_i (days)	v_i (m day^{-1})	s_i (m)	b_i	S_i (m)
Water	0.052	3200	165	1.0	165
Particulates					
CPOM[†]	6.9	0.06	0.40	0.60	0.24
FPOM (fast)[‡]	6.9	7.4	51	0.27	13
FPOM (slow)[‡]	99.0	1.4	141	0.080	11
Epilithon	5.6	0.0	0.0	0.054	0.0
Total particulates	14.0[*]	1.8[*]	25[*]	1.0	25
Consumers					
Snails	150	0.005	0.77	0.024	0.019
Shredders	76	0.13	9.8	0.0002	0.0030
Collectors	105	0.12	13	0.003	0.03
Net spinners	220	0.0	0.0	0.007	0.0
Predators	14	0.007	0.10	0.008	0.0008
Total consumers	150[*]	0.013	2.0[*]	0.028	0.056
Total	18[*]	10	190	1.0	190

[*] Weighted average; [†] coarse particulate organic matter (>1 mm); [‡] fine particulate organic matter (<1 mm). Two turnover rates, 'fast' and 'slow', were resolved from the ^{32}P dynamics of the FPOM compartment.
t_i is average residence time in compartment i; v_i is average downstream velocity of compartment i. Average travel distance in a compartment, $s_i = v_i t_i$. The contribution of compartment i to spiralling length is $S_i = b_i s_i$, where b_i is the probability of passing through compartment i. Particulate and consumer turnover lengths are the sums of S_i values within these categories.

circumstances.) Second, in addition to its role as a fundamental cellular constituent in all organisms, nitrogen is also involved in biologically mediated redox reactions, such as nitrification which yields energy for metabolism and consumes oxygen, and denitrification, in which nitrate serves as a terminal electron acceptor. These processes involve specialized organisms and occur with very different stoichiometric coupling to carbon flow than does cellular assimilation of nitrogen. Finally, there are two major sources of inorganic nitrogen for algal and microbial assimilation (NH_4 and NO_3), rather than one (PO_4) in the phosphorus cycle. Understanding of the nitrogen cycle in rivers is limited not only by the complexity of the cycle, but by the fact that very few isotopic tracer studies have been conducted in rivers or in river-simulating microcosms. The available isotope,

^{15}N, is not radioactive, but is otherwise more difficult and expensive to work with than are the radioisotopes of phosphorus.

Estimates by Meybeck (1982) of worldwide average concentrations of dissolved nitrogen in unpolluted rivers are ($\mu g\ l^{-1}$): dissolved organic nitrogen (DON) 260; nitrate 100; ammonium 15; nitrite 1. In agricultural watersheds in North America, DON averages about 1000 $\mu g\ l^{-1}$, and dissolved inorganic nitrogen typically ranges from ~700 to 5000 $\mu g\ l^{-1}$ (Omernik 1977). Algae and other micro-organisms tend to use ammonium in preference to nitrate, which must be reduced before it can be assimilated (Sprent 1987). Based on results from lacustrine and marine ecosystems (Eppley *et al* 1979; Axler *et al* 1982) it is reasonable to conjecture that in many streams and rivers, ammonium supplies a substantial part, if not the

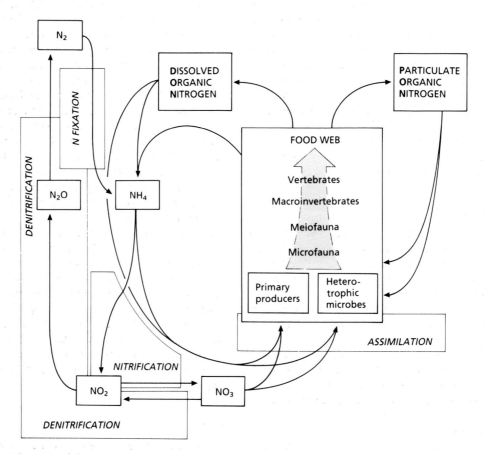

Fig. 7.8 Schematic of the nitrogen cycle in streams and rivers.

majority, of the nitrogen assimilation by algae and heterotrophic micro-organisms. Much of this ammonium may be supplied by excretion from consumers and cycle rapidly (and over short distances) back into the microbial community. Stanley and Hobbie (1981), using ^{15}N, found that uptake by river plankton of ammonium exceeded that of nitrate by a factor of three, while the ammonium concentration averaged only about half that of nitrate. Thus, the ammonium pool turned over approximately six times faster than the nitrate pool. The uptake of dissolved inorganic nitrogen (DIN, consisting of nitrate plus ammonia) averaged about 5 mg m^{-2} h^{-1}, but peaked at near 70 mg m^{-2} h^{-1} in the summer at maximum photoplankton concentrations. On an annual basis, the utilization of DIN within the 60-km study reach exceeded upstream inputs to the reach by a factor of three, corresponding to an uptake length of 20 km, with an average turnover time of the DIN pool of about 10 days. In the summer, under high demand and low flows, turnover times fell to 3–10 h, corresponding to uptake lengths of <1 km. Stanley and Hobbie also concluded that bacteria may have played an important role, in addition to phytoplankton, in taking up the DIN, and that approximately one-third of the DON entering the reach was converted to DIN, presumably through microbial assimilation and regeneration.

Biological uptake

Little is known about the magnitude of ammonium utilization by benthic organisms. Several studies have documented rapid uptake by the benthos of ammonium introduced at enriching levels (McColl 1974; Newbold *et al* 1983b; Richey *et al* 1985; Hill & Warwick 1987). Ammonium, however, is similar to phosphorus in that it is strongly subject to physical sorption and cells can incorporate ammonia temporally in excess of growth requirements (Eppley & Renger 1974; Conway & Harrison 1977). Thus it is difficult to draw inferences from these studies about the utilization of ammonium under natural concentrations.

As with ammonium there are no studies other than that of Stanley and Hobbie (1981) of nitrate

gross uptake under undisturbed conditions. However, there have been a number of mass balance studies observing disappearance of nitrate, which has in some cases been equated to a net incorporation of nitrate by benthic algae and heterotrophic micro-organisms. Fisher *et al* (1982) and Grimm (1987) observed longitudinal declines in nitrate concentrations in a desert stream associated with rapid growth of benthic algae following scouring storms. Maximum calculated uptake rates were 240 mg m^{-2} day^{-1}. Fisher *et al* (1982) noted that comparing net primary production to calculated uptake rates gave a C:N ratio of 4:1, or lower than the Redfield ratio of 6.6:1 (Redfield *et al* 1963), suggesting that some of the net nitrate uptake went to other fates, such as denitrification.

Uptake of experimentally injected nitrate has been observed in several streams (McColl 1974; Sebetich *et al* 1984; Triska *et al* 1989a; Aumen *et al* 1990; Munn & Meyer 1990; Webster *et al* 1991) in which background DIN concentrations were <150 µg l^{-1}. In other streams, all with background concentrations >100 µg l^{-1}, no net uptake was observed (McColl 1974; Richey *et al* 1985; Aumen *et al* 1990). The relation to DIN, however, is not entirely consistent, and background ammonium concentrations (not always reported) may be an important factor. In the most detailed study, where substantial net uptake of added nitrate occurred over 9 days of injection (Triska *et al* 1989a, 1989b), background ammonium concentrations were undetectable, while background nitrate was 25–50 µg l^{-1}.

Kim *et al* (1990) analysed uptake kinetics of nitrate added to artificial channels in which periphyton grew on Plexiglas slides. They obtained half-saturation constants for uptake (K_s) in the range of 50–80 µg l^{-1} from laboratory batch experiments, and then fitted an uptake model (section 7.4) to the dynamics of nitrate injected into their channels. From this, they obtained values for maximum uptake in the range of 0.15–0.65 µg N s^{-1} g^{-1} ash-free-dry-mass (AFDM) of periphyton.

Nitrification

Nitrification, the biological oxidation of ammonium to nitrite and then to nitrate, seems to occur in streams and rivers whenever both ammonium and oxygen are available. It is of particular interest as a water quality issue because it can reduce potentially toxic levels of ammonium discharges. Nitrification, however, consumes dissolved oxygen, and so exacerbates the oxygen depletion associated with oxidation of organic wastes. Much of the dissolved and particulate organic nitrogen in waste discharges may also ultimately be nitrified, after degradation to ammonium. Rates of nitrification in rivers, therefore, are sometimes reported as the rate at which both ammonium and organic nitrogen are converted to nitrate. These rates vary approximately in the range of $0.1-0.5$ day^{-1} depending on temperature, the proportion of the nitrogen source that is ammonium, and other factors (McCutcheon 1987). Nitrification may play a large role in maintaining ammonium levels in unpolluted streams at low levels. Triska et al (1990) found that nitrification occurred in the sediments, reducing relatively high ammonium levels in influent groundwater to low levels observed in the stream. Nitrification appears to be associated largely with the sediments in streams and smaller rivers (Matulewich & Finstein 1978; Cooper 1983; Cooper 1984) but may occur largely in suspension in larger rivers (McCutcheon 1987). Estimated benthic rates of nitrification in streams range from 29 mg m^{-2} day^{-1} (Chatarpaul et al 1980) to 2.5 g m^{-2} day^{-1} (Cooper 1984).

Denitrification

Denitrification, the reduction of nitrate to dinitrogen, is carried out by bacteria using the nitrate as the terminal electron acceptor for oxidative metabolism in the absence of oxygen. Denitrification is included in the general category of dissimilatory nitrate reduction, which can also include reduction of nitrate to ammonium. Denitrification can be a significant pathway for loss of nitrogen from the river and has been implicated in explaining downstream declines in nitrogen concentrations in several watersheds (Kaushik et al 1975; Hill 1979, 1988; Swank & Caskey 1982; Cooper & Cooke 1984). Most estimates of denitrification in streams fall into the range of $10-200$ mg m^{-2} day^{-1} (Chatarpaul &

Robinson 1979; Cooper & Cooke 1984; Duff et al 1984b; Hill & Sanmugadas 1985; Christensen & Sørensen 1988; Seitzinger 1988; Christensen et al 1989). Nearly all of these measurements, however, have been conducted in streams with relatively high nitrate concentrations (i.e. $0.5-10$ mg l^{-1}), and in streams that are not nitrate enriched normal rates of denitrification may be much lower (Duff et al 1984b; S.P. Seitzinger, personal communication).

Although denitrification requires anoxic conditions, it has been observed in aerated sediments and in relatively thin epilithic films (Nakajima 1979; Duff et al 1984b; Duff & Triska 1990). Evidently, denitrification occurs in microzones of anoxia within the sediments and biofilms. Oxygen produced by benthic algae may inhibit denitrification (Duff et al 1984b) resulting in diel variations in denitrification rates (Christensen et al 1990).

Denitrification requires an organic carbon source and proceeds faster where more carbon is available in the water and in the sediments (Hill & Sanmugadas 1985; Duff & Triska 1990). Typical denitrification rates of $10-200$ mg m^{-2} day^{-1} are equivalent to the respiration of $0.03-0.7$ g O$_2$ m^{-2} day^{-1}. By comparison, oxygen consumption in rivers generally ranges from 0.2 to 10 g O$_2$ m^{-2} day^{-1} (Bott et al 1985). Thus, denitrification may contribute a significant portion of the oxidative metabolism in streams where nitrate supply is high and appropriate habitat and carbon resources are available. Denitrification may occur simultaneously with nitrification (Duff et al 1984a; Cooke & White 1987). In Little Lost Man Creek, California, USA, nitrogen enters the deep sediments (hyporheos) as ammonium, mixes with oxygenated interstitial water and nitrifies (Triska et al 1990), yielding a net downstream increase in nitrogen concentrations. Yet significant denitrification also occurs within the hyporheos, presumably in anoxic microzones (Duff & Triska 1990). Chatarpaul et al (1980) found that tubificid worms in sediments enhanced both nitrification and denitrification.

Most of the measurements reported above estimated denitrification by the acetylene block method (Balderston et al 1976; Yoshinari et al 1976), although ^{15}N (e.g. van Kessel 1977;

Chatarpaul & Robinson 1979) and direct measurement of nitrogen evolution (Seitzinger 1988) have also been employed. The acetylene block method is subject to various uncertainties, the most serious of which is that it interferes with nitrification. As a result, the method can greatly underestimate denitrification when it is coupled to nitrification of an ammonia source (SP Seitzinger, CP Nielson, J Caffrey & PB Christenson, unpublished data).

Nitrogen fixation

Blue-green algae and micro-organisms fix nitrogen in streams and rivers, but few measurements have been made. Horne and Carmiggelt (1975) reported nitrogen fixation by the blue-green alga *Nostoc* of $42-360$ mg N m^{-2} year^{-1}, while Francis *et al* (1985) estimated nitrogen fixation in pool sediments of 5.1 g N m^{-2} year^{-1}. Even these higher values, however, are small in comparison with other nitrogen fluxes as noted above.

7.8 THE ROLE OF CONSUMERS

Animals in the food web may account for a considerable portion of the nutrient cycle, but their influence on nutrient cycling involves, in addition: (1) direct influences on prey populations, which may 'cascade' through several trophic levels; (2) indirect influences of regenerated nutrients; and (3) indirect influences of physical transformations and translocations of nutrients and of the physical habitat (Kitchell *et al* 1979).

Algal grazers have been reported to assimilate $30-98\%$ of algal nitrogen uptake (Grimm 1988) and roughly 80% of algal phosphorus uptake (Mulholland *et al* 1983). This is consistent with many studies showing that grazing macroinvertebrates and fish maintain biomass of epilithic algae substantially below levels that would occur in the absence of grazers (e.g. McAuliffe 1984; Jacoby 1985; Stewart 1987; Power *et al* 1988; Feminella *et al* 1989; Hart & Robinson 1990). Indirect, or cascading, control of algal biomass by predators of grazers has also been demonstrated (Power *et al* 1985; Gilliam *et al* 1989; Power 1990). The weight-specific productivity of grazed algae normally exceeds that of ungrazed algae

(Kehde & Wilhm 1972; Summer & McIntire 1982; Gregory 1983; Lamberti & Resh 1983; Lamberti *et al* 1987, 1989; Hill & Harvey 1990), and a similar effect has been observed on phosphorus turnover rate (Mulholland *et al* 1983). However, the net effect of depressing biomass and increasing biomass turnover is normally to decrease productivity (and nutrient uptake) on an areal basis, and only slight, or statistically non-significant, cases of an actual stimulation of areal productivity have been observed.

The possibility that some moderate level of grazing might stimulate, or maximize, areal productivity is of interest because the phenomenon has been observed in lentic systems and microcosms (Cooper 1973; Flint & Goldman 1975; Bergquist & Carpenter 1986; Elser & Goldman 1991), and is usually attributed to the effect of consumers in regenerating nutrients, as predicted by nutrient cycling models (Lane & Levins 1977; Carpenter & Kitchell 1984). Stimulation may have gone undetected in streams because experiments have employed flumes that are too short to observe recycling effects (i.e. much shorter than the spiralling length). However, analysis of a spiralling model (Newbold *et al* 1982b) indicated that the stimulation effect (which would produce a shorter spiralling length) is possible in streams, but unlikely to occur because: (1) streams do not suffer depletion of the limiting nutrient in the same manner as lakes or closed systems (section 7.3); and (2) a simultaneous effect of grazers, which is to dislodge and suspend particles, tends to nullify the benefits of regeneration.

In woodland streams, heterotrophic microbes may take up far more inorganic nutrients from the water than do autotrophs (e.g. Table 7.1). It appears, however, that relatively little of this uptake is assimilated or regenerated by macroinvertebrate consumers. In Walker Branch, Tennessee, for example, where detritus accounts for 95% of the phosphorus uptake, <3% of this is eventually assimilated by macroinvertebrates (Table 7.1). Studies of detritus processing have similarly shown that carbon assimilation by macroinvertebrates is only a small percentage of carbon lost to microbial respiration (Cummins *et al* 1973; Fisher & Likens 1973; Webster 1983; Smock & Roeding 1986; Petersen *et al* 1989). In

marine and lacustrine environments, the protozoans and other very small animals (i.e. the meiofauna) assimilate much of the microbial production
and play a major role in nutrient regeneration
(e.g. Johannes 1965; Andersen *et al* 1986; Caron
& Goldman 1990). Work in riverine environments, although still limited, suggests an equally
strong role for meiofauna (Findlay *et al* 1986; Bott
& Kaplan 1990; Crosby *et al* 1990). In streams
and rivers, the meiofauna may include early-
instar insect larvae, as well as protozoans and
animals that remain microscopic throughout
their life history. Whether energy and nutrients
from the meiofauna can then pass to higher
trophic levels via predation remain essentially
uninvestigated.

Macroinvertebrate consumers exert a substantially larger effect on nutrient cycling than
their energy consumption would suggest. Macroinvertebrate shredders increase significantly the
speed the conversion of large detrital particles to
small particles (e.g. Petersen & Cummins 1974;
Wallace *et al* 1982; Cuffney *et al* 1990) with
consequent effects on particle transport. Cuffney
et al (1990) found that removal of macroinvertebrate populations from a small stream reduced
annual transport of seston by 40%. Webster (1983)
estimated that macroinvertebrate activity accounted for 27% of annual seston transport, and
as much as 83% of transport during low summer
flow. From the standpoint of phosphorus and
nitrogen transport, these effects are probably not
important since, as we have seen, turnover lengths
in headwater streams are short. The transport
does, however, increase the organic carbon turnover length, depleting carbon stocks in headwater
reaches, and so depresses the available substrate
for energy and nutrient utilization.

Newbold *et al* (1982b) hypothesized that, by
increasing the ratio of surface to volume of detrital particles, leaf shredders might enhance
microbial activity and nutrient uptake. Experimental tests, however, showed little evidence of
this effect. Shredders instead decrease phosphorus
utilization both through direct assimilation of
the detrital pool and through increasing downstream particulate losses (Mulholland *et al* 1985a).

Net-spinning caddisfly (Trichoptera) and black-
fly (Diptera: Simuliidae) larvae filter particles

from the water column and therefore actually
reduce transport. Although downstream declines
in seston transport have been observed below
reservoirs and lake outfalls where filter feeders
are abundant (e.g. Maciolek & Tunzi 1968) and
Ladle *et al* (1972) inferred a substantial effect of
Simulium on particle transport in a chalk stream,
it appears that filter feeders in most streams and
rivers exert a small or insignificant effect on total
particle transport (e.g. McCullough *et al* 1979;
Newbold *et al* 1983a). However, they may substantially reduce the transport of particles in
specific high-quality food classes (Georgian &
Thorp 1992).

REFERENCES

Ammerman JW, Azam F. (1985) Bacterial 5'-
nucleotidase in aquatic ecosystems: a novel mechanism of phosphorus regeneration. *Science* **227**:
1338–40. [7.6]

Andersen OK, Goldman JC, Caron DA, Dennett MR.
(1986) Nutrient cycling in a microflagellate food
chain: III. Phosphorus dynamics. *Marine Ecology
Progress Series* **31**: 47–55. [7.8]

Aumen NG, Hawkins CP, Gregory SV. (1990) Influence
of woody debris on nutrient retention in catastrophically disturbed streams. *Hydrobiologia* **190**: 183–92.
[7.6, 7.7]

Axler RP, Gersberg RM, Goldman CR. (1982) Inorganic
nitrogen assimilation in a subalpine lake. *Limnology
and Oceanography* **27**: 53–65. [7.7]

Balderston WL, Sherr B, Payne WJ. (1976) Blockage by
acetylene of nitrous oxide reduction in *Pseudomonas
perfectomarinus*. *Applied and Environmental
Microbiology* **31**: 504–8. [7.7]

Ball RC, Hooper FF. (1963) Translocation of phosphorus
in a trout stream ecosystem. In: Schultz V, Klement
AW Jr (eds) *Radioecology*, pp 217–28. Reinhold
Publishing, New York. [7.2, 7.6]

Barsdate RJ, Prentki RT, Fenchel T. (1974) Phosphorus
cycle of model ecosystems: significance for de-
composer food chains and effect of bacterial grazers.
Oikos **25**: 239–51. [7.6]

Bencala K. (1984) Interactions of solutes and streambed
sediment. 3. A dynamic analysis of coupled hydrologic
and chemical processes that determine solute transport. *Water Resources Research* **20**: 1804–14. [7.4]

Bergquist AM, Carpenter SR. (1986) Limnetic herbivory:
effects on phytoplankton populations and primary
production. *Ecology* **67**: 1351–60. [7.8]

Biggs BJF, Close ME. (1989) Periphyton biomass dynamics in gravel bed rivers: the relative effects of

flows and nutrients. *Freshwater Biology* **22**: 209–31. [7.3]

Bothwell ML. (1985) Phosphorus limitation of lotic periphyton growth rates: an intersite comparison using continuous-flow troughs (Thompson River system, British Columbia). *Limnology and Oceanography* **30**: 527–42. [7.3, 7.6]

Bothwell ML. (1988) Growth rate responses of lotic periphytic diatoms to experimental phosphorus enrichment: the influence of temperature and light. *Canadian Journal of Fisheries and Aquatic Sciences* **45**: 261–70. [7.3]

Bothwell ML. (1989) Phosphorus-limited growth dynamics of lotic periphytic diatom communities: areal biomass and cellular growth rate responses. *Canadian Journal of Fisheries and Aquatic Sciences* **46**: 1293–301. [7.3]

Bott TL, Kaplan LA. (1990) Potential for protozoan grazing of bacteria in streambed sediments. *Journal of the North American Benthological Society* **9**: 336–45. [7.8]

Bott TL, Brock JT, Dunn CS, Naiman RJ, Ovink RW, Petersen RC. (1985) Benthic community metabolism in four temperate stream systems: an inter-biome comparison and evaluation of the river continuum concept. *Hydrobiologia* **123**: 3–45. [7.5, 7.7]

Bowie GL, Mills WB, Porcella DP, Campbell CL, Pagenkopf JR, Rupp GL *et al* (1985) *Rates, constants, and kinetics formulations in surface water quality modeling* 2nd edn. EPA 600/3-85/040. EPA Environmental Research Laboratory, Athens, Georgia. [7.4]

Burkholder-Crecco JM, Bachmann RW. (1979) Potential phytoplankton productivity of three Iowa streams. *Proceedings of the Iowa Academy of Sciences* **86**: 22–5. [7.3]

Cahill TH, Imperato P, Verhoff FH. (1974) Phosphorus dynamics in a watershed. *Journal of the Environmental Engineering Division, Proceedings of the American Society of Civil Engineers* **100**: 439–58. [7.6]

Caron DA, Goldman JC. (1990) Protozoan nutrient regeneration. In: Capriuto GM (ed.) *Ecology of Marine Protozoa*, pp 283–306. Oxford University Press, New York. [7.8]

Carpenter SR, Kitchell JF. (1984) Plankton community structure and limnetic primary production. *American Naturalist* **124**: 159–72. [7.8]

Cembella AD, Antia NJ, Harrison PJ. (1984a) The utilization of inorganic and organic phosphorus compounds as nutrients by eukaryotic microalgae: a multidisciplinary perspective. Part 1. *Critical Reviews in Microbiology* **10**: 317–91. [7.6]

Cembella AD, Antia NJ, Harrison PJ. (1984b) The utilization of inorganic and organic phosphorus compounds as nutrients by eukaryotic microalgae: a multi-disciplinary perspective. Part 2. *Critical Reviews in Microbiology* **11**: 13–81. [7.6]

Chapman BM. (1982) Numerical simulation of the transport and speciation of nonconservative chemical reactants in rivers. *Water Resources Research* **18**: 155–67. [7.4]

Chapra SC, Reckhow KH. (1983) *Engineering Approaches for Lake Management. Vol. 2: Mechanistic Modeling.* Butterworth Publishers, Boston. [7.4]

Chatarpaul L, Robinson JB. (1979) Nitrogen transformations in stream sediments: ^{15}N studies. In: Litchfield CD, Seyfried PL (eds) *Methodology for Biomass Determinations and Microbial Activities in Sediments*, pp 119–27. ASTM STP 673. American Society for Testing and Materials. Philadelphia, USA. [7.7]

Chatarpaul L, Robinson JB, Kaushik NK. (1980) Effects of tubificid worms on denitrification and nitrification in stream sediment. *Canadian Journal of Fisheries and Aquatic Sciences* **37**: 656–63. [7.7]

Christensen PB, Sørensen J. (1988) Denitrification in sediments of lowland streams: Regional and seasonal variation in Gelbæck and Rabis Bæk, Denmark. *FEMS Microbial Ecology* **53**: 335–44. [7.7]

Christensen PB, Nielsen LP, Revsbech NP, Sørensen J. (1989) Microzonation of denitrification activity in stream sediments as studied with a combined oxygen and nitrous oxide microsensor. *Applied and Environmental Microbiology* **55**: 1234–41. [7.7]

Christensen PB, Nielsen LP, Sørensen J, Revsbech NP. (1990) Denitrification in nitrate-rich streams: Diurnal and seasonal variation related to benthic oxygen metabolism. *Limnology and Oceanography* **35**: 640–51. [7.7]

Conway HL, Harrison PJ. (1977) Marine diatoms grown in chemostats under silicate or ammonium limitation. IV. Transient response of *Chaetoceros debilis, Skeletonoma costatum*, and *Thalassiosira gravida* to a single addition of the limiting nutrient. *Marine Biology* **43**: 33–43. [7.7]

Cooke JG, White RE. (1987) The effect of nitrate in stream water on the relationship between denitrification and nitrification in a stream-sediment microcosm. *Freshwater Biology* **18**: 213–26. [7.7]

Cooper AB. (1983) Effect of storm events on benthic nitrifying activity. *Applied and Environmental Microbiology* **46**: 957–60. [7.7]

Cooper AB. (1984) Activities of benthic nitrifiers in streams and their role in oxygen consumption. *Microbiol Ecology* **10**: 317–34. [7.7]

Cooper AB, Cooke JG. (1984) Nitrate loss and transformation in 2 vegetated headwater streams. *New Zealand Journal of Marine and Freshwater Research* **18**: 441–50. [7.7]

Cooper DC. (1973) Enhancement of net primary productivity by herbivore grazing in aquatic laboratory microcosms. *Limnology and Oceanography* **18**: 31–7. [7.8]

Corning KE, Duthie HC, Paul BJ. (1989) Phosphorus and glucose uptake by seston and epilithon in boreal forest streams. *Journal of the North American Benthological Society* **8**: 123–33. [7.6]

Cosser PR. (1989) Nutrient concentration–flow relationships and loads in the South Pine River, southeastern Queensland. 1. Phosphorus loads. *Australian Journal of Marine and Freshwater Research* **40**: 613–30. [7.6]

Crosby MP, Newell RIE, Langdon CJ. (1990) Bacterial mediation in the utilization of carbon and nitrogen from detrital complexes by *Crassostrea virginica*. *Limnology and Oceanography* **35**: 625–39. [7.8]

Cuffney TF, Wallace JB, Lugthart GJ. (1990) Experimental evidence quantifying the role of benthic invertebrates in organic matter dynamics of headwater streams. *Freshwater Biology* **23**: 281–99. [7.8]

Cummins KW, Petersen RC, Howard FO, Wuycheck JC, Holt VI. (1973) The utilization of leaf litter by stream detritivores. *Ecology* **54**: 336–45. [7.8]

Cummins KW, Sedell JR, Swanson FJ, Minshall GW, Fisher SG, Cushing CE et al. (1983) Organic matter budgets for stream ecosystems: problems in their evaluation. In: Barnes JR, Minshall GW (eds) *Stream Ecology. Application and Testing of General Ecological Theory* pp 299–353. Plenum Press, New York. [7.5]

D'Angelo DJ, Webster JR, Benfield EF. (1991) Mechanisms of stream phosphorus retention: an experimental study. *Journal of the North American Benthological Society* **10**: 225–37. [7.6]

Décamps H, Capblancb J, Tourenq JN. (1984) Lot. In: Whitton BA (ed.) *Ecology of European Rivers.* pp 207–35. Blackwell Scientific Publications. [7.3, 7.6]

DePinto JV. (1979) Water column death and decomposition of phytoplankton: an experimental and modeling review. In: Scavia D, Robertson A (eds) *Perspectives on Lake Ecosystem Modeling,* pp 25–52. Ann Arbor Science Publishers, Ann Arbor, Michigan. [7.6]

Dillon PJ, Rigler FH. (1974) The phosphorus–chlorophyll relationship in lakes. *Limnology and Oceanography* **19**: 767–73. [7.3]

Downes MT, Paerl HW. (1978) Separation of two dissolved reactive phosphorus fractions in lakewater. *Journal of the Fisheries Research Board of Canada* **35**: 1635–9. [7.6]

Droop MR. (1973) Some thoughts on nutrient limitation in algae. *Journal of Phycology* **9**: 264–72. [7.6]

Duff JH, Triska FJ. (1990) Denitrification in sediments from the hyporheic zone adjacent to a small forested stream. *Canadian Journal of Fisheries and Aquatic Sciences* **47**: 1140–7. [7.7]

Duff JH, Stanley KC, Triska FJ, Avanzino RJ. (1984a) The use of photosynthesis-respiration chambers to measure nitrogen flux in epilithic algal communities. *Verhandlungen Internationale Vereinigung für Theoretische und Angewandt Limnologie* **22**: 1436–43. [7.7]

Duff JH, Triska FJ, Oremland RS. (1984b) Denitrification associated with stream periphyton: chamber estimates from undisrupted communities. *Journal of Environmental Quality* **13**: 514–18. [7.7]

Dugdale RC, Goering JJ. (1967) Uptake of new and regenerated nitrogen in primary productivity. *Limnology and Oceanography* **12**: 196–206. [7.2]

Edwards RT, Meyer JL. (1987) Metabolism of a subtropical low gradient blackwater river. *Freshwater Biology* **17**: 251–64. [7.5]

Eisenreich SJ, Armstrong DE. (1977) Chromatographic investigation of inositol phosphate esters in lake waters. *Environmental Science and Technology* **11**: 497–501. [7.6]

Elser JJ, Goldman CR. (1991) Zooplankton effects on phytoplankton in lakes of contrasting trophic status. *Limnology and Oceanography* **36**: 64–90. [7.8]

Elser JJ, Kimmel BL. (1985) Nutrient availability for phytoplankton production in a multiple-impoundment series. *Canadian Journal of Fisheries and Aquatic Sciences* **42**: 1359–70. [7.3]

Elser JJ, Marzolf ER, Goldman CR. (1990) Phosphorus and nitrogen limitation of phytoplankton growth in the freshwaters of North-America: a review and critique of experimental enrichments. *Canadian Journal of Fisheries and Aquatic Sciences* **47**: 1468–77. [7.3]

Elwood JW, Nelson DJ. (1972) Periphyton production and grazing rates in a stream measured with a ^{32}P material balance method. *Oikos* **23**: 295–303. [7.6]

Elwood JW, Newbold JD, O'Neill RV, Stark RW, Singley PT. (1981a) The role of microbes associated with organic and inorganic substrates in phosphorus spiralling in a woodland stream. *Verhandlungen Internationale Vereinigung für Theoretische und Angewandt Limnologie* **21**: 850–6. [7.6]

Elwood JW, Newbold JD, Trimble AF, Stark RW. (1981b) The limiting role of phosphorus in a woodland stream ecosystem: effects of P enrichment on leaf decomposition and primary producers. *Ecology* **62**: 146–58. [7.3]

Elwood JW, Newbold JD, O'Neill RV, Van Winkle W. (1983) Resource spiralling: an operational paradigm for analyzing lotic ecosystems. In: Fontaine TD III, Bartell SM (eds) *The Dynamics of Lotic Ecosystems,* pp 3–27. Ann Arbor Science, Ann Arbor, Michigan. [7.2]

Elwood JW, Mulholland PJ, Newbold JD. (1988) Microbial activity and phosphorus uptake on decomposing leaf detritus in a heterotrophic stream.

Verhandlungen Internationale Vereinigung für Theoretische und Angewandt Limnologie **23**: 1198–208. [7.6]

Eppley RW, Renger EH. (1974) Nitrogen assimilation of an oceanic diatom in nitrogen-limited continuous culture. *Journal of Phycology* **10**: 15–23. [7.7]

Eppley RW, Renger EH, Harrison WG, Cullen JJ. (1979) Ammonium distribution in southern California coastal water and its role in the growth of phytoplankton. *Limnology and Oceanography* **24**: 495–509. [18.7]

Feminella JW, Power ME, Resh VH. (1989) Periphyton responses to invertebrate grazing and riparian canopy in three northern California coastal streams. *Freshwater Biology* **22**: 445–57. [7.8]

Findlay S, Carlough L, Crocker MT, Gill HK, Meyer JL, Smith PJ. (1986) Bacterial growth on macrophyte leachate and fate of bacterial production. *Limnology and Oceanography* **31**: 1335–41. [7.8]

Fisher SG. (1977) Organic matter processing by a stream-segment ecosystem: Fort River, Massachusetts, USA. *Internationale Revue der Gesamten Hydrobiologie* **62**: 701–27.

Fisher SG, Likens GE. (1973) Energy flow in Bear Brook, New Hampshire: an integrative approach to stream ecosystem metabolism. *Ecological Monographs* **43**: 421–39. [7.5, 7.8]

Fisher SG, Gray LJ, Grimm NB, Busch DE. (1982) Temporal succession in a desert stream ecosystem following flash flooding. *Ecological Monographs* **52**: 93–110. [7.7]

Flint RW, Goldman CR. (1975) The effects of a benthic grazer on the primary productivity of the littoral zone of Lake Tahoe. *Limnology and Oceanography* **20**: 935–44. [7.8]

Fox AM. (1994) Macrophytes. In: Calow P, Petts GE (eds) *The Rivers Handbook*, vol. 1, pp 216–33. Blackwell Scientific Publications, Oxford. [7.2]

Francis MM, Naiman RJ, Melillo JM. (1985) Nitrogen fixation in subarctic streams influenced by beaver (*Castor canadensis*). *Hydrobiologia* **121**: 193–202. [7.7]

Francko DA. (1986) Epilimnetic phosphorus cycling: influence of humic materials and iron on coexisting major mechanisms. *Canadian Journal of Fisheries and Aquatic Sciences* **43**: 302–10. [7.6]

Freeman MC. (1986) The role of nitrogen and phosphorus in the development of *Cladophora glomerata* (L.) Kutzing in the Manawatu River, Zew Zealand. *Hydrobiologia* **131**: 23–30. [7.3]

Furumai H, Kondo T, Ohgaki S. (1989) Phosphorus exchange kinetics and exchangeable phosphorus forms in sediments. *Water Research* **23**: 685–91. [7.6]

Georgian T, Thorp JH. (1992) Effects of microhabitat selection on feeding rates of net-spinning caddisfly

larvae. *Ecology* **73**: 229–40. [7.8]

Gilliam JF, Fraser DF, Sabat AM. (1989) Strong effects of foraging minnows on a stream benthic invertebrate community. *Ecology* **70**: 445–52. [7.8]

Gloss SP, Reynolds RC Jr, Mayer LM, Kidd DE. (1981) Reservoir influences on salinity and nutrient fluxes in the arid Colorado River basin. In: Stefan HG (ed.) *Symposium on Surface Water Impoundments*, 2–5 June 1980, Minneapolis, pp 1618–29. American Society of Civil Engineers, New York. [7.6]

Green DB, Logan TJ, Smeck NE. (1978) Phosphate adsorption–desorption characteristics of suspended sediments in the Maumee River Basin of Ohio. *Journal of Environmental Quality* **7**: 208–12. [7.6]

Gregory SV. (1978) Phosphorus dynamics on organic and inorganic substrates in streams. *Verhandlungen Internationale Vereinigung für Theoretische und Angewandt Limnologie* **20**: 1340–6. [7.6]

Gregory SV. (1983) Plant–herbivore interactions in stream systems. In: Minshall GW, Barnes JR (eds) *Stream Ecology. Application and Testing of General Ecological Theory*, pp 157–89. Plenum Press, New York. [7.8]

Grimm NB. (1987) Nitrogen dynamics during succession in a desert stream. *Ecology* **68**: 1157–70. [7.3, 7.7]

Grimm NB. (1988) Role of macroinvertebrates in nitrogen dynamics of a desert stream. *Ecology* **69**: 1884–93. [7.8]

Grimm NB, Fisher SG. (1986) Nitrogen limitation in a Sonoran Desert stream. *Journal of the North American Benthological Society* **5**: 2–15. [7.3]

Grobbelaar JU. (1983) Availability to algae of N and P adsorbed on suspended solids in turbid waters of the Amazon River. *Archiv für Hydrobiologie* **96**: 302–16. [7.6]

Gromiec MJ, Loucks DP, Orlob GT. (1983) Stream quality modeling. In: Orlob GT (ed.) *Mathematical Modeling of Water Quality: Streams, Lakes, and Reservoirs*, pp 176–226. John Wiley and Sons, New York. [7.5]

Harms LL, Vidal PH, McDermott TE. (1978) Phosphorus interactions with stream-bed sediments. *Journal of the Environmental Engineering Division of the American Society of Civil Engineers* **104**: 271–88. [7.6]

Hart DD, Robinson CT. (1990) Resource limitation in a stream community: phosphorus enrichment effects on periphyton and grazers. *Ecology* **71**: 1494–502. [7.3, 7.8]

Herbes SE, Allen HE, Mancy KH. (1975) Enzymatic characterization of soluble organic phosphorus in lake water. *Science* **187**: 432–4. [7.6]

Hill AR. (1979) Denitrification in the nitrogen budget of a river ecosystem. *Nature* **281**: 291–2. [7.7]

Hill AR. (1982) Phosphorus and major cation mass

balances for two rivers during low summer flows. *Freshwater Biology* **12**: 293–304. [7.6]

Hill AR. (1988) Factors influencing nitrate depletion in a rural stream. *Hydrobiologia* **160**: 111–22. [7.7]

Hill AR, Sanmugadas K. (1985) Denitrification rates in relation to stream sediment characteristics. *Water Research* **19**: 1579–86. [7.7]

Hill AR, Warwick J. (1987) Ammonium transformations in springwater within the riparian zone of a small woodland stream. *Canadian Journal of Fisheries and Aquatic Sciences* **44**: 1948–56. [7.7]

Hill WR, Harvey BC. (1990) Periphyton responses to higher trophic levels and light in a shaded stream. *Canadian Journal of Fisheries and Aquatic Sciences* **47**: 2307–14. [7.8]

Hill WR, Knight AW. (1988) Nutrient and light limitation of algae in two northern California streams. *Journal of Phycology* **24**: 125–32. [7.3]

Hino S. (1989) Characterization of orthophosphate release from dissolved organic phosphorus by gel filtration and several hydrolytic enzymes. *Hydrobiologia* **174**: 49–55. [7.6]

Horne AJ, Carmiggelt CJW. (1975) Algal nitrogen fixation in California streams: seasonal cycles. *Freshwater Biology* **5**: 461–70. [7.7]

Horner RR, Welch EB. (1981) Stream periphyton development in relation to current velocity and nutrients. *Canadian Journal of Fisheries and Aquatic Sciences* **38**: 449–57. [7.3]

Horner RR, Welch EB, Veenstra RB. (1983) Development of nuisance periphytic algae in laboratory streams in relation to enrichment and velocity. In: Wetzel RG (ed.) *Periphyton of Freshwater Ecosystems*, pp 121–34. Junk Publishers, The Hague. [7.3]

Horner RR, Welch EB, Seeley MR, Jacoby JM. (1990) Responses of periphyton to changes in current velocity, suspended sediment and phosphorus concentration. *Freshwater Biology* **24**: 215–32. [7.3]

Hullar MA, Vestal JR. (1989) The effects of nutrient limitation and stream discharge on the epilithic microbial community in an oligotrophic Arctic stream. *Hydrobiologia* **172**: 19–26. [7.3]

Ittekot V. (1988) Global trends in the nature of organic matter in river suspensions. *Nature* **332**: 436–8. [7.5]

Jacoby JM. (1985) Grazing effects on periphyton by *Theodoxus fluviatilis* (Gastropoda) in a lowland stream. *Journal of Freshwater Ecology* **3**: 265–74. [7.8]

Johannes RE. (1965) Influence of marine protozoa on nutrient regeneration. *Limnology and Oceanography* **10**: 434–42. [7.8]

Johnson AH, Bouldin DR, Goyette EA, Hedges AM. (1976) Phosphorus loss by stream transport from a rural watershed: quantities, processes, and sources. *Journal of Environmental Quality* **5**: 148–57. [7.6]

Jordan C, Dinsmore P. (1985) Determination of biologically available phosphorus using a radiobioassay technique. *Freshwater Biology* **15**: 597–603. [7.6]

Karlsson G, Löwgren M. (1990) River transport of phosphorus as controlled by large scale land use changes. *Acta Agriculturae Scandinavica* **40**: 149–62. [7.6]

Kaushik NK, Hynes HBN. (1968) Experimental study on the role of autumn-shed leaves in aquatic environments. *Journal of Ecology* **56**: 229–43. [7.6]

Kaushik NK, Robinson JB, Sain P, Whiteley HR, Stammers WN. (1975) A quantitative study of nitrogen loss from water of a small spring-fed stream. In: *Water Pollution Research in Canada*, pp 110–17. Proceedings of the 10th Canadian Symposium, Toronto, February 1974. Institute for Environmental Studies, University of Toronto, Toronto. [7.7]

Kaushik NK, Robinson JB, Stammers WN, Whiteley HR. (1981) Aspects of nitrogen transport and transformation in headwater streams. In: Lock MA, Williams DD (eds) *Perspectives in Running Water Ecology*, pp 113–39. Plenum Press, New York. [7.7]

Kehde PM, Wilhm JL. (1972) The effects of grazing by snails on community structure of periphyton in laboratory streams. *The American Midland Naturalist* **87**: 8–24. [7.8]

Keithan ED, Lowe RL, Deyoe HR. (1988) Benthic diatom distribution in a Pennsylvania stream: role of pH and nutrients. *Journal of Phycology* **24**: 581–5. [7.3]

Keup LE. (1968) Phosphorus in flowing waters. *Water Research* **2**: 373–86. [7.6]

Kim BK, Jackman AP, Triska FJ. (1990) Modeling transient storage and nitrate uptake kinetics in a flume containing a natural periphyton community. *Water Resources Research* **26**: 505–15. [7.4, 7.7]

Kitchell JF, O'Neill RV, Webb D, Gallepp GW, Bartell SM, Koonce JF, Ausmus BS. (1979) Consumer regulation of nutrient cycling. *BioScience* **29**: 28–34. [7.8]

Klotz RL. (1985) Factors controlling phosphorus limitation in stream sediments. *Limnology and Oceanography* **30**: 543–53. [7.3, 7.6]

Klotz RL. (1988) Sediment control of soluble reactive phosphorus in Hoxie Gorge Creek, New York. *Canadian Journal of Fisheries and Aquatic Sciences* **45**: 2026–34. [7.6]

Knorr DF, Fairchild GW. (1987) Periphyton, benthic invertebrates and fishes as biological indicators of water quality in the East Branch Brandywine Creek. *Proceedings of the Pennsylvania Academy of Science* **61**: 61–6. [7.3]

Krogstad T, Løvstad Ø. (1989) Erosion, phosphorus and phytoplankton response in rivers of South-Eastern Norway. *Hydrobiologia* **183**: 33–41. [7.3, 7.6]

Kunishi HM, Taylor AW, Heald WR, Gburek WJ, Weaver RN. (1972) Phosphate movement from an agricultural watershed during two rainfall periods. *Journal of Agricultural and Food Chemistry* **20**:

900–5. [7.6]

Kuwabara JS, Helliker P. (1988) Trace contaminants in streams. In: Cheremisinoff PN, Cheremisinoff NP, Cheng SL (eds) *Civil Engineering Practice*, Vol. 5, pp 739–65. Technomic Publishing, Lancaster, Pennsylvania. [7.4]

Ladle M, Bass JAB, Jenkins WR. (1972) Studies on the production and food consumption by the larval Simuliidae (Diptera) of a small chalk stream. *Hydrobiologia* 39: 429–48. [7.8]

Lamberti GA, Resh VH. (1983) Stream periphyton and insect herbivores: an experimental study of grazing by a caddisfly population. *Ecology* 64: 1124–35. [7.8]

Lamberti GA, Ashkenas LR, Gregory SV, Steinman AD. (1987) Effects of three herbivores on periphyton communities in laboratory streams. *Journal of the North American Benthological Society* 6: 92–104. [7.8]

Lamberti GA, Gregory SV, Ashkenas LR, Steinman AD, McIntire CD. (1989) Productive capacity of periphyton as a determinant of plant herbivore interactions in streams. *Ecology* 70: 1840–56. [7.8]

Lane P, Levins R. (1977) The dynamics of aquatic systems. 2. The effects of nutrient enrichment on model plankton communities. *Limnology and Oceanography* 22: 454–71. [7.8]

Lean DRS. (1973) Movements of phosphorus between its biologically important forms in lake water. *Journal of the Fisheries Research Board of Canada* 30: 1525–36. [7.6]

Lean DRS, Nalewajko C. (1976) Phosphate exchange and organic phosphorus excretion by freshwater algae. *Journal of the Fisheries Research Board of Canada* 33: 1312–23. [7.6]

Leonard RL, Kaplan LA, Elder JF, Coats RN, Goldman CR. (1979) Nutrient transport in surface runoff from a subalpine watershed, Lake Tahoe Basin, California. *Ecological Monographs* 49: 281–310. [7.6]

Li WC, Armstrong DE, Williams JDH, Harris RF, Syers JK. (1972) Rate and extent of inorganic phosphate exchange in lake sediments. *Soil Science Society of America, Proceedings* 36: 279–85. [7.6]

Lock MA, Ford TE, Hullar MAJ, Kaufman M, Vestal JR, Volk GS, Ventullo RM. (1990) Phosphorus limitation in an arctic river biofilm – a whole ecosystem experiment. *Water Research* 24: 1545–9. [7.3]

Maltby L. (1994) Heterotrophic microbes. In: Calow P, Petts GE (eds) *The Rivers Handbook*, vol. 1, pp 165–94. Blackwell Scientific Publications, Oxford. [7.2]

Maltby L. (1994) Detritus processing. In: Calow P, Petts GE (eds) *The Rivers Handbook*, vol. 1, pp 331–53. Blackwell Scientific Publications, Oxford. [7.2]

McAuliffe JR. (1984) Resource depression by a stream herbivore – effects on distributions and abundances of other grazers. *Oikos* 42: 327–33. [7.8]

McCallister DL, Logan TJ. (1978) Phosphate adsorption–desorption characteristics of soils and bottom sediments in the Maumee River Basin of Ohio. *Journal of Environmental Quality* 7: 87–92. [7.6]

McColl RHS. (1974) Self-purification of small freshwater streams: phosphate, nitrate, and ammonia removal. *New Zealand Journal of Marine Freshwater Research* 8: 375–88. [7.6, 7.7]

McCullough DA, Minshall GW, Cushing CE. (1979) Bioenergetics of a stream 'collector' organism, *Tricorythodes minutus* (Insecta: Ephemeroptera). *Limnology and Oceanography* 24: 45–58. [7.8]

McCutcheon S. (1987) Laboratory and instream nitrification rates for selected streams. *Journal of Environmental Engineering* 113: 628–46. [7.7]

McCutcheon SC. (1989) *Water Quality Modeling. Vol. 1. Transport and Surface Exchange in Rivers.* CRC Press, Boca Raton, Florida. [7.4]

Maciolek JA, Tunzi MG. (1968) Microseston dynamics in a simple Sierra Nevada lake-stream system. *Ecology* 49: 60–75. [7.8]

Marker AFH, Gunn RJM. (1977) The benthic algae of some streams in southern England. III. Seasonal variations in chlorophyll *a* in the seston. *Journal of Ecology* 65: 223–34. [7.5]

Mason JW, Wegner GD, Quinn GI, Lange EL. (1990) Nutrient loss via groundwater discharge from small watersheds in southwestern and south central Wisconsin. *Journal of Soil and Water Conservation* 45: 327–31. [7.6]

Matulewich VA, Finstein MS. (1978) Distribution of autotrophic nitrifying bacteria in a polluted river (the Passaic). *Applied and Environmental Microbiology* 35: 67–71. [7.7]

Mayer LM, Gloss SP. (1980) Buffering of silica and phosphate in a turbid river. *Limnology and Oceanography* 25: 12–22. [7.6]

Megard RO. (1981) Effects of planktonic algae on water quality in impounds of the Mississippi River in Minnesota. In: Stefan HG (ed.) *Symposium on Surface Water Impoundments*, 2–5 June 1980, Minneapolis, pp 1575–84. American Society of Civil Engineers, New York. [7.3]

Meybeck M. (1982) Carbon, nitrogen, and phosphorus transport by world rivers. *American Journal of Science* 282: 401–50. [7.6, 7.7]

Meyer JL. (1979) The role of sediments and bryophytes in phosphorus dynamics in a headwater stream ecosystem. *Limnology and Oceanography* 24: 365–75. [7.6]

Meyer JL. (1980) Dynamics of phosphorus and organic matter during leaf decomposition in a forest stream. *Oikos* 34: 44–53. [7.6]

Meyer JL, Edwards RT. (1990) Ecosystem metabolism and turnover of organic carbon along a blackwater river continuum. *Ecology* 71: 668–77. [7.5]

Chapter 7

Meyer JL, Johnson C. (1983) The influence of elevated nitrate concentration on rate of leaf decomposition in a stream. *Freshwater Biology* 13: 177–83. [7.3]

Meyer JL, Likens GE. (1979) Transport and transformation of phosphorus in a forest stream ecosystem. *Ecology* 60: 1255–69. [7.6]

Meyer JL, McDowell WH, Bott TL, Elwood JW, Ishizaki C, Melack JM et al. (1988) Elemental dynamics in streams. *Journal of the North American Benthological Society* 7: 410–32. [7.1]

Minear RA. (1972) Characterization of naturally occurring dissolved organophosphorus compounds. *Environmental Science & Technology* 6: 431–7. [7.6]

Minshall GW, Petersen RC, Cummins KW, Bott TL, Sedell JR, Cushing CE, Vannote RL. (1983) Interbiome comparison of stream ecosystem dynamics. *Ecological Monographs* 53: 1–25. [7.5]

Minshall GW, Petersen RC, Bott TL, Cushing CE, Cummins KW, Vannote RL, Sedell JR. (1992) Stream ecosystem dynamics of the Salmon River Idaho: an 8th order drainage system. *Journal of the North American Benthological Society* 11: 111–37. [7.5]

Mulholland PJ. (1981) Organic carbon flow in a swamp–stream ecosystem. *Ecological Monographs* 51: 307–22. [7.5]

Mulholland PJ, Watts JA. (1982) Transport of organic carbon to the oceans by rivers of North America: a synthesis of existing data. *Tellus* 34: 176–86. [7.5]

Mulholland PJ, Newbold JD, Elwood JW, Hom CL. (1983) The effect of grazing intensity on phosphorus spiralling in autotrophic streams. *Oecologia* 58: 358–66. [7.8]

Mulholland PJ, Elwood JW, Newbold JD, Webster JR, Ferren LA, Perkins RE. (1984) Phosphorus uptake by decomposing leaf detritus: effect of microbial biomass and activity. *Verhandlungen Internationale Vereinigung für Theoretische und Angewandt Limnologie* 22: 1899–905. [7.6]

Mulholland PJ, Elwood JW, Newbold JD, Ferren LA. (1985a) Effect of a leaf-shredding invertebrate on organic matter dynamics and phosphorus spiralling in heterotrophic laboratory streams. *Oecologia* 66: 199–206. [7.8]

Mulholland PJ, Newbold JD, Elwood JW, Ferren LA, Webster JR. (1985b) Phosphorus spiralling in a woodland stream: seasonal variations. *Ecology* 66: 1012–23. [7.6]

Mulholland PJ, Minear RA, Elwood JW. (1988) Production of soluble, high molecular weight phosphorus and its subsequent uptake by stream detritus. *Verhandlungen Internationale Vereinigung für Theoretische und Angewandt Limnologie* 23: 1190–7. [7.6]

Mulholland PJ, Steinman AD, Elwood JW. (1990) Measurement of phosphorus uptake length in streams: comparison of radiotracer and stable PO_4 releases. *Canadian Journal of Fisheries and Aquatic Sciences* 47: 2351–7. [7.4, 7.6]

Munn MD, Osborne LL, Wiley MJ. (1989) Factors influencing periphyton growth in agricultural streams of central Illinois. *Hydrobiologia* 174: 89–97. [7.3]

Munn NL, Meyer JL. (1990) Habitat-specific solute retention in two small streams: an intersite comparison. *Ecology* 71: 2069–82. [7.6, 7.7]

Munn N, Prepas E. (1986) Seasonal dynamics of phosphorus partitioning and export in two streams in Alberta, Canada. *Canadian Journal of Fisheries and Aquatic Sciences* 43: 2464–71. [7.6]

Naiman RJ, Melillo JM, Lock MA, Ford TE, Reice SR. (1987) Longitudinal patterns of ecosystem processes and community structure in a subarctic river continuum. *Ecology* 68: 1139–56. [7.5]

Nakajima T. (1979) Denitrification by the sessile microbial community of a polluted river. *Hydrobiologia* 66: 57–64. [7.7]

Newbold JD. (1987) Phosphorus spiralling in rivers and river-reservoir systems: implications of a model. In: Craig JF, Kemper JB (eds) *Regulated Streams: Advances in Ecology*, pp 303–27. Plenum Press, New York. [7.3]

Newbold JD, Elwood JW, O'Neill RV, Van Winkle W. (1981) Measuring nutrient spiralling in streams. *Canadian Journal of Fisheries and Aquatic Sciences* 38: 860–3. [7.2]

Newbold JD, Mulholland PJ, Elwood JW, O'Neill RV. (1982a) Organic carbon spiralling in stream ecosystems. *Oikos* 38: 266–72. [7.5]

Newbold JD, O'Neill RV, Elwood JW, Van Winkle W. (1982b) Nutrient spiralling in streams: implications for nutrient limitation and invertebrate activity. *American Naturalist* 120: 628–52. [7.3, 7.8]

Newbold JD, Elwood JW, O'Neill RV, Sheldon AL. (1983a) Phosphorus dynamics in a woodland stream ecosystem: a study of nutrient spiralling. *Ecology* 64: 1249–65. [7.2, 7.6, 7.8]

Newbold JD, Elwood JW, Schulze MS, Stark RW, Barmeier JC. (1983b) Continuous ammonium enrichment of a woodland stream: uptake kinetics, leaf decomposition, and nitrification. *Freshwater Biology* 13: 193–204. [7.3, 7.7]

O'Connor DJ. (1988a) Models of sorptive toxic substances in freshwater systems. I. Basic equations. *Journal of Environmental Engineering* 114: 507–32. [7.4]

O'Connor DJ. (1988b) Models of sorptive toxic substances in freshwater systems. II. Lakes and reservoirs. *Journal of Environmental Engineering* 114: 533–51. [7.4]

O'Connor DJ. (1988c) Models of sorptive toxic substances in freshwater systems. III. Streams and rivers. *Journal of Environmental Engineering* 114: 552–74.

[7.4]

O'Neill RV. (1979) A review of linear compartmental analysis in ecosystem science. In: Matis JH, Patten BC, White GC (eds) *Compartmental Analysis of Ecosystem Models*, pp 3–28. International Co-operative Publishing House, Fairland, Maryland. [7.4]

Omernik JM. (1977) *Nonpoint source-stream nutrient level relationships: a nationwide survey.* EPA-600/3-77-105, Ecological Research Series, United States Environmental Protection Agency, Washington, DC. [7.6, 7.7]

Orlob GT (ed.). (1983) *Mathematical Modeling of Water Quality: Streams, Lakes, and Reservoirs.* John Wiley and Sons, New York. [7.4]

Paerl HW, Downes MT. (1978) Biological availability of low versus high molecular weight reactive phosphorus. *Journal of the Fisheries Research Board of Canada* 35: 1639–43. [7.6]

Paul BJ, Duthie HC. (1989) Nutrient cycling in the epilithon of running waters. *Canadian Journal of Botany* 67: 2302–9. [7.6]

Paul BJ, Corning KE, Duthie HC. (1989) An evaluation of the metabolism of sestonic and epilithic communities in running waters using an improved chamber technique. *Freshwater Biology* 21: 207–17. [7.6]

Peters RH. (1978) Concentrations and kinetics of phosphorus fractions in water from streams entering Lake Memphremagog. *Journal of the Fisheries Research Board of Canada* 35: 315–28. [7.6]

Peters RH. (1981) Phosphorus availability in Lake Memphremagog and its tributaries. *Limnology and Oceanography* 26: 1150–61. [7.6]

Petersen RC, Cummins KW. (1974) Leaf processing in a woodland stream. *Freshwater Biology* 4: 343–68. [7.8]

Petersen RC, Cummins KW, Ward GM. (1989) Microbial and animal processing of detritus in a woodland stream. *Ecological Monographs* 59: 21–39. [7.8]

Peterson BJ, Hobbie JE, Hershey AE, Lock MA, Ford TE, Vestal JR, McKinley VL *et al.* (1985) Transformation of a tundra river from heterotrophy to autotrophy by addition of phosphorus. *Science* 229: 1383–6. [7.3, 7.6]

Pomeroy LR. (1970) The strategy of mineral cycling. *Annual Review of Ecology and Systematics* 1: 171–90. [7.2]

Pomeroy LR, Smith EE, Grant CM. (1965) The exchange of phosphate between estuarine water and sediments. *Limnology and Oceanography* 10: 167–72. [7.6]

Power ME. (1990) Effects of fish in river food webs. *Science* 250: 811–14. [7.8]

Power ME, Matthews WJ, Stewart AJ. (1985) Grazing minnows, piscivorous bass, and stream algae: dynamics of a strong interaction. *Ecology* 66: 1448–56. [7.8]

Power ME, Stewart AJ, Matthews WJ. (1988) Grazer control of algae in an Ozark Mountain stream: effects of short-term exclusion. *Ecology* 69: 1894–8. [7.8]

Prairie YT, Kalff J. (1988) Particulate phosphorus dynamics in headwater streams. *Canadian Journal of Fisheries and Aquatic Sciences* 45: 210–15. [7.6]

Pringle CM. (1987) Effects of water and substratum nutrient supplies on lotic periphyton growth: an integrated bioassay. *Canadian Journal of Fisheries and Aquatic Sciences* 44: 619–29. [7.3]

Pringle CM, Bowers JA. (1984) An *in situ* substratum fertilization technique: diatom colonization on nutrient-enriched, sand substrata. *Canadian Journal of Fisheries and Aquatic Sciences* 41: 1247–51. [7.3]

Pringle CM, Paaby-Hansen P, Vaux PD, Goldman CR. (1986) *In situ* nutrient assays of periphyton growth in a lowland Costa Rican stream. *Hydrobiologia* 134: 207–13. [7.3]

Redfield AC, Ketchum BH, Richards FA. (1963) The influence of organisms on the composition of sea water. In: Hill MN (ed.) *The Sea*, Vol. 2, pp 26–77. Interscience, New York. [7.7]

Rekolainen S. (1989) Effect of snow and soil frost melting on the concentrations of suspended solids and phosphorus in two rural watersheds in western Finland. *Aquatic Sciences* 51: 211–23. [7.6]

Reynolds CS. (1994) Algae. In: Calow P, Petts GE (eds) *Th Rivers Handbook*, vol. 1, pp 195–215. Blackwell Scientific Publications, Oxford. [7.2]

Rhee G-Y. (1974) Phosphate uptake under nitrate limitation by *Scenedesmus* sp. and its ecological implications. *Journal of Phycology* 10: 470–5. [7.6]

Richey JE, Hedges JI, Devol AH, Quay PD, Victoria R, Martinelli L, Forsberg BR. (1990) Biogeochemistry of carbon in the Amazon River. *Limnology and Oceanography* 35: 352–71. [7.5]

Richey JS, McDowell WH, Likens GE. (1985) Nitrogen transformations in a small mountain stream. *Hydrobiologia* 124: 129–39. [7.7]

Rigler FH. (1966) Radiobiological analysis of inorganic phosphorus in lakewater. *Verhandlungen Internationale Vereinigung für Theoretische und Angewandt Limnologie* 16: 465–70. [7.5, 7.6]

Ruttner F. (1963) *Fundamentals of Limnology* 3rd edn. University of Toronto Press, Toronto. [7.3]

Saunders JF III, Lewis WM Jr. (1988) Transport of phosphorus, nitrogen, and carbon by the Apure River, Venezuela. *Biogeochemistry* 5: 323–42. [7.6]

Sayler GS, Puziss M, Silver M. (1979) Alkaline phosphatase assay for freshwater sediments: application to perturbed sediment systems. *Applied and Environmental Microbiology* 38: 922–7. [7.6]

Schlessinger WH, Melack JM. (1981) Transport of organic carbon in the world's rivers. *Tellus* 33: 172–87. [7.5]

Sebetich MJ, Kennedy VC, Zand SM, Avanzino RJ,

Zelweger GW. (1984) Dynamics of added nitrate and phosphate compared in a northern California woodland stream. *Water Resources Bulletin* **20**: 93–101. [7.7]

Seitzinger SP. (1988) Denitrification in freshwater and coastal marine ecosystems: ecological and geochemical significance. *Limnology and Oceanography* **33**: 702–24. [7.7]

Simmons BL, Cheng DMH. (1985) Rate and pathways of phosphorus assimilation in the Nepean River at Camden, New South Wales. *Water Research* **19**: 1089–96. [7.6]

Smart MM, Jones JR, Sebaugh JL. (1985) Streamwatershed relations in the Missouri Ozark Plateau Province. *Journal of Environmental Quality* **14**: 77–82. [7.6]

Smith REH, Harrison WG, Harris L. (1985) Phosphorus exchange in marine microplankton communities near Hawaii. *Marine Biology* **86**: 75–84. [7.6]

Smock LA, Roeding CE. (1986) The trophic basis of production of the macroinvertebrate community of a southeastern USA blackwater stream. *Holarctic Ecology* **9**: 165–74. [7.8]

Søballe DM, Kimmel BL. (1987) A large-scale comparison of factors influencing phytoplankton abundance in rivers, lakes, and impoundments. *Ecology* **68**: 1943–54. [7.3]

Sprent JI. (1987) *The Ecology of the Nitrogen Cycle.* Cambridge University Press, Cambridge. [7.7]

Stabel H-H, Geiger M. (1985) Phosphorus adsorption to riverine suspended matter. Implications for the P-budget of lake Constance. *Water Research* **19**: 1347–52. [7.6]

Stainton MP. (1980) Errors in molybdenum blue methods for determining orthophosphate in freshwater. *Canadian Journal of Fisheries and Aquatic Sciences* **37**: 472–8. [7.6]

Stanley DW, Hobbie JE. (1981) Nitrogen recycling in a North Carolina coastal river. *Limnology and Oceanography* **26**: 30–42. [7.7]

Stewart AJ. (1987) Responses of stream algae to grazing minnows and nutrients: a field test for interactions. *Oecologia* **72**: 1–7. [7.8]

Stockner JG, Shortreed KRS. (1978) Enhancement of autotrophic production by nutrient addition in a coastal rainforest stream on Vancouver Island. *Journal of the Fisheries Research Board of Canada* **35**: 28–34. [7.3]

Stream Solute Workshop. (1990) Concepts and methods for assessing solute dynamics in stream ecosystems. *Journal of the North American Benthological Society* **9**: 95–119. [7.1, 7.6]

Streeter HW, Phelps EB. (1925) A study of the pollution and natural purification of the Ohio River. III. Factors concerned in the phenomena of oxidation and reacration. Bulletin 146, US Public Health Service, Washington DC. pp 1–75. [7.5]

Strickland JDH, Parsons TR. (1972) *A Practical Handbook of Seawater Analysis.* Bulletin 167, 2nd edn. Fisheries Research Board of Canada. Ottawa. [7.6]

Sumner WT, McIntire CD. (1982) Grazer–periphyton interactions in laboratory streams. *Archives in Hydrobiology* **93**: 135–57. [7.8]

Swank WT, Caskey WH. (1982) Nitrate depletion in a second-order mountain stream. *Journal of Environmental Quality* **11**: 581–4. [7.7]

Tarapchak SJ. (1983) Soluble reactive phosphorus measurements in lake water: evidence for molybdate-enhanced hydrolysis. *Journal of Environmental Quality* **12**: 105–8. [7.6]

Tarapchak SJ, Herche LR. (1988) Orthophosphate concentrations in lake water: analysis of Rigler's radiobioassay method. *Canadian Journal of Fisheries and Aquatic Sciences* **45**: 2230–7. [7.6]

Taylor AW, Kunishi HM. (1971) Phosphate equilibria on stream sediment and soil in a watershed draining an agricultural region. *Journal of Agricultural and Food Chemistry* **19**: 827–31. [7.6]

Thomann RV. (1984) Physico-chemical and ecological modeling the fate of toxic substances in natural water systems. *Ecological Modelling* **22**: 145–70. [7.4]

Thomann RV, Mueller JA. (1987) *Principles of Surface Water Quality Modeling and Control.* Harper and Row Publishers, New York. [7.4]

Triska FJ, Kennedy VC, Avanzino RJ, Zellweger GW, Bencala KE. (1989a) Retention and transport of nutrients in a third-order stream: channel processes. *Ecology* **70**: 1877–92. [7.3, 7.7]

Triska FJ, Kennedy VC, Avanzino RJ, Zellweger GW, Bencala KE. (1989b) Retention and transport of nutrients in a third-order stream in northwestern California: hyporheic processes. *Ecology* **70**: 1893–905. [7.3]

Triska FJ, Duff JH, Avanzino RJ. (1990) Influence of exchange flow between the channel and hyporheic zone on nitrate production in a small mountain stream. *Canadian Journal of Fisheries and Aquatic Sciences* **47**: 2099–111. [7.7]

van Kessel JF. (1977) Factors affecting the denitrification rate in two water-sediment systems. *Water Research* **11**: 259–67. [7.7]

Vannote RL, Minshall GW, Cummins KW, Sedell JR, Cushing CE. (1980) The river continuum concept. *Canadian Journal of Fisheries and Aquatic Sciences* **37**: 130–7. [7.5]

Verhoff FH, Melfi DA. (1978) Total phosphorus transport during storm events. *Journal of the Environmental Engineering Division, Proceedings of the American Society of Civil Engineers.* **104**: 1021–6. [7.6]

Verhoff FH, Melfi DA, Yaksich SM. (1979) Storm travel distance calculations for total phosphorus and suspended materials in rivers. *Water Resources Research*

15: 1354–60. [7.6]

Viner AB. (1988) Phosphorus on suspensoids from the Tongariro River (North Island, New Zealand) and its potential availability for algal growth. *Archiv für Hydrobiologie* **111**: 481–89. [7.6]

Vollenweider RA. (1976) Advances in defining critical loading levels for phosphorus in lake eutrophication. *Memorie dell'Istituto Italiano di Idrobiologia* **33**: 53–83. [7.3]

Wallace JB, Webster JR, Woodall WR. (1977) The role of filter feeders in flowing waters. *Archives in Hydrobiologie* **79**: 506–32. [7.1]

Wallace JB, Webster JR, Cuffney TF. (1982) Stream detritus dynamics: regulation by invertebrate consumers. *Oecologia* **53**: 197–200. [7.5, 7.8]

Walling DE, Webb BW. (1994) Waterquality I. physical characteristics. In: Calow F, Petts GE (eds) *The Rivers Handbook*, vol. 1, pp 48–72. Blackwell Scientific Publications, Oxford. [7.1]

Waters TF. (1961) Standing crop and drift of stream bottom organisms. *Ecology* **42**: 532–7. [7.5]

Waters TF. (1965) Interpretation of invertebrate drift in streams. *Ecology* **46**: 327–34. [7.5]

Webster JR. (1975) *Analysis of potassium and calcium dynamics in stream ecosystems on three southern Appalachian watersheds of contrasting vegetation.* PhD thesis, University of Georgia, Athens. [7.1]

Webster JR. (1983) The role of benthic macroinvertebrates in detritus dynamics of streams: a computer simulation. *Ecological Monographs* **53**: 383–404. [7.8]

Webster JR, Benfield EF. (1986) Vascular plant breakdown in freshwater ecosystems. *Annual Review of Ecology and Systematics* **17**: 567–94. [7.5]

Webster JR, Patten BC. (1979) Effects of watershed perturbation on stream potassium and calcium dynamics. *Ecological Monographs* **49**: 51–72. [7.1, 7.2]

Webster JR, D'Angelo DJ, Peters GT. (1991) Nitrate and phosphate uptake in streams at Coweeta Hydrologic Laboratory. *Verhandlungen Internationale Vereinigung für Theoretische und Angewandt Limnologie* **24**: 1681–6. [7.7]

Wetzel RG, Ward AK. (1994) Primary production. In: Calow P, Petts GE (eds) *The Rivers Handbook*, vol. 1, pp 354–69. Blackwell Scientific Publications, Oxford. [7.2]

White E, Payne G, Pickmere S, Pick FR. (1981) Orthophosphate and its flux in lake waters. *Canadian Journal of Fisheries and Aquatic Sciences* **38**: 1215–19. [7.6]

White RE, Beckett PHT. (1964) Studies on the phosphate potentials of soils. Part I. The measurement of phosphate potential. *Plant and Soil* **20**: 1–16. [7.6]

Whitford LA, Schumacher GJ. (1961) Effect of current on mineral uptake and respiration by a fresh-water alga. *Limnology and Oceanography* **6**: 423–5. [7.3]

Whitford LA, Schumacher GJ. (1964) Effect of a current on respiration and mineral uptake in *Spirogyra* and *Oedogonium*. *Ecology* **45**: 168–70. [7.3]

Wiley MJ, Osborne LL, Larimore RW. (1990) Longitudinal structure of an agricultural prairie river system and its relationship to current stream ecosystem theory. *Canadian Journal of Fisheries and Aquatic Sciences* **47**: 373–84. [7.3]

Wong SL, Clark B. (1976) Field determination of the critical nutrient concentrations for *Cladophora* in streams. *Journal of the Fisheries Research Board of Canada* **33**: 85–92. [7.3]

Wuhrmann K, Eichenberger E. (1975) Experiments on the effects of inorganic enrichment of rivers on periphyton primary production. *Verhandlungen Internationale Vereinigung für Theoretische und Angewandt Limnologie* **19**: 2028–34. [7.3]

Yoshinari T, Hynes R, Knowles R. (1976) Acetylene inhibition of nitrous oxide reduction and measurement of denitrification and nitrogen fixation in soil. *Soil Biology and Biochemistry* **9**: 177–83. [7.7]

8: In-stream Hydraulics and Sediment Transport

P. A. CARLING

8.1 INTRODUCTION

Rivers are linear features and so it might be supposed that the flow hydraulics would be unidirectional and relatively simple to describe. To some extent this is true. Except for the largest of the world rivers, the effects of wind-waves and geostrophic circulation can be ignored, as can periodically reversing flow which dominates tidal reaches. However, to complicate matters, rivers are characterized by relatively rapid changes in flow rate and consequently can be regarded as stationary in behaviour only over short time spans. The relative shape and roughness of the channel may change as water levels fluctuate, and the twisting and turning of meandering channels means only short reaches can be considered as linear conduits.

Despite complexities, the relatively two-dimensional nature of many rivers has meant that engineers have been able to describe the flow structure present at any one moment by two-dimensional or even one-dimensional flow models with sufficient accuracy for most practical applications. However, all rivers are characterized by turbulent flow, and large rivers in particular have a complex three-dimensional structure that can be well described only by sophisticated mathematical modelling, scaled by laboratory data obtained using delicate apparatus. The latter are usually not suitable or are difficult and expensive to use in real rivers.

It is clear, then, that there are a number of levels of increasing complexity at which the natural river system can be described. It is very important to decide at which level of complexity observations should be made to obtain answers to

problems and to consider whether genuine understanding can be obtained at any one level. For example, at the most basic level a good correlation may be obtained between the behaviour of an organism and mean current speed when in fact it is the level of turbulence intensity that is really controlling behaviour. However, moving to a higher level of explanation may not be productive if too many assumptions are required to produce a 'working' model of hydraulic structure. For many applications, sophisticated explanation is neither necessary nor cost effective. As Peters and Goldberg (1989) observed, average data may well describe the environment, and standards exist that assume temporal stability of simple phenomena (Gore 1978; Newbury 1984). Peters and Goldberg (1989) further noted that the field scientist frequently 'has to rely on rough, robust apparatus, often lacking sensitivity' and consequently it is important to consider what can realistically be achieved when designing any field monitoring or experimental programme. The tools available must be capable of providing at least a degree of insight into the problem of interest.

With these limitations in mind, a considerable degree of understanding can be obtained by those who appreciate the complexities of natural flow structure even if they are unable to describe fully many phenomena mathematically. To this end, this chapter aims to provide a practical approach to dealing with the intricacy of hydraulics and sediment transport within the context of recent research related to real rivers, and to emphasize those methods that are likely to be fruitful in practical applications. It is not possible within the space available to describe all procedures fully but details can be found in the references cited.

8.2 THE NEAR-BED BOUNDARY LAYER

Fluvial currents are driven by gravitational gradients, either imposed by the nature of the terrain or as modified by fluvial erosion and deposition and the quantity of water delivered to the channel. In turn, the structure of the flow is mediated by the friction induced by the channel boundary. In deep rivers the region where the frictional effects are felt, the *boundary layer*, may occupy only a small proportion of the total depth whilst in shallow rivers it may extend to the water surface. Laminar (non-turbulent) flow never occurs throughout the full depth in natural rivers, so that it is turbulence which transfers frictional forces throughout the fluid and redistributes suspended particles. Further, turbulence intensity mediates the momentary level of shear stress exerted on the boundary which may result in the movement of bed and bank sediments so modifying the shape and capacity of the river channel. A consideration of the velocity structure is therefore of prime importance.

Turbulent flow may be divided into smooth, transitional and rough hydrodynamic regimes. Such a consideration is required to select appropriate equations to describe the velocity structure. For flat sand-beds the division may be given by considering the roughness Reynolds number, a non-dimensional ratio defined by the shear velocity (u_*, defined below), the grain size (D) and the kinematic viscosity (v):

Smooth turbulent: $u_* D/v < 3.5$ (1a)

Transitional: $3.5 < u_* D/v < 68$ (1b)

Rough turbulent: $u_* D/v > 68$ (1c)

The ratio expresses the balance of inertial and viscous forces. Where bed roughness is due to gravel or ripples, for example, then D (the grain size) needs replacing by some other characteristic roughness length (k_s; Table 8.1). The Reynolds number, with the length defined by flow depth, will be referred to again in section 8.3.

The time-averaged velocity (denoted as U, ignoring the usual over-bar) usually increases from zero at the bed to the free-stream velocity (U_x) at the edge of the boundary layer where the water is sufficiently deep to exceed the boundary layer thickness (Fig. 8.1(a)). However, in shallow flow the boundary layer may extend close to the surface (Fig. 8.1(b)). The theoretical structure of the flow can be divided into sublayers. The layer closest to the bed, the *bed layer*, is usually thin, often only a few millimetres thick. In slow flow over smooth beds (such as clay) it may be termed the *laminar sublayer*. In natural flows, however, the laminar nature may be disrupted. Local distortions in the velocity profile and turbulence levels then exist within the bed layer. The theoretical thickness of the laminar layer (δ) in *ideal* flow can be estimated using the relationship:

$$\delta = 11.5 v/u_*$$ (2)

The value of the constant can vary between 8 and 20 (Chriss & Caldwell 1983) but in rough turbulent flow over a gravel bed, where the calculated thickness is only millimetres, the layer is

Table 8.1 Assessment of probable hydrodynamic regime

Bed type	k_s (cm)	u_* (cm s^{-1})	$u_* k_s/v$	Regime
Smooth mud	0.006	1.2	0.5	Smooth
Smooth sand	0.03	2.2	5	Transitional
Dunes	15	2.8	3000	Rough
Flat gravel bed	1.5	3.6	400	Rough
Rocks	>30	4.6	>10^4	Rough

k_s is the equivalent roughness, u_* is the shear velocity, and v is the kinematic viscosity (0.013 cm^2 s^{-1}, 10°C, freshwater). Reproduced with modification from Soulsby (1983).

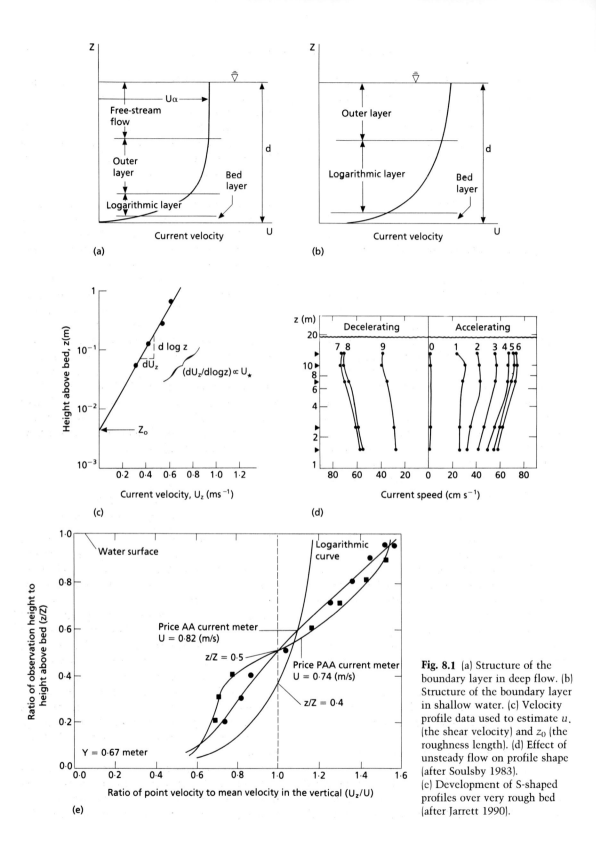

Fig. 8.1 (a) Structure of the boundary layer in deep flow. (b) Structure of the boundary layer in shallow water. (c) Velocity profile data used to estimate u_* (the shear velocity) and z_0 (the roughness length). (d) Effect of unsteady flow on profile shape (after Soulsby 1983). (e) Development of S-shaped profiles over very rough bed (after Jarrett 1990).

disrupted and often absent. In general, if the bed roughness value is greater than the calculated thickness of δ, then the latter is absent.

Many ecological texts argue that invertebrates are adapted morphologically to live within a laminar layer (see references in Statzner & Holm 1982), but Carling (1991) has argued that in many streams invertebrates that venture out of the interstitial environment are subject to low current speed but *high shear stress and turbulence levels*, a point made by Décamps and co-workers (1972, 1975) but largely ignored in contemporary literature.

Above the bed layer is the *logarithmic layer* (Figs 8.1(a) & 8.1(b)), the basic form of which is neither affected by the local roughness of the bed nor by the free-stream flow structure. As its name implies, the structure can be described by logarithmic functions. This layer is very important, as measurements taken within it allow estimation of the shear stress acting on the bed. Usually it extends over 10–15% of the depth and may extend to the surface. To estimate the boundary shear stress, it is important not to obtain current readings in the outer layer (Figs 8.1(a) & 8.1(b)), where the flow may be non-logarithmic. The usual relationship for the log-layer in the rough turbulent regime is:

$$U_z = (u_* / \kappa) \ln (z/z_0) \qquad (3)$$

where von Kármán's constant (κ) equals 0.40, and the roughness length (z_0) scales with the roughness of the river bed (Table 8.1). An estimate of z_0 can be obtained assuming $z_0 = D/30$ for sand or $D/15$ for well-sorted gravel, whilst the roughness of periodic bed-forms depends on height and spacing (e.g. Wooding *et al* 1973). Estimates of u_* and z_0 can be obtained from a regression analysis of current speed (U_z) against the height (z) above the bed (Fig. 8.1(c)). The shear stress (τ_0) is related to u_* as $\tau_0 = \rho u_*^2$. However, profiles should be plotted to ensure that the data conform to the logarithmic model; otherwise incorrect values of u_* and z_0 may be obtained.

The structure of the outer layer may be influenced by the free-stream velocity and consequently may not be described by any universal relationship. It is worth noting that the common practice of assuming that the depth-averaged velocity in the profile exists at 0.6 of the depth assumes that the logarithmic profile extends to the surface (Walker 1988).

Although Fig. 8.1(c) is a true representation of an idealized velocity profile, the logarithmic profile may be distorted by such effects as acceleration or deacceleration (Fig. 8.1(d)), extreme bed roughness (dunes or large rocks; Jarrett 1990), or bank drag which may suppress the filament of maximum velocity to below the water surface (Fig. 8.2). In the case of large-scale roughness an S-shaped profile may be present with a logarithmic section close to the bed and a further log-profile at some distance from the bed (Fig. 8.1(e)). If this reflects the influence of two scales of roughness (Dyer 1971), respectively that induced by the size of the bed material and that owing to the size of larger projections such as dunes, then each section of the log-profile may be treated separately to estimate the frictional effect of both roughness scales (Paris 1989). However, such an approach is open to criticism and where the height of the bed roughness is of the same order as the water depth no accepted theory exists to describe the vertical current structure.

8.3 BULK FLOW IN STRAIGHT CHANNELS

Theoretical consideration of bulk flow divides the flow properties into three categories: uniform, gradually varied and rapidly varied flow. Uniform flow is constant in depth, so the pressure distribution remains constant. Consequently the streamlines, water surface and bed profile are all parallel. Gradually varied flow is typified by gradual changes in cross-section and hence depth, so that streamlines may diverge or converge. The water surface and bed may not be parallel everywhere and the pressure distribution varies. Rapidly varied flow is typified by such phenomena as hydraulic drops and is common locally in steep, rough-bedded mountain streams (Whittaker 1987).

In many rivers, section properties change only slowly so that locally the flow can be treated as uniform. In addition, flow may be considered as steady (no change in discharge or velocity with time) or unsteady. Examples are the constant compensation flow immediately downstream of a reservoir which is dependent on a steady release of water from a reservoir, and the varying discharge owing to the passage of a storm-generated flood wave through the channel. If flow does not change too rapidly with respect to the time necessary to sample it, then it is often acceptable to treat it as if steady conditions pertain.

There are a number of fundamental non-dimensional ratios in basic hydraulics. Two important ones describe the state of flow. These are the Reynolds number (*Re*) and the Froude number (*Fr*). A form of the *Re* has already been defined in equation 1. Often the characteristic length is expressed as *d* (the depth) or as *R* (the hydraulic radius, i.e. sectional area/section perimeter) which is appropriate when width to mean depth ratios are less than 13. Using this definition of length, in natural rivers laminar flow exists when *Re* < 500 and fully turbulent flow exists when *Re* > 2500.

Fr expresses the relationship between the inertial force and the swiftness of a gravity wave:

$$Fr = U/\sqrt{gd} \qquad (4)$$

where *g* is the acceleration due to gravity. If *Fr* < 1.0 the flow is regarded as subcritical (tranquil) and where *Fr* > 1.0 the flow is supercritical (rapid). In the latter case surface waves appear and break repeatedly in an upstream direction. Supercritical flow in most rivers only occurs locally. Even in mountain torrents the reach-averaged Froude numbers rarely exceed 0.6 (Jarrett 1984).

If uniform flow exists, the driving force (the component of the weight of the water acting downslope) is balanced by the resisting force (the bed resistance). If the resisting force varies as the square of velocity:

$$\tau_0 = kU^2 \qquad (5)$$

then the balance of forces may be written as:

$$U^2 = (\rho g/k)RS \qquad (6)$$

(where τ_0 is the shear stress, *U* is the depth-averaged velocity, ρ is the density of water, *g* is the acceleration due to gravity, *R* is the hydraulic radius, *S* is the energy slope and the square root

of the term in parentheses is often termed the Chezy coefficient) or as:

$$\tau_0 = \rho gRS = \gamma RS \qquad (7)$$

where γ is the specific weight of water. This indicates that the bed shear stress is balanced by the product of the unit weight of water, the hydraulic radius and the bed slope.

The friction associated with the flow resistance can be expressed by the Darcy–Weisbach friction factor:

$$f = 8(\sqrt{gRS}/U)^2 = 8(u_*/U)^2 \qquad (8)$$

which is dimensionless. In equation 8, u_* is the shear velocity. The Chezy C coefficient can be regarded as a form of resistance coefficient and alternative formulations may be used such as the Manning number. These may be related one to another (e.g. Richards 1982).

Equations 7 and 8 express the total shear resistance imposed by the boundary over a given reach and give no information as to how this is distributed locally around the perimeter. The local shear resistance associated with individual velocity profiles may differ across the bed and needs to be integrated to yield an average term. Variation in roughness across the bed and between bed and banks can induce significant variation across the section such that the velocity structure of straight channels is distorted.

This observation highlights the fact that natural flow is three dimensional. In natural channels with width to depth ratios greater than about 13, the effects of the banks are not felt in the central two-thirds of the flow. However, as channels narrow and deepen the importance of bank resistance cannot be ignored. Roughening the walls of a hydraulic flume alters the flow structure, depressing the maximum filament of velocity to below the surface. The same effect is seen in natural channels (Fig. 8.2) and strong upwelling and secondary flow cells develop along the banks (Bhomik 1982). Once the river goes over-bank the effect may intensify with momentum transfer distorting the flow structure in the vicinity of the banks. In wide channels, the flow may be characterized by more than one flow cell (Fig. 8.2), each exhibiting a degree of secondary 'corkscrew' spiralling flow (Fig. 8.3).

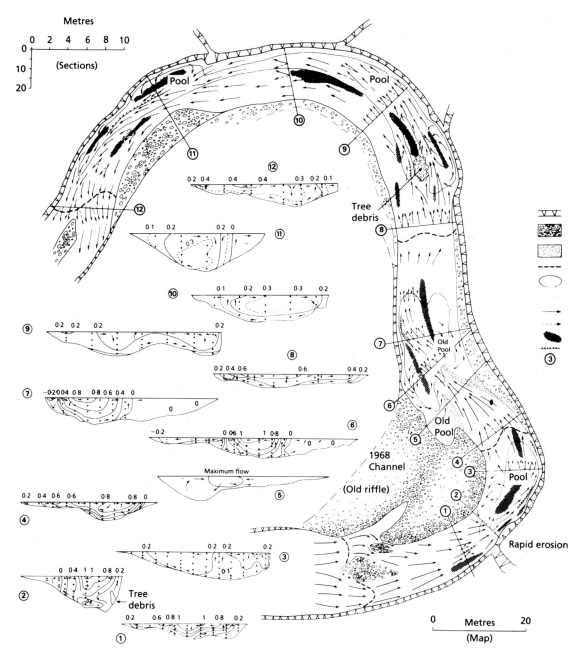

Fig. 8.2 Channel morphology, secondary flows and cross-section isovels through a meander loop. Velocities in m s^{-1} (after Hooke & Harvey 1983).

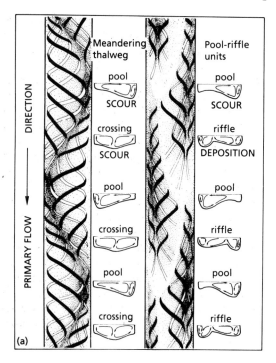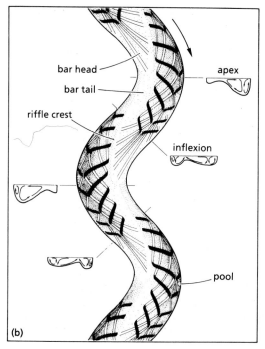

Fig. 8.3 Models of flow structure in (a) straight and (b) meandering channels (after Thompson 1986).

The resistance to flow is generated by the nature and scale of the boundary material (Table 8.2). The relationship is not simple, however, and large-scale distortion of the channel alignment, such as bank protrusions or large bars on the bed, will add resistance to flow over that associated with a flat bed of sand or gravel. Where the resulting flow distortion is gross then energy dissipation will be intense, resulting in additional flow resistance.

It is usual to recognize four kinds of flow resistance: the *skin* (or grain) *resistance* (f') associated with the individual sand or gravel particles making up the bed; the *form resistance* (f'') associated with the presence of ripples or dunes; the *internal distortion resistance* associated with banks and changes in channel alignment; the *spill resistance*, a localized phenomenon associated with rapid flow, for example between and over boulders in a mountain stream. In this latter case Froude numbers are high (>0.5) and the roughness coefficient is a function of the Froude number. For low Froude numbers, the resistance

factors sum to give the total roughness, i.e.:

$$f = f' + f'' + \ldots$$

It is possible to determine the relative contributions of skin and form roughness by considering pressure distributions (Shen *et al* 1990), but in the field it is usually accomplished by analysing compound velocity profiles or by estimating the skin roughness (Prestegaard 1983) using a relative roughness relationship such as proposed by Limerinos (1970):

$$1/\sqrt{f'} = 1.16 + 2 \log (R/D_{84}) \qquad (9)$$

where D_{84} is the 84th percentile at the coarse end of a size distribution of the bed material. In principle, in straight, wide channels the difference between this estimate of f' and a field measure of the total resistance is an estimate of the form of drag contribution. The skin friction factor is of fundamental importance when estimating the threshold shear stress required to initiate sediment transport.

In straight uniform channels there is important

Table 8.2 Typical values of the roughness length (z_0) (see text for more details)

Bed type	z_0 (cm)
Mud	0.02
Mud/sand	0.07
Silty sand	0.005
Flat-bed sand	0.04
Rippled sand	0.60
Small dunes	0.12
Large gravelly dunes	<0.50
Sand/gravel	0.03
Mud/sand/gravel	0.03
Fine gravel	0.30
Medium gravel	0.65
Coarse gravels	1.64
Slumped banks	2.42
Moss bed	<5.00
Prone ranunculus	<20.00

Data are from a variety of sources including Soulsby (1983), Bridge and Jarvis (1977), Dyer (1970, 1972) and unpublished data of author. Values are means. Standard errors of estimates are typically up to 1.0, indicating wide variation induced, for example, in mixtures where fines can infill surface void space smoothing rougher textures.

variation in the velocity and flow resistance across the section (Fig. 8.3(a)). If the flow is competent to move sediment (scour and fill), then a degree of spatial sorting may occur, changing the local roughness characteristic. This in turn can be exaggerated by aquatic vegetation growing in the more suitable habitats, and by the presence of discontinuities in bank alignment or the presence of tributary junctions. These areas may have well-developed secondary flow cells with reverse flow flowing up-stream along the bank (see Fig. 8.2). Where the river is wide enough, remote sensing can readily demonstrate this variability (Plate 8.1, facing p. 170). The instigation of scour and fill, altering the topography, will inevitably result in a gradual variation in flow from one section to another.

A good example of gradually varied flow is provided by pool and riffle sequences. Straight channels are frequently characterized by deeper sections called *pools* and shallow sections termed *riffles* (Fig. 8.4(a)). There is some debate about how these are initiated but, once formed, they are hydrodynamically stable and regularly spaced along the river (typically every three to ten widths; Ferguson 1981). The surface sediment in riffles is often coarser than that in pools, which may have a blanket of fines on the bottom (Hirsch & Abrahams 1981). During low stage, riffles have faster currents over them than the pools owing to the reduced cross-sectional area (Carling 1991). In some streams, riffles are wider than pools so that the flow diverges. A central shallow bar may develop with deep channels impinging against and eroding the banks (Hooke 1986). This divergence and convergence may induce secondary spiralling flow and complex shear stress distribution patterns on the bed. As the discharge increases the relative difference in pool and riffle cross-sectional area reduces so that the mean velocities and mean shear stress in each section begin to converge (Fig. 8.4(b)). Convergence usually occurs close to bankfull, but local variations in the shear stress in pool or riffle may persist.

8.4 BULK FLOW IN CURVED CHANNELS

In a straight channel, variation in the strength of secondary circulation and the cross-channel velocity component results in a meandering zone of bedload transport (Hey & Thorne 1975) which can lead to the deposition of alternate channel-side bars (see Fig. 8.3(a)) and the instigation of meandering (Leopold 1982). Flow through a bend is three dimensional (Okoye 1989), and although a useful description of flow structure can be obtained using impeller current meters (Bridge & Jarvis 1977), for a full description the secondary flow cells need to be defined by using electromagnetic flowmeters. Typically, the flow is dominated by a large cell directing fast-flowing surface water towards the outer bank and slower near-bed water towards the inner bank (point bar) (Figs 8.2 & 8.3(b)). A small inner cell of opposite polarity may occur close to the outer bank (Thorne & Hey 1979; Thorne & Rais 1984). Such complex flow results in spatial variation in the shear stress on the bed. Peak values obtain where the isovels are compressed near the bed owing to high velocities or downwelling, whilst low shear stresses

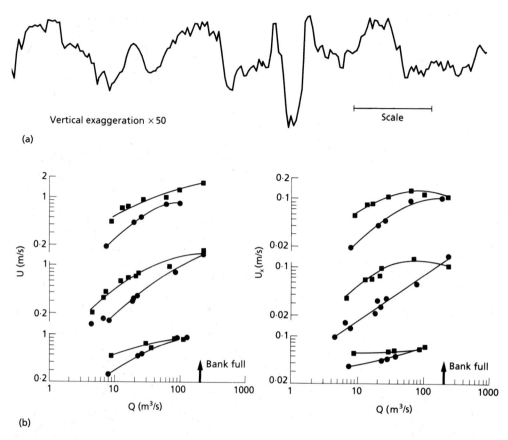

Fig. 8.4 (a) Longitudinal bed profile of the River Severn near Welshpool, UK, showing distinct pool and riffle structure. (b) Convergence of mean current velocity and shear velocity in riffles (squares) and pools (dots) as discharge increases to bankfull.

obtain where there is upwelling (Hey 1984). Peak values associated with the main flow line are typically 1.5–2.5 times the sectionally averaged value, but local downwelling may result in factors of four applying.

The locus of the maximum velocity depends on the discharge (Fig. 8.5), but at the entrance to the bend the swiftest flow is usually located close to the inner bank, whilst near the apex of the bend the flow has moved to close the outer bank. Here the flow lines are highly distorted. Near the surface, upwelling and flow away from the bank results in low applied shear on the bank, but near the bed the two cells flow towards the bank and the stress on the bank-foot may be high, resulting in basal scour and bank recession. The track of

the bed-load zone tends to mimic the variation in the locus of the maximum velocity filament (Dietrich *et al* 1984) and, as the outward flow away from the inside of the meander results in decreased shear stress close to the inside of the bend, deposition occurs there (Richards 1982; Dietrich & Smith 1984; Parker & Andrews 1985). The flow in nature can be shown to compare well with recent theoretical models (Bridge 1984; Dietrich & Smith 1984; Lee & Hsieh 1989), but improvement in the models will require a proper understanding of the turbulent structure associated with highly variable topography.

Turbulence in rivers has been little studied (McQuivey 1973; Heslop & Allen 1989) and much more is known from laboratory studies and from

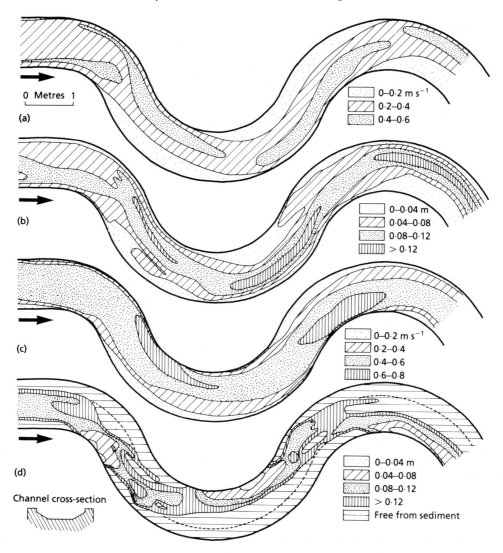

0–0.2 m s^{-1}
0.2–0.4
0.4–0.6

0–0.04 m
0.04–0.08
0.08–0.12
> 0.12

0–0.2 m s^{-1}
0.2–0.4
0.4–0.6
0.6–0.8

0–0.04 m
0.04–0.08
0.08–0.12
> 0.12
Free from sediment

0 Metres 1

(a)

(b)

(c)

(d)

Channel cross-section

Fig. 8.5 Patterns of surface velocity and flow depth observed by Martvall and Nilsson (1972) in a laboratory meandering sand-bedded channel. (a) Surface velocity:discharge = 0.014 m^3 s^{-1}; (b) flow depth:discharge = 0.014 m^3 s^{-1}; (c) surface velocity:discharge = 0.05 m^3 s^{-1}; (d) flow depth:discharge = 0.05 m^3 s^{-1} (after Allen 1984).

the shallow marine environment. If x is the downstream direction, y the transverse direction and z the vertical direction, then the quantities of interest are the time-averaged variances of the velocities in these respective directions; u^2, v^2 and w^2. The Reynolds shear stresses, averaged over a few minutes, are also frequently used to summarize turbulence information; these are written as $-\rho uw$, $-\rho vw$, $-\rho uv$. The term $-\rho vw$,

for example, represents the rate of turbulent transfer across the x–y plane in the y (cross-channel) direction. Intermittent fluctuations over longer timescales give rise to unsteady flow events in the near-bed region, known as 'bursting' (Kline *et al* 1967). Accelerated flow events acting towards the bed are termed 'sweeps' and rapid fluid movements away from the bed are called 'ejections'. In the laboratory such behaviour has a profound influence on the entrainment of non-cohesive sediments (Fig. 8.6).

Although it is not clear whether the phenomena observed in the field are directly analogous to the laboratory experience, unsteady bedload transport in the sea is driven by 'bursting-type' cycles (Thorne *et al* 1989), whilst suspension of sand may be periodic and related to 'ejection' events (Soulsby *et al* 1987). Unsteady transport of gravel bedload is now known to occur widely in rivers and in some cases may be related to the inherent (turbulent) structure of river flow. Consequently, further effort will be directed to understand the

intermittence in turbulence structure (Williams *et al* 1989).

Secondary currents and turbulence induce mixing across the river, but it must not be implied that complete mixing always results. Many dispersion models often underestimate the residence time of a tracer (Day & Wood 1976; Bencala & Walters 1983), in part because of inadequate consideration of the effect of slow-flowing water close to the banks. However, more extensive patches of slow-flowing or 'still' water can occur (see Fig. 8.2) which mix only slowly across a shear zone with the main advective flow (Chatwin &

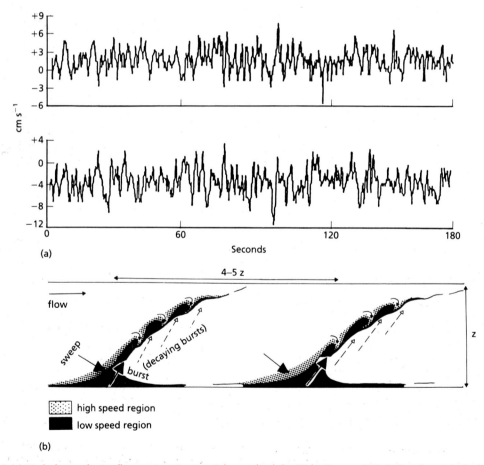

Fig. 8.6 (a) Turbulent velocity fluctuations in a straight reach of the River Severn, UK, 0.5 m above the bed. Upper trace shows vertical velocities and the lower trace shows the cross-stream component (from Heslop & Allen 1989). (b) Schematic diagram depicting high speed turbulent ejection of fluid and suspended solids away from the sediment bed (after Leeder 1983).

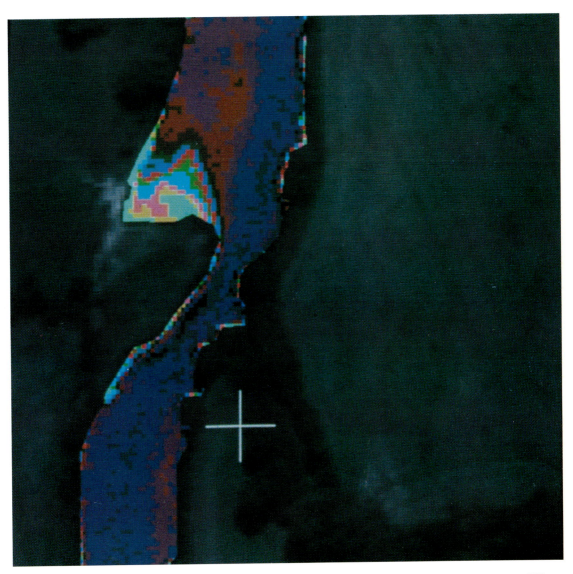

Plate 8.1 False colour remote sensing image of algal concentrations in 'dead zones' along the River Severn, UK, during low summer flow. Dark blue represents low concentrations in very turbulent flow; red represents intermediate concentrations; and the light colours represent high concentrations in almost stagnant water. River is up to 60 m wide. Flow from bottom to top.

Allen 1985). The effect of these 'dead zones' is to delay the passage through a reach of solutes and suspended solids (Young & Wallis 1987) and so they are important in the context of chemical dispersion processes. They also may act as refuges for organisms during high flows and as temporary sinks for fine sediments during low flows. Recently, *in situ* fluorimetry and airborne remote sensing have been used to detect the spatial variation in flow structure (Reynolds *et al* 1991).

Similar complex flow patterns can be expected at channel confluences (Best 1987; Roy & Bergeron 1990) and in braided stream environments with multiple channels (Ashmore & Parker 1983), where flow separation and slack-water areas in cut-off channel reaches are common.

8.5 INITIAL MOTION OF LOOSE BED SEDIMENT

The stability of a particle depends primarily upon its immersed weight and the drag and lift forces imposed by the flow (Fig. 8.7(a)). The actual value of the forces depends on fluctuations in the current speed as well as the variation in particle

shape, density and degree of exposure upon the surface (see James 1990 for a review). Consequently, equilibrium can be expressed as a force balance (e.g. White 1940; Komar & Li 1988) between the shear stress promoting entrainment and the particle size, density and gravity resisting entrainment, expressed as a non-dimensional ratio:

$$\theta = \tau_0 / (\rho_s - \rho) g D \qquad (10)$$

where θ is Shields parameter and ρ_s is the density of sediment; all other terms are defined above. Variation in θ with the particle Reynolds number (Fig. 8.7(b)) produces a curve known as the Shields function. There are few data to validate the curve for fine silts at low Reynolds numbers (<2) where cohesive forces (Mantz 1978; 1980) and negative lift forces tend to augment the resistance force. At high Reynolds numbers (>200), viscous forces become unimportant and the ratio tends to a constant value that depends on such factors as particle shape, degree of protrusion from the bed, and the overall degree of particle sorting and bed roughness. Traditionally this value has been 0.06 for well-sorted sediment and 0.047 for poorly sorted material (Miller *et al* 1977). The actual

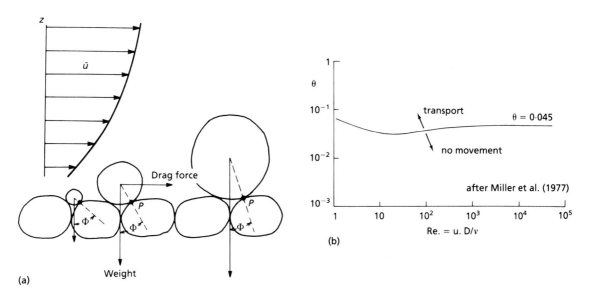

Fig. 8.7 (a) Particles in pivoting position for entrainment, illustrating that the larger the grain the greater its exposure to the flow and the smaller the pivoting angle (ϕ) (after Komar 1989). (b) Shields entrainment function (equation 10) plotted as a function of the particle Reynolds number.

value depends in part on how initial motion is defined, so that a region of uncertainty exists, but the function reaches a minimum ($\theta \sim 0.03$) in the smooth turbulent regime, which could be representative of fine sand in natural rivers.

Considerable interest has centred recently on the correct function for mixtures of coarse gravel of irregular shape in natural rivers (Komar 1987, 1989). Flattened pebbles may become imbricated (i.e. overlapped like shingling on a roof), increasing resistance to entrainment, but usually variation in particle exposure (Fig. 8.7(a)) and shape, for example, result in differing sizes of rounded pebbles becoming entrained for a given value of θ. Consequently for some mixtures many susceptible pebbles may be regarded as equally mobile, whilst others may be entrained only as the force on the bed increases (Ashworth & Ferguson 1989). The importance of equal mobility or selective entrainment depends upon the nature of particle sorting, packing and bed roughness (Komar & Carling 1991; Carling *et al* 1992). For example, gravel particles can roll freely over a flat sand-bed and so it should not be assumed axiomatically that small particles will be entrained before larger ones.

The critical shear stress needed to move a well-mixed sediment is usually related to the $D_{50\%}$ of the grading curve, but many coarse gravel beds become size segregated in the vertical so that a coarse layer, devoid of fines, overlies finer gravels or sand beneath the surface. This may be up to three times as coarse as the subsurface, and is usually called the *armour layer*, because a high shear stress is needed to destroy it and so (during moderate discharges) it protects the finer sediments from entrainment. To add an element of confusion, in the USA especially, this layer is often referred to as *pavement* whilst the term armour is reserved for surfaces that cannot readily be eroded (Parker *et al* 1982). Where segregated surfaces are present, initial motion of the armour can be calculated using either surface samples (Wolman 1954) or the $d_{84\%}$ of a bulk bed–sediment sample. Despite the interest in defining the threshold for initial motion, little attention has been given to the cessation of transport. Reid and Frostick (1984) and Lisle (1989) noted that gravel bedload transport continued until the shear

stress on the bed was only a fifth of that required to initiate motion.

The cohesive materials most likely to be found in rivers are muds and clays. These may be entrained by fluid shear or by corrasion, that is erosion owing to the scouring action of gravels or blocks of cohesive sediment swept over the surface (Allen 1984). As with non-cohesive materials, once a threshold has been exceeded, erosion increases as the applied stress increases. Factors that affect the yield strength of the mud include particle size, mineralogy, chemistry, compaction history and water temperature, so that satisfactory universal relationships for initial entrainment have not been devised. In a similar vein, hydraulic patterns and sediment transport have been little studied in bedrock channels. Erodibility depends on the nature of the imposed bedload, as well as weathering and solution rates (Howard 1987).

8.6 BED-FORM PHASE EXISTENCE FIELDS

A unidirectional flow of water over loose sediment may deform the bed to create a variety of bedforms (Fig. 8.8(a)). The most extensively studied material is fine sand, but bedforms may also be generated in silt and gravels. Starting with a flat planar bed of fine sand and low competent shear stress, some sediment transport may take place over the flat surface (low-stage plane bed). *Ripples* develop in fine sand (up to 0.6 mm in size) once a critical shear stress has been exceeded. Ripples are sharp crested, roughly triangular in profile, with gentle upstream (stoss) slopes and steeper downstream (lee) slopes close to the angle of repose of granules in water. Ripple crestlines are more or less transverse to the flow direction but may be sinuous and bifurcate. Migration downstream is by erosion on the stoss slope and deposition over the lee slope. Heights are less than 0.04 m and wavelengths are below 0.6 m. The literature concerning ripples is vast and complex, but a useful summary has been provided by Allen (1984).

At higher velocities ripples are replaced by superficially similar but much larger forms called *dunes*. These may reach metres in height with

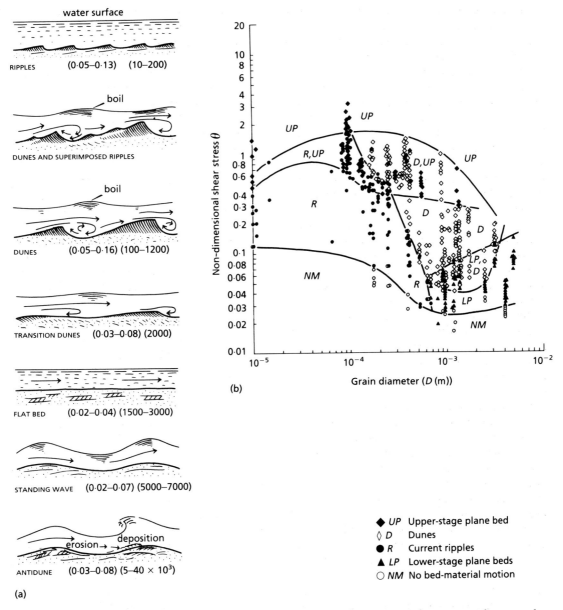

Fig. 8.8 (a) The sequence of sandy bed-forms generated as current speed increases. Values in parentheses are the friction factor (*f*) and the bedload transport rate (p.p.m.) (after Richards 1982). (b) Bed-form existence field constructed using laboratory data for predominantly sandy sediment (after Allen 1984).

wavelengths of tens of metres. Unlike ripples, the scale of which is related to grain size, dunes scale with depth of flow. Consequently, they may be related to large-scale turbulent eddies with dimensions comparable to the flow depth or boundary-

layer thickness. The transition from ripples to dunes is usually abrupt, although conditions can exist where ripples form on the stoss-side of dunes. With increasing velocity, the dunes become more rounded and lower (so-called *hump-*

back dunes; Saunderson & Lockett 1983) until the bed-forms are washed out and an upper-stage flat bed forms, characterized by intense bedload transport. A tremendous variety of names has been applied to various dune-like features (Ashley 1990): some of the confusion has arisen because of the variety and scales of non-equilibrium geometries that can occur in natural accelerating and deaccelerating flows (Allen 1983).

As the velocity increases further ($Fr > 0.8$), standing waves form at the water surface, and the bed is remolded into a train of sediment waves (*anti-dunes*) in phase with the water waves. As Fr increases, so the amplitude of the waves increases. These bed-forms may migrate upstream with the upstream slope steepening until the geometry is unstable and the features are washed out catastrophically only to begin to form again (Hand 1974; Allen 1984; Yagishita & Taira 1989).

A great variety of bed-form stability or existence fields have been proposed and some of these are appropriate for particular purposes (reviewed by Allen 1984). One approach is to consider the grain size of the bed material and the mean flow velocity (e.g. Southard & Boguchwal 1973), but the most successful diagrams, in terms of adequately delineating the existence fields, consider the shear velocity, non-dimensional shear stress (equation 10) or the streampower ($\tau_0 U_z$) and particle size (Fig. 5.8(b)). These diagrams are constructed largely from laboratory data collected under idealized conditions.

Little is known of the stability fields of large bed-forms in rivers (with scales equivalent to river width), especially where sediments are coarser than 1 mm (gravels). At the smallest scale, a few individual gravel particles may be grouped into *clusters* (Brayshaw 1983, 1984) which resist entrainment, whilst at larger scales (equivalent to the dune scale) a variety of low-amplitude *bars* develops. The larger features are frequently affected by the proximity of the channel margins and subunits may be 'welded on' as flows fluctuate, resulting in a variety of sediment associations (Brierley 1989). Consequently an alarming proliferation of nomenclature has resulted to describe features that may not be hydrodynamically distinct (Smith 1978). Nomenclature often refers to the position of the bars in the channel, for example

alternating bar, medial bar, transverse bar, point bar (Richards 1982). It is not certain how these features relate to flow dynamics but they are long lived relative to the period of formative discharges. Crowley (1983), for example, considered that such features formed in sandy sediment are not hydrodynamically equivalent to the regime bed-forms considered in Fig. 8.8(b), which generally respond rapidly to hydrodynamic regime. Ashley (1990) noted that as these bars are channel-scale features they should scale with bankfull (formative) discharge, but this observation is not particularly illuminating. Examples of gravel bars to compare with Crowley's sand-bed features are provided by Bluck (1976).

Bluck (1976) considered that the surface facies arrangement of bars can be divided into bar-head (upstream) and bar-tail (downstream) deposits. The bar-head gravels form during high-flow stages, whilst as flow drops the focus for deposition shifts to the bar tail, and this has its own suite of deposits. Sediment is delivered to the bar head by either the migration of low-amplitude gravel dunes (Carling 1990), which leave distinct cross-bedded deposits over the pre-existing sediment surface, or by the progression of mobile diffuse gravel sheets (Kuhnle & Southard 1988; Whiting *et al* 1988), which leave indistinct stratigraphic structures (Wells & Dohrenwend 1985).

Bed-forms also occur in cohesive materials such as muds (Allen 1984). These include ripples and corrasion marks that can develop into distinctive flutes and scallops. Allen (1971) noted a distinct relationship between the wavelength of mud ripples and the hydraulic radius, but importantly found no relationship with flow velocity.

8.7 SUSPENSION DYNAMICS

The study of suspended sediments is vital for understanding siltation processes, but also the control of sediment-associated contamination (Golterman *et al* 1983). Suspended grains are held above the bed by turbulent fluctuations in the fluid (Brush *et al* 1962). Although turbulent structure cannot be regarded as isotropic, as a general rule the variances of the turbulent fluctuations of velocity in the x, y and z planes are of similar magnitude and close to the bed are roughly equivalent to u_*. Consequently, a general criterion for suspension is that the shear velocity should exceed the velocity (w) at which the grain would settle in still clear water:

$$u_* > w \tag{11}$$

or in terms of the shear stress:

$$\tau_0 > \rho w^2 \tag{12}$$

Middleton (1976) gave support to this general function from the results of hydraulic measurements in natural rivers. Sometimes more conservative estimates have been applied to high concentration turbulent flows (Richardson & Zaki 1954), in which case u_* is multiplied by a factor of up to 1.6.

An alternative rough approximation was suggested by Graf (1971), who argued that as the upward velocity component and the Reynolds number are indicative of the turbulence level, then a function relating these should act as a guideline for suspension. When calibrated using empirical data, this reduced to:

$$u_z = 0.17(Ud)^{0.46} \tag{13}$$

If the settling velocity is known (see Vanoni 1977) then this can be compared with u_z. Clearly, turbulence will still bring some suspended grains into close contact with the bed, and if there are open pore spaces, as in a gravel bed, then suspended particles may be entrapped even if quiescent deposition is not possible (Carling 1984).

The actual prediction of sediment fluxes requires measured suspended sediment profiles (for techniques see Jansen *et al* 1979; Mangelsdorf *et al* 1990) or a prediction of the vertical distribution, together with velocity profile data. Inadequate sampling designs can result in variable results (Horowitz *et al* 1990). Coarse sediments tend to be found close to the bed with only the finest sediment in the outer layer. For any given grain size, the distribution throughout the total depth (d) can be approximated if a reference concentration (C_a) is known at a given height (a) above the bed:

$$\frac{C_z}{C_a} = \left[\frac{d-z}{z} \frac{a}{d-a} \right]^{\beta} \tag{14}$$

The exponent $\beta \sim w/0.4u_*$ for fine sediment but

takes larger values for coarse sediments. For low β values the distribution is nearly uniform, whereas for large values little sediment is found near the surface (Graf 1971). A comprehensive treatment of suspension mechanics is given by Vanoni (1977) whilst simpler introductions are provided by Henderson (1966) and Graf (1971).

Once the stress on the bed (τ_0) falls below a critical value (τ_c) suspended sediments can begin to settle out. Not all the sediment will fall out at once because of its distribution by size in the vertical and because there is still a turbulent current flowing. The rate of deposition (R_d) for any size fraction can be estimated where the near-bed concentration (C_a) is known as follows:

$$R_d = C_a w[1 - (\tau_0/\tau_c)] \tag{15}$$

The above overview is extremely simplified but provides a practical means to make reasonable assessments of the likelihood of resuspension or deposition.

8.8 TRANSPORT OF BED MATERIALS

Once a particle resting on the bed is entrained it may go into suspension. However, if it is too heavy for suspension it may travel along the bed surface. Particles may roll or slide along the bed (the *traction load*) or hop as they rebound on impact with the bed. In the latter case, ballistic trajectories occur and the material is said to move by *saltation*. Many particles moving by either of the mechanisms are collectively referred to as *bedload* and it is of considerable importance in many investigations to be able to estimate or measure this load.

A great number of bedload transport equations have been proposed over the years. Many of the more sophisticated are calibrated only with flume data and are difficult to apply in the field. Where equations have been calibrated using field data, reasonable results can be expected when the equations are applied to rivers very similar to those for which they were devised, but performance can be very poor in other environments. A major problem is that the equations assume that the flow will transport as much material as it is *competent* to do so, so that a *capacity* load is assumed. However, because of the armouring effect in many rivers and vagaries in supply of sediment to rivers from the land, sediment loads are often lower than predicted. Of the simpler formulae for gravel-bed rivers that of Meyer-Peter and Mueller (1948) gives reliable results:

$$\eta \, \rho \, g \, R \, S - 0.047 \, (\rho - \rho_s) \, g \, D_m = 0.25 \, \rho^{1/3} \, I_b^{2/3} \tag{16}$$

$$\eta = (n'/n)^{1.5} \tag{17}$$

$$n' = 0.038 \, D_{84}^{1/6} \tag{18}$$

$$n = \frac{R^{2/3} \, S^{1/2}}{U} \tag{19}$$

$$I_b = \left[\frac{\text{LHS}}{0.25\rho^{1/3}} \right]^{3/2} \tag{20}$$

where I_b is the bedload transport rate, n is Manning's roughness; all other terms have been defined above.

Equation 16 has three terms. The first on the left represents total shear stress acting on the bed grains; the second, shear at initial motion; and the right-hand side, the transport rate. The correction factor, η (equation 17), is the ratio of the grain roughness coefficient to the total roughness coefficient, raised to a power of 1.5. For gravel-bed rivers where bed-forms are absent it will be close to unity. The grain roughness is calculated from a Strickler-type function (equation 18) whilst the total roughness is calculated from stream characteristics (equation 19). The second term is the critical shear stress in terms of Shields entrainment function (equation 10). Thus for total shear stress less than the value of the second term, the left-hand side (LHS) of the equation is negative indicating zero transport, whilst for a greater total stress it is positive and gives increasing transport rates with increasing flow. The weight per unit width and time, I_b, can be calculated using equation 20, and dividing the result by g yields the mass per unit width and time.

Alternative formulae are available, such as that of Ackers and White (1973). The Ackers–White approach was developed from analysis of about 1000 sets of flume data, based on dimensionless expressions for sediment transport and derived from the streampower $(\tau_0 U_z)$ concept of Bagnold

(1966). More recently Parker and colleagues (1982) introduced a method for gravel-bed rivers with well-developed armour layers. Analysis showed that once the armour was broken up, all grain sizes were equally mobile. This latter approach has not been tested widely, but when applied to several gravel-bed rivers in north-west England gave results consistently greater (by a factor of four) than were obtained using the Meyer–Peter and the Ackers–White formulations. The latter two approaches yielded almost identical results of a magnitude commensurate with transport rates which have been sampled directly in similar British rivers. Although somewhat more complex than the Meyer–Peter approach, the Ackers–White model can also be applied to sand-bedded rivers, and deserves attention.

Many transport equations have been proposed for the calculation of bedload transport in sand-bedded channels. A major difference between these expressions and those for gravel-bed rivers is in the expression of the shear stress available for transporting material. In gravel rivers the total roughness depends largely on grain roughness, but in sand-bed streams the total roughness is influenced by form resistance, limiting the shear available for transport. For sand-bed rivers the formula of Yang (1973), based on the concept of unit stream power, has been widely compared with data-sets from both field and laboratory (Stevens & Yang 1989) and shown to be one of the most reliable available. Other formulae suitable for sand-bed rivers are those of Toffaleti (1968) and Ackers and White (1973).

An alternative to estimating the transport rate is to measure it directly. This can be done by trapping sediment in pits (Reid *et al* 1980; Bathurst 1987; Carling 1989), or by electro-magnetic detection (Reid *et al* 1984; Custer *et al* 1987). Another alternative is to lower a sampler directly on to the bed from a boat or gantry and to trap material for a given time period. The problems here include representative sampling of a spatially and time-variable load across the complete bed-width, together with the risk of the sampler not sampling efficiently, either by disrupting the flow field and altering the local transport rate or indeed by digging into the sediment surface or not sitting evenly on the bed. Today,

the most popular sampler in the USA is the Helley–Smith pressure difference sampler (Helley & Smith 1971) with a uniform entrance section (7 × 7 cm or 17 × 15 cm). Extensively calibrated (see Hubbell 1987) for sand and fine gravel, it is of limited value for very coarse gravels. For the latter, other samplers, notably the full-size VUV (Novak 1957; Gibbs & Neill 1973), with a large entrance section (20 cm high × 50 cm wide) are required (Carling 1989). The particular sampling requirements needed to map bedload transport vectors are given by Dietrich and Smith (1984).

The use of a sampling or estimation procedure assumes that a unique relationship exists between the characteristics of the flow at the time of sampling and the mass of bedload in transit past that point, although this is rarely the case (Church 1985). It has already been noted that supply limitations may reduce the amount actually transported. In addition it is now accepted that temporal variations in transport rates occur (Fig. 8.9(a)) at a variety of timescales even when the flow is uniform and steady (Gomez *et al* 1989). Instantaneous fluctuations (seconds) in transport rates may be related to the inherent stochastic nature of the entrainment and transport process (e.g. bursting processes), whilst periodicities scaled in minutes and hours might be related to the controlling influence of the bed armour (Gomez 1983) or to clusters of particles forming or breaking up (Naden & Brayshaw 1987) or to the periodic passage of bed-forms. Reid and colleagues (1985), for example, noted that bedload transport was more intense on the falling limb of a hydrograph because, after the discharge peak, the bed surface armour had been disrupted. Longer-term periodicity or variation can result from intra-event (Reid *et al* 1985) or seasonal exhaustion of supply (Carling & Hurley 1987), or longer-term controls on sediment availability (e.g. Meade 1985; Roberts & Church 1986).

Given the unsteady nature of the sediment transport process, it is not surprising that the nature of the particle sorting and depositional processes over bed-forms is not well understood. Although considerable attention has been given to the flow structure (Nelson & Smith 1989), particle sorting and the formation of sedimentary structures over sandy ripples and dunes (e.g. Bridge & Best 1988), the same cannot be said for coarser sediments. Carling (1990) described a sorting process over gravel dunes, whilst Iseya and Ikeda (1987) and Dietrich *et al* (1989) demonstrated that the longitudinal sorting processes can result in rhythmic fluctuations in the bedload and characteristic depositional facies.

Better studied is the sorting of particles through meander bends in association with the formation of point bars (Dietrich 1987). At the bend entrance, where shoaling is significant, the largest particles in motion are near the inside of the bank and the finest are near the outer bank (Fig. 8.9(b)). However, because of the interaction of the near-bed flow and bar slope, the direction that particles move relative to the cross-stream plane through the bend depends on the magnitudes of the lift and drag forces relative to frictional resistance of the bed. The net result is that large particles generally exhibit a tendency to roll outwards obliquely across the bar surface towards deeper water whilst finer particles move into shallow waters on the upper bar surface. There are two views on how equilibrium is reached. The traditional one is that equilibrium is achieved when the cross-stream component of the particles' weight and lift force are balanced by the inward component of the drag force owing to secondary circulation. In this model particles travel along lines defined by equal depth. The alternative view is that equilibrium is achieved when the outward zone of maximum shear stress is balanced by convergent sediment transport owing to a net outward flux of bedload. The two models are summarized by Richards (1982) and Dietrich (1987), respectively. In either case the processes result in the characteristic fining inwards sequence of deposits observed on point bars (Fig. 8.9(c); Allen 1971).

Increasingly, the field investigator and the modeller are turning their attention to the transport dynamics of mixtures of particle sizes, including distinctly bimodal mixes of sand and gravel (Ferguson *et al* 1989). Close attention to the dynamics of the complete particle size distribution will be required to elucidate complex sorting patterns and entrainment and deposition processes (Shih & Komar 1990). A process that is

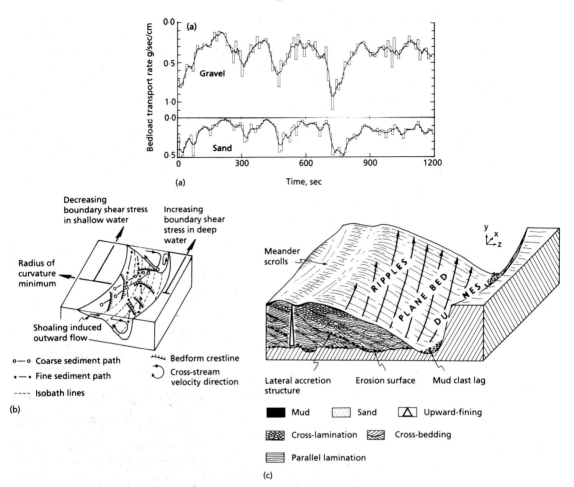

Fig. 8.9 (a) Unsteady transport of sand and gravel fractions of the bedload in steady fluid flow (from Iseya & Ikeda 1987). (b) Sketch showing the primary transport pathways of fine and coarse bedload through a channel bend (from Dietrich 1987). (c) Cross-section through sandy point-bar deposits, showing general direction of fine particle transport, the locus of bed-forms and the characteristic sedimentary deposits (after Allen 1984).

still not well understood is the one by which fine sediments are deposited and infiltrate down into the void space in gravel beds. This is not only important in understanding downstream sorting and the construction of sedimentary bodies, but also has practical significance for fish spawning habitat (Everest *et al* 1987). The effect on the latter has become prominent because human activities in many environments have increased the runoff of fine sediments into rivers which previously transported mainly coarse bed-material (Scrivener & Brownlee 1989), or river regulation has precluded high flows which previously

flushed fines away so that silt accumulates. Salmonid fish in particular deposit their eggs in gravels by excavating and then backfilling a depression in the bed (Gustafson-Marjanen & Moring 1984; Crisp & Carling 1989). A flow of oxygenated water through the hummocky structure occurs (Thibodeaux & Boyle 1987; Jobson & Carey 1989) and this is needed to ensure successful development to hatching. The young fish (alevins) need to be able to swim up through the interstitial spaces to emerge. Consequently, heavy siltation (>20% 1 mm sand in bed) will reduce the recruitment of young fish to the population (Lisle 1989). The depth to which the fines will settle depends on their size relative to the size of the bed material (Einstein 1968; Carling 1984; Diplas & Parker 1985) and to its vertical grading (Frostick *et al* 1984). This is because the particle size of the bed material mediates the size of the intervening pores. However, scour and fill will alter the active bed surface level so that sand lenses may occur at depth (Lisle 1989). Some particles trickling down will inevitably be larger than the gaps between pebbles. Consequently, lodgement occurs, hindering the passage of other particles, and as a result a sand-seal may develop at any depth. The finest materials may infiltrate much deeper into the bed, although fine clays (<3 μm) can become attached to and coat bed particles, a process influenced by water chemistry (Matlack *et al* 1989). The inner point-bar environment is particularly typified by clay infiltration deposits (Fig. 8.9(c)) because not only does the bedload transport process move the finest sediments into this region, but fine suspended sediments are delivered to this area only on high flows when concentrations are high. Fluctuating water levels also induce intra-gravel flow which aids the infiltration process, and as the point-bar is rarely subject to extensive reworking, deposits are persistent.

Empirical criteria have been developed to describe the likelihood of fine particles filtering down into gravel beds. These methods are based on the relative sizes of the two populations of sediment and have been reviewed by Carey (1991). Should lodgement occur, so that a sand-seal develops, the thickness of the seal can be estimated using a relationship proposed by Lisle (1989).

$$T = 0.1A[\rho_s \lambda_g (1 - \lambda_m)]^{-1} \tag{21}$$

where T is the seal thickness, A is the area considered, ρ_s is the sediment density, and λ_g and λ_m are gravel and sand porosities, respectively.

In addition, flow rates through fish spawning beds can be measured using standpipes and dilution techniques (Carling & Boole 1986) whilst the nature of the silted sediments can be studied using a freeze-coring technique, which retrieves undisturbed cores of bed gravel complete with the sandy infill (Crisp & Carling 1989). The method is also useful for investigating the depo-

sitional history of polluted sediments (Petts *et al* 1989). However, in regulated streams which are not subject to intensive periodic scour by high flows, fines are rarely winnowed from below the surface layer (one or two grain-diameters thick). This is because laboratory (Carling 1984) and field experience (Kondolf *et al* 1987) have demonstrated that the bed sediments need to be disturbed to remove the fines (Reiser *et al* 1985). In many cases this can be achieved only by mechanical cleansing (see collected papers in Cassidy 1981).

REFERENCES

Ackers P, White WR. (1973) Sediment transport: new approach and analysis. *Journal of the Hydraulics Division of American Society of Civil Engineers* **99**: 2041–60. [8.8]

Allen, JRL. (1971) *Physical Processes of Sedimentation.* Allen and Unwin, London. [8.6, 8.8]

Allen JRL. (1983) River bedforms: progress and problems. In: Collinson JD, Lewin J (eds) *Modern and Ancient Fluvial Systems.* Special Publication of the International Association of Sedimentologists, **6**: 19–33. [8.6]

Allen JRL. (1984) Sedimentary structures: their character and physical basis. *Developments in Sedimentology 30.* Elsevier, Amsterdam. p. 1256 [8.5, 8.6, 8.8]

Ashley GM. (1990) Classification of large-scale subaqueous bedforms: a new look at an old problem. *Journal of Sedimentary Petrology* **60**: 160–72. [8.6]

Ashmore P, Parker G. (1983) Confluence scour in coarse braided streams. *Water Resources Research* **19**: 392–402. [8.4]

Ashworth PJ, Ferguson RI. (1989) Size-selective entrainment of bed load in gravel bed streams. *Water Resources Research* **25**: 627–34. [8.5]

Bagnold RA. (1966) An approach to the sediment transport problem from general physics. *Professional Paper, US Geological Survey 4221.* Washington, DC. [8.8]

Bathurst JC. (1987) Measuring and modelling bedload transport in channels with coarse bed materials. In: Richards KS (ed.) *River Channels: Environment and Process,* pp 272–94. Blackwell Scientific Publications, Oxford. [8.8]

Bencala KE, Walters RA. (1983) Simulation of solute transport in a mountain pool-and-riffle stream: a transient storage model. *Water Resources Research* **9**: 718–24. [8.4]

Best JL. (1987) Flow dynamics at river channel confluences: implications for sediment transport and bed morphology. In: Ethridge FG, Flores RM, Harvey MD (eds). *Recent Developments in Fluvial Sedimentology*, pp 27–35. Society of Economic Paleontologists and Mineralogists Publication No. 39. Tulsa, Oklahoma. [8.4]

Bhomik NG. (1982) Shear stress distribution and secondary currents in straight open channels. In: Hey RD, Bathurst JC, Thorne CR (eds) *Gravel-bed Rivers*, pp 31–61. John Wiley & Sons, Chichester. [8.3]

Bluck BJ. (1976) Sedimentation in some Scottish rivers of low sinuosity. *Transactions of the Royal Society of Edinburgh* **69**: 425–56. [8.6]

Brayshaw AC. (1983) The hydrodynamics of particle clusters and sediment entrainment in coarse alluvial channels. *Sedimentology* **30**: 137–43. [8.6]

Brayshaw AC. (1984) Characteristics and origin of cluster bedforms in coarse-grained alluvial channels. *Canadian Society of Petroleum Geologists Memoir* **10**: 77–85. [8.6]

Bridge JS. (1984) Flow and sedimentary processes in river bends: comparisons of field observations and theory. In: Elliot CM (ed). *River Meandering*, pp 857–72. *American Society of Civil Engineers.*

Bridge JS, Best JL. (1988) Flow, sediment transport and bedform dynamics over the transition from dunes to upper-stage plane beds: implications for the formation of planar laminae. *Sedimentology* **35**: 753–63. [8.8]

Bridge JS, Jarvis J. (1977) Velocity profiles and bed shear stress over various bed configurations in a river bend. *Earth Surface Processes* **2**: 281–94 [8.3, 8.4]

Brierley G. (1989) River planform facies models: the sedimentology of braided, wandering and meandering reaches of the Squamish River, British Columbia. *Sedimentary Geology* **6**: 17–35. [8.6]

Brush LM, Ho H-W, Singamsetti R. (1962) A study of sediment in suspension. In: *Symposium of the International Association of Hydrological Sciences. Commission on Land Erosion.* IAHS Publication No. 59, pp 293–310. [8.7]

Carey WP. (1991) Evaluating filtration processes in alluvial channels. *Proceedings of the Fifth Federal Interagency Sedimentation Conference.* (in press). [8.8]

Carling PA. (1984) Deposition of fine and coarse sand in an open-work gravel bed. *Canadian Journal of Fisheries and Aquatic Sciences* **41**: 263–70. [8.7, 8.8]

Carling PA. (1989) Bedload transport in two gravel-bedded streams. *Earth Surface Processes and Landforms* **14**: 27–39. [8.8]

Carling PA. (1990) Particle over-passing on depth-limited gravel bars. *Sedimentology* **37**: 345–55. [8.8]

Carling PA. (1991) An appraisal of the velocity reversal hypothesis for stable pool-riffle sequences in the River Severn, England. *Earth Surface Processes and Landforms* **16**: 19–31. [8.2, 8.3, 8.6]

Carling PA. (In Press) The nature of the fluid boundary –

layer and the selection of parameters for benthic ecology. *Freshwater Biology.*

Carling PA, Boole P. (1986) An improved conduction-metric standpipe technique for measuring interstitial seepage velocities. *Hydrobiologia* **135**: 3–8. [8.8]

Carling PA, Hurley MA. (1987) A time-varying stochastic model of the frequency and magnitude of bedload transport events in two small trout streams. In: Thorne CR, Bathurst JC, Hey RD (eds) *Sediment Transport In Gravel-bed Rivers*, pp 897–920. Wiley, Chichester. [8.8]

Carling PA, Kelsey A, Glaister MS. (1992) Effect of bed roughness, particle shape and orientation on initial motion criteria. In: Billi P *et al* (eds) *Dynamics of Gravel-bed Rivers*, pp 23–37. John Wiley & Sons, Chichester. [8.5]

Cassidy JJ (ed.) (1981) *Salmon-spawning gravel: a renewable resource in the Pacific Northwest?* Conference held in Seattle, Washington 6–7 October 1980. Washington Water Research Center, Pullman, Washington. [8.8]

Chatwin PD, Allen CM. (1985) Mathematical models of dispersion in rivers and estuaries. *Annual Review of Fluid Mechanics* **17**: 119–49. [8.4]

Chriss TM, Caldwell DR. (1983) Universal similarity and the thickness of the viscous sublayer at the ocean floor. *Journal of Geophysical Research* **89** (part C4): 6403–14. [8.2]

Church M. (1985) Bedload in gravel-bed rivers: observed phenomena and implications for computation. *Proceedings of the Annual Meeting of the Canadian Society for Civil Engineering, Saskatoon, 1985.* **2**: 17–37. [8.8]

Crisp DT, Carling PA. (1989) Observations on siting, dimensions and structure of salmonid redds. *Journal of Fish Biology* **34**: 119–34. [8.8]

Crowley KD. (1983) Large-scale bed configurations (macroforms), Platte River Basin, Colorado and Nebraska: primary structures and formative processes. *Geological Society of America Bulletin* **94**: 117–33. [8.6]

Custer SG, Bugosh N, Ergenzinger PE, Anderson BC. (1987) Electromagnetic detection of pebble transport in streams: a method for measurement of sediment-transport waves. In: Ethridge F, Flores R. (eds) *Recent Developments in Fluvial Sedimentology*, pp 21–6. Society of Economic Paleontologists and Mineralogists Publication No. 39. Tulsa, Oklahoma. [8.8]

Day TJ, Wood LR. (1976) Similarity of the mean motion of fluid particles dispersing in a natural channel. *Water Resources Research* **12**: 655–66. [8.4]

Décamps H, Capblancq J, Hirigoyen JP. (1972) Etude des conditions découlement près du substrat en canal experimental. *Verhandlungen Internationale Vereinigung für Theoretische und Angewandt Limnologie* **18**: 718–25. [8.2]

Décamps H, Larrony G, Trivellato D. (1975) Approche hydrodynamique de la microdistribution d'invertebrates benthiques en eau courante. *Annales de Limnologie* **11**: 79–100. [8.2]

Dietrich WE. (1987) Mechanics of flow and sediment transport in river bends. In: Richards KS (ed.) *River Channels: Environment and Process*, pp 179–227. Blackwell Scientific Publications, Oxford. [8.8]

Dietrich WE, Smith JD. 1984. Bedload transport in a river meander. *Water Resources Research* **20**: 1355–80. [8.4, 8.8]

Dietrich WE, Smith JD, Dunne T. (1984) Boundary shear stress, sediment transport and bed morphology in a sand-bedded river meander during high and low flow. In: Elliot CM (ed.) *River Meandering*, pp 632–9. ASCE, New York. [8.4]

Dietrich WE, Kirchner JW, Ikeda H, Iseya F. (1989) Sediment supply and the development of the coarse surface layer in gravel bedded rivers. *Nature* 340, 215–17. [8.8]

Diplas P, Parker G. (1985) Pollution of gravel spawning grounds due to fine sediment. *St. Anthony Falls Hydraulic Laboratory Project Report 240*. University of Minnesota, Minneapolis. [8.8]

Dyer KR. (1970) Current velocity profiles in a tidal channel. *Geophysical Journal of the Royal Astronomical Society* **22**: 153–61. [8.3]

Dyer KR. (1971) Current velocities in a tidal channel. *Geophysical Journal of the Royal Astronomical Society* **22**: 153–61. [8.2]

Dyer KR. (1972) Bed shear stresses and the sedimentation of sandy gravels. *Marine Geology* **13**: M31–M36. [8.3]

Einstein HA. (1968) Deposition of suspended particles in a gravel bed. *Journal of the Hydraulics Division of American Society of Civil Engineers* **94**: 1197–205. [8.8]

Everest FH, Beschta RL, Scrivener JC, Koshi KV, Sedell JR, Cederholm CJ. (1987) Fine sediment and salmonid production: a paradox. In: Salo EO, Cundy TW (eds) *Streamside Management: Forestry and Fisheries Interactions*, pp 98–142. Symposium Proceedings, University of Washington, February, 1986. Institute of Forest Resources, University of Washington. [8.8]

Ferguson RI. (1981) Channel forms and channel changes. In: Lewin J (ed.) *British Rivers*, pp 90–125. Allen and Unwin, London. [8.3]

Ferguson RI, Prestegaard KL, Ashworth PJ (1989) Influence of sand on hydraulics and gravel transport in a braided gravel bed river. *Water Resources Research* **25**: 635–43. [8.8]

Frostick LE, Lucas PM, Reid I. (1984) The infiltration of fine matrices into coarse-grained alluvial sediments and its implications for stratigraphical interpretation. *Journal of the Geological Society of London* **141**: 955–65. [8.8]

Gibbs CJ, Neill CR. (1973) Laboratory testing of model VUV bedload sampler. Research Council of Alberta Report REH./72/2. Alberta, Canada..[8.8]

Golterman HL, Sly PG, Thomas RL. (1983) Study of the relationship between water quality and sediment transport. *Technical Paper in Hydrology No. 26*. Unesco, Paris. [8.7]

Gomez B. (1983) Temporal variations in bedload transport rates: the effect of progressive armouring. *Earth Surface Processes and Landforms* **8**: 41–54. [8.8]

Gomez B, Naff RL, Hubbell DW. (1989) Temporal variations in bedload transport rates associated with the migration of bedforms. *Earth Surface Processes and Landforms* **14**: 135–56. [8.8]

Gore JA. (1978) A technique for predicting in-stream flow requirements of benthic macroinvertebrates. *Freshwater Biology* **8**: 141–51. [8.1]

Graf WH. (1971) *Hydraulics of Sediment Transport*. McGraw-Hill, New York. [8.7]

Gustafson-Marjanen KI, Moring JR. (1984) Construction of artificial redds for evaluating survival of Atlantic salmon eggs and alevins. *North American Journal of Fish Management* **4**: 455–6. [8.8]

Hand BM. (1974) Supercritical flow in density currents. *Journal of Sedimentary Petrology* **44**: 637–48. [8.6]

Helley EJ, Smith W. (1971) Development and calibration of a pressure-difference bedload sampler. *US Geological Survey Open-File Report*. Washington, DC. [8.8]

Henderson FM. (1966) *Open Channel Flow*. Macmillan, New York. [8.7]

Heslop SE, Allen CM. (1989) Turbulence and dispersion in larger UK rivers. In: *Proceedings of the International Association of Hydraulic Research Congress*, Ottawa, Canada, pp D75–D82. [8.4]

Hey RD. (1984) Plan geometry of river meanders. In: Elliot CM (ed.) *River Meandering*, pp 30–43. ASCE, New York. [8.4]

Hey RD, Thorne CR. (1975) Secondary flows in river channels. *Area* **7**: 191–6. [8.4]

Hirsch PJ, Abrahams AD. (1981) The properties of bed sediments in pools and riffles. *Journal of Sedimentary Petrology* **51**: 757–76. [8.3]

Hooke JM. (1986) The significance of mid-channel bars in an active meandering river. *Sedimentology* **33**: 839–50. [8.3]

Hooke JM, Harvey AM. (1983) Meander changes in relation to bend morphology and secondary flows. In: Collinson JD, Lewin J (eds) *Modern and Ancient Fluvial Systems*, pp 121–32. Special Publication No. 6, International Association of Sedimentologists. Blackwell Scientific Publications, Oxford. [8.2]

Horowitz AJ, Rinella FA, Lamothe P *et al* (1990) Variations in suspended sediment and associated trace element concentrations in selected riverine cross-sections. *Environmental Science and Technology* **24**: 1313–20. [8.7]

Howard AD. (1987) Modelling fluvial systems: rock-, gravel- and sand-bed channels. In: Richards KS (ed.) *River channels: Environment and Process*, pp 69–94. Blackwell Scientific Publications, Oxford. [8.5]

Hubbell DW. (1987) Bedload sampling and analysis. In: Thorne CR, Bathurst JC, Hey RD (eds) *Sediment Transport in Gravel-bed Rivers*, pp 89–120. Wiley, Chichester. [8.8]

Iseya F, Ikeda H. (1987) Pulsations in bedload transport rates induced by a longitudinal sediment sorting: a flume study using sand and gravel mixtures. *Geografiska Annaler* **69A**: 15–27. [8.8]

James CS. (1990) Prediction of entrainment conditions for nonuniform, noncohesive sediments. *Journal of Hydraulic Research* **28**: 25–41. [8.5]

Jansen PP, van Bendegom L, van den Berg J, de Vries M, Zanen A. (1979) *Principles of River Engineering: The Non-tidal Alluvial River*. Pitman, London. [8.7]

Jarrett RD. (1984) Hydraulics of high gradient streams. *Journal of Hydraulic Engineering* **110**: 1517–18. [8.3]

Jarrett RD. (1990) Hydrological and hydraulic research in mountain streams. *Water Resources Research* **26**: 419–29. [8.2]

Jobson HE, Carey WP. (1989) Interaction of fine sediment with alluvial streambeds. *Water Resources Research* **25**: 135–40. [8.8]

Kline SJ, Reynolds WC, Schraub FA, Rundstadler PW. (1967) The structure of the turbulent boundary layer. *Journal of Fluid Mechanics* **30**: 741–73. [8.4]

Komar PD. (1987) Selective gravel entrainment and the empirical evaluation of flow competence. *Sedimentology* **34**: 1165–76. [8.5]

Komar PD. (1989) Flow-competence evaluations of the hydraulic parameters of floods: an assessment of the technique. In: Beven K, Carling P (eds) *Floods: Hydrological, Sedimentological and Geomorphological Implications*, pp 107–34. Wiley, Chichester. [8.5]

Komar PD, Carling PA. (1991) Grain sorting in gravel-bed streams and the choice of particle sizes for flow-competence evaluations. *Sedimentology* **38**: 489–502. [8.5]

Komar PD, Li Z. (1988) Applications of grain pivoting and sliding analysis to selective entrainment of gravel and to flow competence evaluations. *Sedimentology* **35**: 681–95. [8.5]

Kondolt GM, Cada GF, Sale MJ. (1987) Assessing flushing-flow requirements for brown trout spawning gravels in steep streams. *Water Resources Bulletin* **23**: 927–35. [8.8]

Kuhnle RA, Southard JB. (1988) Bedload transport fluctuations in a gravel bed laboratory channel. *Water Resources Research* **24**: 247–60. [8.6]

Lee H-J, Hsieh K-C. (1989) Flow characteristics in an alluvial river bend. In: *Proceedings of Technical Session B. Fluvial Hydraulics*, pp B25–B31. IAHR, Ottawa, Canada, 21–25 August 1989. [8.4]

Leeder MR. (1983) On the interactions between turbu-lent flow, sediment transport and bedform mechanics in channelized flows. In: Collinson JD, Lewin J (eds) *Modern and Ancient Fluvial Systems*, pp 121–32. Special Publication No. 6, International Association of Sedimentologists. Blackwell Scientific Publications, Oxford. [8.4]

Leopold LD. (1982) Water surface topography in river channels and the implications for meander develop-ment: In: Hey RD, Bathurst JC, Thorne CR (eds) *Gravel-bed Rivers*, pp 359–88. John Wiley & Sons, Chichester. [8.4]

Limerinos JT. (1970) Determination of the Manning coefficient from measured bed roughness in natural channels. *US Geological Survey Water Supply Paper 1898-B*. Washington, DC. [8.3]

Lisle TE. (1989) Sediment transport and resulting depo-sition in spawning gravels, North Coastal California. *Water Resources Research* **25**: 1303–19. [8.5, 8.8]

McQuivey RS. (1973) Summary of turbulence data from rivers, conveyance channels and laboratory flumes. *US Geological Survey Professional Paper 802B*. Washington, DC. [8.4]

Mangelsdorf J, Scheurmann K, Weiss F-H. (1990) *River Morphology*. Springer-Verlag, Berlin. [8.7]

Mantz PA. (1978) Bedforms produced by fine, cohesion-less, granular and flaky sediments under subcritical water flows. *Sedimentology* **25**: 83–103. [8.5]

Mantz PA. (1980) Laboratory flume experiments on the transport of cohesionless silica silts by water streams. *Proceedings of the Institution of Civil Engineers* **69**: 977–94. [8.5]

Martvall S, Nilsson G. (1972) *Experimental Studies of Meandering*. University of Uppsala, UNGI Report No. 20. [8.4]

Matlack KS, Houseknecht DW, Applin KP. (1989) Emplacement of clay into sand by infiltration. *Journal of Sedimentary Petrology* **59**: 77–87. [8.8]

Meade R. (1985) Wave-like movement of bedload sedi-ment, East Fork River, Wyoming. *Environmental Geology and Water Science* **7**: 215–25. [8.8]

Meyer-Peter E, Mueller R. (1948) Formulas for bed-load transport. In: *Proceedings of the International Association of Hydraulic Research*, Third Annual Conference, Stockholm, pp 39–64. IAHR, Delft, Ottawa, Canada. [8.8]

Middleton GV. (1976) Hydraulic interpretation of sand size distributions. *Journal of Geology* **84**: 405–26. [8.7]

Miller MC, McCave IN, Komar PD. (1977) Threshold of sediment motion under unidirectional currents. *Sedimentology* **24**: 507–27. [8.5]

Naden PS, Brayshaw AC. (1987) Small and medium-scale bedforms in gravel-bed river. In: Richards KS (ed.) *River Channels: Environment and Process*, pp 249–71. Blackwell Scientific Publications, Oxford. [8.8]

Nelson JM, Smith JD. (1989) Mechanics of flow over

ripples and dunes. *Journal of Geophysical Research* **94**: 8146–62. [8.8]

Newbury RW. (1984) Hydrologic determinants of aquatic insect habitats. In: Resh VH, Rosenberg DM (eds) *The Ecology of Aquatic Insects*, pp 323–57. Praeger, New York. [8.1]

Novak P. (1957) Bedload meters—development of a new type and determination of their efficiency with the aid of scale models. In: *Transactions of the International Association of Hydraulic Research*, Vol. 1, pp A9-1–A-11. Seventh General Meeting, Lisbon. [8.8]

Okoye KG. (1989) Mean flow structure in model alluvial channel bends. In: *Proceedings of Technical Session B. Fluvial Hydraulics*, pp B-81–B-90. IAHR, Ottawa, Canada, 21–25 August 1989. Delft, Ottama, Canada. [8.4]

Paris E. (1989) Velocity distribution over macroscale roughness: preliminary results. In: *Fourth International Symposium on River Sedimentation*, pp 625–32. Water Resources and Electric Power Press, Beijing. [8.2]

Parker G, Andrews ED. (1985) Sorting of bedload sediment by flow in meander bends. *Water Resources Research* **21**: 1361–73. [8.4]

Parker G, Klingeman PC, McLean DG. (1982) Bedload and size distribution in paved gravel-bed streams. *Journal of the Hydraulics Division, American Society of Civil Engineers* **108**: 544–71. [8.5, 8.8]

Peters JJ, Goldberg A. (1989) Flow data in large alluvial channels. In: Malesimović C, Radojković M (eds) *Computational Modelling and Experimental Methods in Hydraulics*, pp 75–85. Elsevier, London. [8.1]

Petts GE, Thoms MC, Brittan K, Atkin B. (1989) A freeze-coring technique applied to pollution by fine sediment in gravel-bed rivers. *The Science of the Total Environment* **84**: 259–72. [8.8]

Prestegaard KL. (1983) Bar resistance in gravel bed streams at bankfull stage. *Water Resources Research* **19**: 472–6. [8.3]

Reid I, Frostick LE. (1984) Particle interaction and its effect on the thresholds of initial and final bedload motion in coarse alluvial channels. In: Koster EH, Steel RJ (eds) *Sedimentology of Gravels and Conglomerates*, pp 61–8. Canadian Society of Petroleum Geologists Memoir 10. [8.5]

Reid I, Layman JT, Frostick LE. (1980) The continuous measurement of bedload discharge. *Journal of Hydraulics Research* **18**: 243–9. [8.8]

Reid I, Brayshaw AC, Frostick LE. (1984) An electromagnetic device for automatic detection of bedload motion and its field applications. *Sedimentology* **31**: 269–76. [8.8]

Reid I, Frostick LE, Layman JT. (1985) The incidence and nature of bedload transport during flood flows in coarse-grained alluvial channels. *Earth Surface Processes and Landforms* **10**: 33–44. [8.8]

Reiser DW, Ramey MP, Lambert TR. (1985) Review of flushing flow requirements in regulated streams. In: *Report to Pacific Gas and Electric Company No. Z19-5-120-84*. San Roman, California. [8.8]

Reynolds CS, Carling PA, Beven KJ. (1991) Flow in river channels: new insights into hydraulic retention. *Archiv fuer Hydrobiologie* **121**: 171–9. [8.4]

Richards KS. (1982) *Rivers: Form and Process in Alluvial Channels*. Methuen, London. [8.3, 8.4, 8.6, 8.8]

Richardson JF, Zaki WN. (1954) Sedimentation and fluidization—Part 1. *Transactions of the Institute of Chemical Engineers* **32**: 35–53. [8.7]

Roberts RG, Church M. (1986) The sediment budget of severely disturbed watersheds, Queen Charlotte Ranges, British Columbia. *Canadian Journal of Forest Research* **16**: 1092–6. [8.8]

Roy AG, Bergeron N. (1990) Flow and particle patterns at a natural river confluence with coarse bed material. *Geomorphology* **3**: 99–112. [8.4]

Saunderson HC, Lockett FP. (1983) Flume experiments on bedforms and structures at the dune–plane bed transition. In: Collinson JD, Lewin J (eds) *Modern and Ancient Fluvial Systems*, pp 49–58. Special Publication No. 6, International Association of Sedimentologists. Blackwell Scientific Publications, Oxford. [8.6]

Scrivener JC, Brownlee MJ. (1989) Effects of forest harvesting on spawning gravel and incubation survival of chum (*Oncorhynchus beta*) and coho salmon (*D. Kisutch*) in Carnation Creek, British Columbia. *Canadian Journal of Fisheries and Aquatic Sciences* **46**: 681–96. [8.8]

Shen HW, Fehlman HM, Mendoza C. (1990) Bedform resistance in open channel flows. *Journal of Hydraulic Engineering* **116**: 799–815. [8.3]

Shih S-M, Komar PD. (1990) Differential bedload transport rates in a gravel-bed stream: a grain-size distribution approach. *Earth Surface Processes and Landforms* **15**: 539–52. [8.8]

Smith ND. (1978) Some comments on terminology for bars in shallow rivers. In: Miall AD (ed.) *Fluvial Sedimentology*, pp 85–8. Canadian Society of Petroleum Geologists Memoir 5. [8.6]

Soulsby (RL). (1983) The bottom boundary layer of shelf seas. In: Johns B (ed.) *Physical Oceanography of Coastal and Shelf Seas*, pp 189–266. Elsevier, Amsterdam. [8.2, 8.3]

Soulsby RL, Atkins R, Salkield AP. (1987) Observations of the turbulent structure of a suspension of sand in a tidal current. In: *Euromech 215, Mechanics of Sediment Transport in Fluvial and Marine Environments*, pp 88–91. Balkema, Rotterdam. [8.4]

Southard JB, Boguchwal LA. (1973) Flume experiments on the transition from ripples to lower flat bed with increasing size. *Journal of Sedimentary Petrology* **43**:

1114–21. [8.4]

Statzner B, Holm TF. (1982) Morphological adaptation of benthic invertebrates to stream flow—an old question studied by means of a new technique (Laser A Doppler Anemometry). *Oecologia* **53**: 290–2. [8.2]

Stevens HH, Yang CT. (1989) Summary and use of selected fluvial sediment-discharge formulae. *US Geological Survey Water Resources Investigations Report 89–4026.* Washington, DC. [8.8]

Thibodeaux LJ, Boyle JD. (1987) Bedform-generated convective transport in bottom sediment. *Nature* **325**: 341–3. [8.8]

Thompson A. (1986) Secondary flows and the pool—riffle unit: a case study of the processes of meander development. *Earth Surface Processes and Landforms* **11**: 631–41. [8.3]

Thorne CR, Hey RD. (1979) Direct measurements of secondary currents at a river inflexion point. *Nature* **280**: 226–8. [8.4]

Thorne CR, Rais S. (1984) Secondary current measurements in a meandering river. In: Elliot CM (ed.) *River Meandering*, pp 675–86. ASCE, New York. [8.4]

Thorne PD, Williams JJ, Heathershaw AD. (1989) *In situ* acoustic measurements of marine gravel threshold and transport. *Sedimentology* **36**: 61–74. [8.4]

Toffaleti FB. (1968) *A procedure for computation of the total river sand discharge and detailed distribution bed to surface.* Committee on Channel Stabilization, US Army Corp of Engineers Waterways Experiment Station Technical Report 5. Vicksburg. [8.8]

Vanoni VA. (1977) *Sedimentation Engineering.* ASCE, New York. [8.7]

Walker JF. (1988) General two-point method for determining velocity in open channel. *Journal of Hydraulic Engineering* **114**: 801–5. [8.2]

Wells SG, Dohrenwend JC. (1985) Relief sheetflood bedforms on Late Quaternary alluvial-pan surfaces in the southwestern United States. *Geology* **13**: 512–13. [8.6]

White CM. (1940) The equilibrium of grains on the bed of a stream. *Proceedings of the Royal Society of London* **A174**: 332–8. [8.5]

Whiting PJ, Dietrich WE, Leopold LB, Drake TG, Shreve RL. (1988) Bedload sheets in heterogeneous sediments. *Geology* **16**: 105–8. [8.6]

Whittaker JG. (1987) Sediment transport in step-pool streams. In: Thorne CR, Bathurst JC, Hey RD (eds) *Sediment Transport in Gravel-bed Rivers*, pp 545–79. John Wiley & Sons, Chichester. [8.3]

Williams JJ, Thorne PD, Heathershaw AD. (1989) Measurements of turbulence in the benthic boundary layer over a gravel bed. *Sedimentology* **36**: 959–71. [8.4]

Wolman MG. (1954) A method of sampling coarse river-bed material. *Transactions of the American Geophysical Union* **35**: 951–6. [8.5]

Wooding RA, Bradley EF, Marshall JK. (1973) Drag due to arrays of roughness elements of varying geometry. *Boundary-Layer Meteorology* **5**: 285–308. [8.2]

Yagishita K, Taira A. (1989) Grain fabric of a laboratory antidune. *Sedimentology* **36**: 1001–5. [8.6]

Yang CT. (1973) Incipient motion and sediment transport. *Journal of the Hydraulics Division of American Society of Civil Engineers* **99**: 1679–704. [8.8]

Young PC, Wallis SG. (1987) The aggregated dead zone model for dispersion. In: *BHRA: Proceedings of a conference on water-quality modelling in the inland natural environment*, pp 421–33. BHRA, Cranfield. [8.4]

9: Channel Morphology and Typology

M. CHURCH

9.1 INTRODUCTION

Several factors govern the physical processes in rivers and hence their morphology. The primary ones are: the volume and time distribution of water supplied from upstream; the volume, timing and character of sediment delivered to the channel; the nature of the materials through which the river flows; the local geological history of the riverine landscape. Secondary factors that can be important determinants of channel morphology include local climate (particularly the occurrence of a freezing winter or an extended dry season), the nature of riparian vegetation, and land use in the drainage basin. Direct modification of the channel by humans is a further important factor in many streams in settled regions.

The size of the channel is determined by the water flow through it, particularly by flood peak flows that effect erosion and channel-shaping sediment transport. Many investigators have advocated floods that recur once in about 1.5–2.5 years (i.e. approximately the mean annual flood) as the bankfull, 'channel-forming flow', but there appears to be no universally consistent correlation between flow frequency and bankfull, nor between flood frequency and effectiveness in creating morphological change. In fact, large rivers flowing in relatively fine, easily mobilized sediments may be shaped predominantly by frequently recurring flows, whereas headwater boulder or cobble-gravel channels may be subjected to major disturbance only during much more extreme events. Certainly, regionally consistent combinations of these circumstances may occur.

The duration of inundation — and possibly the season when inundation occurs — may be signifi-
cant in determining the character of channel edge and riparian habitats. Figure 9.1 presents a simple classification of channel sections that reflects qualitatively flow depth and duration. Correlations have been demonstrated between riparian plant communities and elevation above some reference water level; hence between plant species and duration of inundation (e.g. Teversham & Slaymaker 1976). A significant channel boundary is the 'lower limit of continuous terrestrial vegetation' — the limit of the 'active channel' in Fig. 9.1 — which is more or less well defined on most stream banks. In some jurisdictions, this forms the limit of the river channel for legal purposes.

River morphology reflects the concentration and calibre of sediment moving down the channel. When sediment delivered to the channel is predominantly fine-grained it is largely carried in suspension, and much is deposited in slack water overbank during floods. This builds relatively high, cohesive banks. The result is a relatively narrow, single-thread channel that habitually meanders. Coarse sediment is transported on or near the bed. Consequently, it is deposited in bars which fill the channel and deflect the river in a less regular pattern of lateral activity. According to the supply of sediment, such channels may meander irregularly, wander, or become braided. They are characteristically wide and shallow, with non-cohesive lower banks formed in the coarse material. The division between 'fine' and 'coarse' sediment depends in some measure on the energy of the stream, but for purposes of this discussion it can be placed within the range 0.3–1.0 mm, that is, in the medium to coarse sand range.

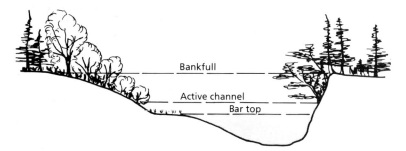

Fig. 9.1 Sketch of channel sections according to morphology (after Osterkamp & Hedman 1982). Within the 'active channel', vegetation is restricted to herbs and to species able to survive extended periods of inundation.

Channels that flow through sediments which they have previously deposited are termed *alluvial channels*. They certainly are competent to modify their form, since they previously moved the sediment that makes up their bed and banks. Here, we find consistent relations between flow and the width, depth and velocity. Such relations are termed *hydraulic geometry*. The actual form of the channel is, of course, influenced by the strength of the material that makes up the bed and banks, so distinctive hydraulic geometries occur for different materials. In fact, the strength of granular sediments has only a moderate range. In comparison, the range of flows may vary over many orders of magnitude as one moves down a river system, and over several orders of magnitude at one place through the seasons. The depth of a stable channel is limited by material strength — hence its ability to withstand the shearing force of the flow — so as rivers become larger they chiefly become wider. The scale of alluvial river channels can be summarized conveniently in the relation between channel width and characteristic flow (Fig. 9.2).

Channel size increases systematically through a river system as the increasing drainage area contributes larger flows to the trunk channel. The morphological scale of the channel changes accordingly (Fig. 9.3). Rivers flowing through non-alluvial material (bedrock or non-alluvial sediments) may depart from the scale relations implied in Figs 9.2 and 9.3.

Because of the significant correlation between channel scale and position in a drainage system, classification of river channels on the basis of their position in the drainage system is of some interest. For this purpose rivers are defined as sequences of *links*, an individual link comprising the channel between two successive tributary confluences (*internal link*), or between the channel source and the first tributary confluence (*external link*). The most commonly cited link classification system defines external links as 'order 1' streams, and defines a link of next higher order whenever two links of *equal* order join. Because of the irregular structure of drainage networks, order is not a reliable indicator of channel size. A more consistent index is link *magnitude* (Shreve 1967), each link being defined as having magnitude equal to the sum of all external links draining to it. The magnitude of a drainage basin is a useful surrogate of its discharge potential. (However, drainage networks are usually defined from maps, which often do not depict the smallest headward tributaries, so that consistent definition often remains a problem; cf. Dunne & Leopold 1978). In small drainage basins, there is also a regionally reliable correlation between drainage area and discharge characteristics.

Geological history and physiographic setting impose constraints on river morphology and behaviour. Most rivers flow in valleys that exert some lateral and vertical control over the channel. Rivers may be confined or entrenched in narrow valleys. Conversely, on broad plains — especially river-constructed deltas or alluvial fans — there are few constraints on lateral movement of the channel. The gradient is theoretically self-adjusted in a purely alluvial channel, but very few rivers conform with this ideal. In most cases, rivers flow in topographic defiles determined by a long geological history, and valley gradient is more or less imposed. River adjustment in the medium term is restricted to the possibility to reduce gradient locally by adoption of a meandered habit.

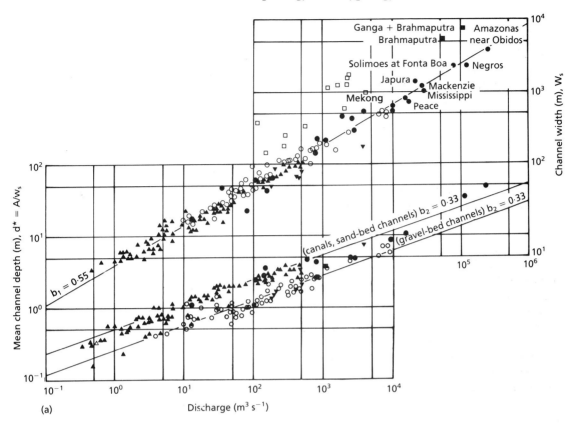

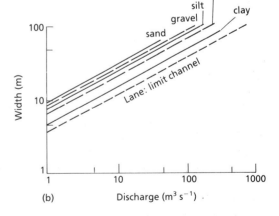

Fig. 9.2 Scale relations for channel width and depth versus flow. (a) Channels that are dynamic models of each other scale as $Q^{0.4}$ in both width and depth, indicating that similarity is preserved. However, alluvial channels are distorted toward relatively greater width as flow increases; that is, the ratio of width to depth increases as channels become larger. ● sand-bed rivers (Leopold & Wolman 1957; Neill 1973); ○ gravel-bed rivers (Bray 1973); ▼ lake outlet channels (gravel) (Kellerhals 1967); ■ □ braided channels (sand, gravel); ▲ flumes (sand). (b) Variation in channel width due to material properties. The 'limit channel' is the narrowest mechanically stable channel for a given flow.

This constraint often affects even major 'alluvial' rivers in those parts of the world that were subject to Pleistocene glaciation, and is generally true of headwaters.

Human activity can affect river channels through direct structural interference, as when artificial channels are constructed or weirs placed to control water levels, or it may result from effects on runoff, sediment calibre and sediment yield produced by the land use or change thereof

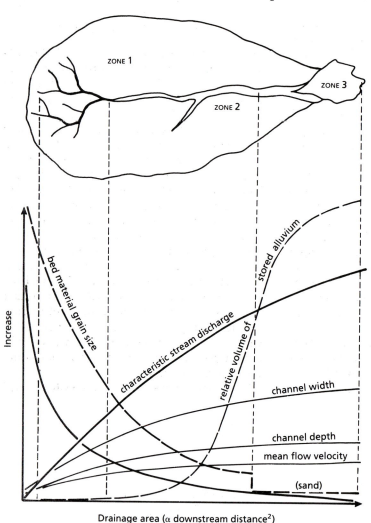

Fig. 9.3 Schematic representation of the variation in channel properties through a drainage basin (based on a concept of Schumm 1977).

in the contributing drainage basin (Kellerhals & Church 1989). Human activities manifest a characteristic intensity on the landscape; it is a reasonable generalization that the impacts of land use are most severely visited upon smaller, headwater channels. None the less, the cumulative impact of widespread human activity can be observed in even the greatest rivers (*cf.* Fremling & Claflin 1984, on the Upper Mississippi River), and individual engineering projects directly control some of the world's largest channels.

The controls described above change along a river. Some changes may be subtle, as sediment load is modified in a long, slowly aggrading channel. Others are abrupt, as at major tributary junctions. Before a river can be classified successfully it must be divided into homogeneous units, so-called *reaches*, within which the controlling factors do not change appreciably. This normally can be done *a priori* on air photographs and maps, at least for larger channels. However, certain reach limits, associated perhaps with changes in geological history or with the history of human activity, may become obvious only after substantial study.

The granular materials that make up the river

bed and banks customarily vary downstream according to the ease with which they can be transported by the river. Accordingly, in headwaters we find cobble and boulder gravels (Fig. 9.3), the larger fractions of which cannot be moved very far, whereas the far downstream reaches of large rivers are formed in sand and silt. In headwaters, then, channel scale converges toward the size of individual boundary elements, whereas downstream the channel becomes very much larger than the characteristic sediment grains. This is the major determinant of the behaviour and morphology of river channels, and provides a framework for classification. In such a scheme, the most fundamental division is between 'small channels', ones in which channel scale is comparable with the scale of individual sediment grains, which therefore are individually significant elements of the channel boundary, and larger channels in which the boundary is made up of aggregate structures of grains.

Channels in bedrock do not easily lend themselves to definition in this scheme. Generally, small channels would be ones defined by individual bedding planes or joint partings.

9.2 SMALL CHANNELS

In coarse material, the scale of small channels is of the order of one to ten particles only. Small channels flowing through cobble (>64 mm) to boulder (>256 mm) sized material are significant in a management context. Channel depth characteristically is of the order of 0.1−1 large grain, so the relative roughness (the ratio, D/d, between grain diameter and flow depth) is usually greater than 1.0. Individual clasts constitute significant form elements of the channel (Fig. 9.4). (Note that definition on the basis of relative size of channel and bounding sediment removes any absolute scale; a sand channel on a beach over which one could step with impunity would not, in this view, be a 'small channel'. In fine materials, small channels are restricted to hillside rills, which are not considered here.) The smallest headwater channels in forested environments are often difficult to identify because the streams flow under vegetation mats and roots. Here and in moorlands, large 'pipes' often feed the channel from concentrated subsurface flow.

The channel forms a sequence of pools dammed

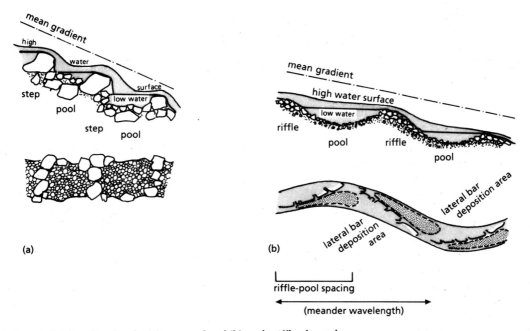

(a)

(b)

Fig. 9.4 Definition sketches for (a) step-pool and (b) pool−riffle channels.

behind individual rocks, or a line of rocks, and falls over the same features. In channels with surface width $w_s \geq 3D$, lines or cells of boulders form, characteristically held by a keystone (Fig. 9.5(a)). In forests, major pieces of wood may be incorporated into the steps. Banks are composed of bedrock, boulders, turf, or heavy root development. Such channels are restricted to headwater positions in drainage systems, so they are characteristically steep. Gradients fall in the range 2° to >20° (4% to >35%). Consequently, steps are frequent and the pools are short; pool length often is of the order of 3–4 w_s only. Gradients are commonly bedrock or debris controlled, and

streambed material may be only one to a few grains deep. Not infrequently, bedrock-controlled cascades or falls occur.

Such channels do not often re-form. Exceptional floods, characteristically with recurrence intervals measured in decades, may break the steps. Experimental observations (Whittaker & Jaeggi 1982) have demonstrated that step-pool features form in extreme flows ($D_{max}/d \approx 1.0$). Initial bed deformation in the flume tends to produce wave-like (antidune) features, but the positions of the largest stones, which move only sporadically, anchor the pattern and promote the development by local scour and deposition of step-pool features

Fig. 9.5 Morphological features in small and intermediate channels. (a) Details of a step; the keystone is under the water fountain. (b) Riffle–pool morphology; the pool upstream of the riffle is filled with cobbles and would be classified as a 'glide' for habitat purposes. (c) Step-pool channel forming a boulder cascade at low flow. (d) Channel-spanning log jam (vertical aerial view); the jam is highly permeable and would be classified as a mesojam in the typology of Fig. 9.7 despite the relatively large volume of debris present.

with characteristic roughness spacing (height to length ratio) of the order of 0.1–0.15. These are much steeper than ordinary anti-dunes and represent a configuration that presents maximum flow resistance. It is obvious from field observation that the locking together of the largest boulders is an important mechanism to stabilize them and control the step-pool formation.

If the height of the steps is determined by the diameter of the largest stones, then roughness spacing of 0.1–0.15 implies that a gradient of 6–9° can be contributed by the dominant grains alone. Steeper gradients can be achieved by aggregate structures. A large portion of the total drop in step-pool channels is controlled by the placement of the largest stones.

In channels with gradients >10–12°, extreme flows may mobilize the entire mass of sediment in the channel and create a debris flow (review in Costa 1984). Afterwards normal flows re-establish the step-pool sequence by moving the larger stones into stable, interlocked positions. Boulders are moved mainly by being undermined so that they roll forward under the influence of gravity.

The frequency of channel re-formation is determined not only by flood history but also by sediment supply, channels with abundant debris influx being subject to frequent reorganization. Many headwater channels are confined in steep gulleys or narrow valleys (so they are directly coupled with hillslope sources of debris). Vertical aggradation and degradation form the major style of instability here. Channels on alluvial cones characteristically become choked with debris and then avulse (Kellerhals & Church 1990); the entire channel shifts abruptly to a new, usually steeper line of descent.

Many small channels, as defined here, have no direct fishery value because they are too steep to be colonized (and often lie beyond impassable barriers). In the lower end of the range, pool-dwelling salmonids occur. Such channels may, however, be important for production of invertebrates and for recruitment of organic material that may drop into the water, both of which may subsequently be transported downstream. Conditions in these headwater channels also influence water quality downstream. Finally, the frequently steep, moist banks of such streams often provide sites for survival of otherwise endangered plants.

9.3 INTERMEDIATE CHANNELS

Intermediate channels have $w_s \gg D$, but they still may be influenced by blockage across a substantial portion or all of their cross-section by unusual sediment accumulations or, more usually, by fallen trees or branches. This criterion places an absolute but approximate upper limit on the width, which in most forest regions is in the range of 20–30 m. In open country, the limit may be observed in much smaller channels, depending on the character of the sediments and sediment transport. Relative roughness usually falls in the range $1.0 > D/d > 0.1$, so that flows are wake dominated or transitional toward deep shear flows. Individual large boulders may still represent significant elements on the boundary of intermediate channels, but they are subordinate to structures formed by many grains, including accumulations into bars (Church & Jones 1982), to elements formed from large parts or root boles of trees (Keller & Swanson 1979), and accumulations of large organic material termed 'jams'. In the absence of significant log jams, the pool–riffle–bar unit (see Fig. 9.4) forms the major physical element in such channels, and it has a characteristic but not invariant spacing of about five to seven channel widths. Significant variation of this basic unit has important implications for habitat quality.

Riffles take several forms (the following terminology is from Grant *et al* 1990). Riffles *sensu stricto* are zones of relatively shallow, rapid flow in comparison with pools. $D/d < 1.0$, in general, and flow remains subcritical or near critical. Local channel gradient characteristically remains less than 1° (2%). Riffles are the dominant rapid in larger intermediate channels with average gradient less than about 0.5° (1%) and occur where the channel is dominated by a sequence of alternating bars with intervening crossovers on the riffles (Fig. 9.5(b)).

Rapids exhibit irregular boulder lines and cells, have $D/d \approx 1.0$, and flow near critical. Some boulders are emergent at low flows. Gradient is typically in the range of 1–2° (2–4%) and flow occurs as a series of poorly defined filaments or discrete *chutes* through the bed structures.

Cascades are steep reaches in which flow occurs over a sequence of emergent boulders, often organized into steps (Fig. 9.5(c)). Flow occurs as supercritical jets between the boulders or falls over them. Gradient is greater than 2°. Cascades may be defined by boulders or by bedrock. When cascades dominate the channel, it is usual for the reach to be classified as a 'small channel' (but see below).

Glides (also 'runs') are extended riffles, often replacing a pool that has been filled by sediment (Fig. 9.5(b)). Pool—riffle or pool—rapid transition zones have a similar character. Glides are not usually distinguished in strictly morphological classifications, but they represent significant habitat units (Bisson *et al* 1981).

Pools can be classified in two major groups (Bisson *et al* 1981; Sullivan 1986). *Backwater pools* occur where major obstructions occur in the channel. These include boulders, bars and large organic debris. Flow diverges into these pools, with declining velocity and energy gradient. Consequently, sediment tends to be deposited in these locations and so they are commonly relatively shallow, with finer than average substrate. Within this class are *separation zones* on the lee side of boulders, logs, or emergent debris accumulations (backwater pools of Bisson; eddy pools of Sullivan). Separation zones also occur at channel confluences and at abrupt bankline angles. *Dammed pools* occur upstream of boulder lines in rapids and cascades and, on a larger scale, upstream of channel-spanning debris jams or gravel accumulations. *Scour pools* (drawdown pools of Sullivan) occur where flow converges over or around a constriction. These are the deepest places in the channel, but velocities are relatively high during high flows even though they may become backwater units at low flow. A number of variants have been described, including *plunge pools* on the downstream side of steps, *lateral scour pools* where flow is deflected by obstructions such as root boles, logs and channel bars, *vertical scour pools* under hanging logs or roots, and *trench pools* where flow is confined for some distance against erosion resistant substrate, such as bedrock.

Lateral scour pools adjacent to sediment ac-

cumulations are the dominant feature in pool—riffle sequences. Most of the other features defined above are distinctive mainly by virtue of local arrangements of boulders, logs or sediment accumulations. However, they constitute significant individual habitat units. For example, certain of the scour pool types provide hiding places under overhanging banks or hanging logs. Figure 9.6 summarizes riffle and pool features in a hierarchical diagram which emphasizes that the major channel units are composed of groupings of the smaller features.

Finally, it is worth emphasizing that the hydraulic character of individual units, and their morphological appearance, changes dramatically as flow varies. Although they invariably are observed and classified at low flow, they are of course formed at flows competent to move sediment and debris.

The third element of the pool—riffle—bar unit is the sediment accumulation zone termed a bar. These occur in areas of flow divergence where the sediment transporting competence of the stream declines, typically upon riffles, or downstream from scour pools (Church & Jones 1982). Bars strictly defined as accumulations of sediment grains require $D/\langle d \rangle < 0.5$, where $\langle d \rangle$ indicates mean depth, in order that the grains may pile up. Typically $D/\langle d \rangle < 0.1$. Since material must be moved by traction on to the bar, we require, following Shield's criterion for competence (see Chapter 5) $\tau_c \rho_s' g D_{90} \approx 0.03$, where $\tau_c = \rho g d S$ is the critical tractive force to move sediment of size D, ρ_s' is submerged sediment density, ρ is the density of water, g is the acceleration of gravity, S is the energy gradient, and D is a relatively large grain size (here D_{90}). If $\rho_s'/\rho = 1.65$, we have $S \leq 0.05 D_{90}/d$. For $D_{90}/d \approx 1.0$, $S \leq 3°$ (5%). In fact, bars become well developed at gradients of less than 2° in association with rapids and riffles. Riffles often constitute the front faces of bars. Bars are often anchored at fixed locations by the patterns of flow divergence and convergence determined by the overall channel form (Lisle 1986).

Like other channel elements, bars exhibit a variety of morphologies (see following section on large channels). In intermediate channels, a variety of relatively ill-defined bar features occurs in association with debris accumulations.

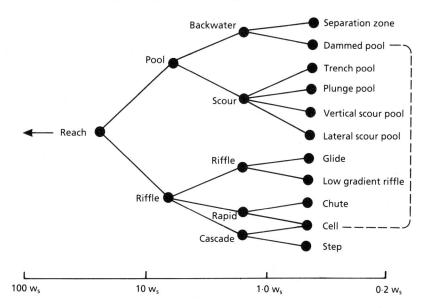

Fig. 9.6 A hierarchical representation of channel morphological units (partly after Sullivan 1986).

Channel morphology is modified along intermediate channels by normal fluvial sediment transport, most of the sediment being moved from bar to bar. However, in reaches containing large amounts of organic debris, steps may be common over individual, transversely oriented logs, or through debris accumulations up to several metres in height (Fig. 9.6(d)). Log steps and jams in forest streams typically account for between 5% and 15% of the total drop (Marston 1982; Hogan 1986) (some studies have reported much higher proportions — to more than 50% — but there remain some questions about equivalence of measurements). The formation and decay of log jams regulates the transfer of bed material downstream. The upstream side of a jam typically consists of a sediment-filled backwater reach. Downstream may be an extended, sediment-starved riffle or rapid. D.L. Hogan (personal communication) has defined a characteristic process/morphology history for log jams (Fig. 9.7). Since jams develop at random along the channel, they often obscure or serve to relocate the normal sequence of bars. Jams also encourage the occurrence of channel avulsions where the channel is not confined, so promoting relocation of the stream channel and the development of blind side-channels and backwaters.

Channels that flow along the base of cliffs or steep hillslopes often become filled with large boulders which the stream is not capable of moving at any stage. Such a channel may technically be a 'small channel', but where the overall gradient is less than about 2° and flow remains continuous between the boulders, the channel is more appropriately classified as intermediate. Similarly, deep, narrow channels are common in alluvial floodplains and deltas. Whilst their scale may qualify them as intermediate channels, their overall morphology and hydraulic conditions identify them as scaled-down large channels, with which they are more appropriately grouped. The introduction of 'regime groups' of channels in the next section of this chapter will provide a hydraulic basis for these assignments.

Intermediate channels constitute the optimum spawning and rearing habitat for a range of migratory and resident fishes, particularly salmonids. Riffles with clean gravels provide spawning sites, whilst pools and side-channels contribute rearing habitat. Microhabitat units within the channel are particularly important in rearing inasmuch as they provide hiding and resting places. Intermediate channels in forest environments may still be dominated by overhanging riparian vegetation which is important as shelter, as a source of drop-in food, and as an agent of streambank stability.

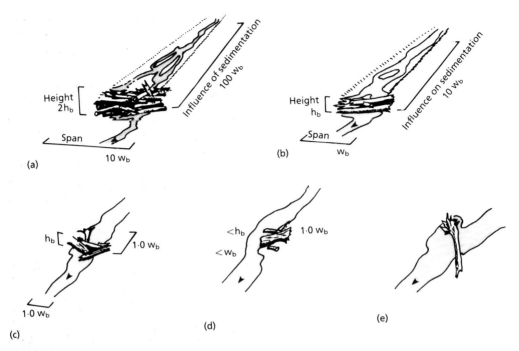

Fig. 9.7 Schematic diagram of log jam morphologies and their effect on channel morphology (courtesy of D.L. Hogan). (a) Megajam; (b) macrojam; (c) mesojam; (d) microjam; (e) individual log pieces. h_b, W_b, are bankfull depth, width of the channel (for scale).

9.4 LARGE CHANNELS

Large rivers are ones in which purely fluvial processes and geological constraints determine the morphology. Riparian effects do not dominate the channel, although laterally unstable channels in forests may recruit large volumes of wood debris which accumulate locally to influence bar and channel developments. (The transition to large channels occurs in many environments somewhere near $w \approx 20-30$ m and bankfull discharge, $Q_b \approx 20-50\,\mathrm{m^3\,s^{-1}}$, but much smaller channels can exhibit the characteristics of 'large' channels.) $D/d < 0.1$ in most cases, and flow is a deep shear flow with a well-defined velocity profile. Bed structure is dominated by pool−riffle sequences amongst major bar-forms, or by meandered bend−pool and crossover sequences. Large rivers are familiar from textbook descriptions. However, most textbooks present a very poor, one-dimensional classification of them.

Because the calibre and volume of sediment supply have an important influence on channel morphology (section 9.1), channel pattern and the style of within-channel sediment storage are two useful bases for morphological classification (Fig. 9.8). Since the style of lateral instability may also be important in determining the character of riverine habitat, this provides a third useful dimension for classification. A classification that uses these dimensions has been developed by Kellerhals *et al* (1976; see also Kellerhals & Church 1989). It is presented in outline in Figs 9.9−9.11.

The hydraulic relations underlying the systematic variation of river channel morphology have been only partly quantified (see Ackers 1988). The most important control of channel morphology clearly is the water flow regime. Scale relations for river channels (as illustrated in Fig. 9.2) are empirically summarized in the equations of hydraulic geometry:

$$P = a_p Q_n^{b_1}$$

$$R = a_r Q_n^{b_2}$$

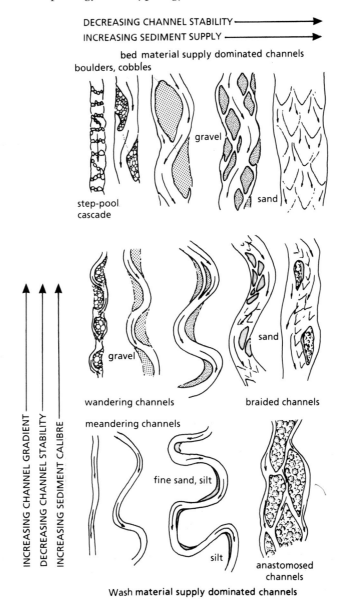

DECREASING CHANNEL STABILITY ⟶

INCREASING SEDIMENT SUPPLY ⟶

bed material supply dominated channels

boulders, cobbles

gravel

step-pool cascade

sand

gravel

sand

wandering channels

braided channels

meandering channels

fine sand, silt

silt

anastomosed channels

Wash material supply dominated channels

INCREASING CHANNEL GRADIENT ⟶

DECREASING CHANNEL STABILITY ⟶

INCREASING SEDIMENT CALIBRE ⟶

Fig. 9.8 Conceptual pattern of morphological types of large channels (based on concepts of Mollard 1973 and Schumm 1985). (a) Bed material supply-dominated channels; (b) wash material supply-dominated channels.

in which P is the wetted perimeter of the channel, $R = A/P$ is the hydraulic radius, A is channel cross-section, and Q_n is the flow for some fixed return period (such as mean annual flow, mean annual flood, etc.). Because of measurement constraints, P is often replaced by w_s, the surface width of the river, and R then becomes $d. = A/w_s$, the 'hydraulic mean depth'.

The coefficients a_p and a_r (or a_w and a_d) are known to depend upon the sedimentary materials that make up the bed and lower banks of the channel, since they control the resistance of the bed and banks to erosion, hence determine the channel shape (*cf.* Fig. 9.2, inset). Therefore, channels flowing in the same materials conform to the same hydraulic geometry and form a *regime*

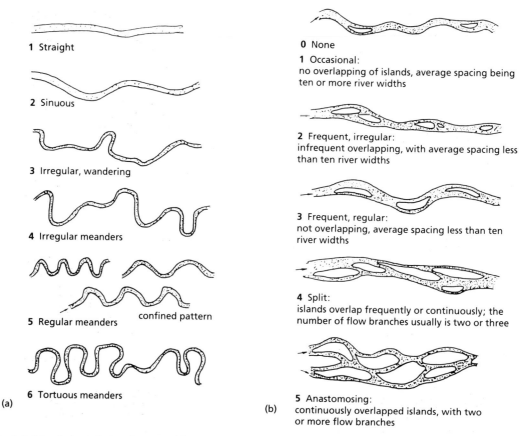

1 Straight

2 Sinuous

3 Irregular, wandering

4 Irregular meanders

5 Regular meanders   confined pattern

6 Tortuous meanders

(a)

0 None

1 Occasional:
no overlapping of islands, average spacing being
ten or more river widths

2 Frequent, irregular:
infrequent overlapping, with average spacing less
than ten river widths

3 Frequent, regular:
not overlapping, average spacing less than ten
river widths

4 Split:
islands overlap frequently or continuously; the
number of flow branches usually is two or three

(b)

5 Anastomosing:
continuously overlapped islands, with two
or more flow branches

Fig. 9.9 Morphological classification of large channels: plan-form pattern (Kellerhals & Church 1989). (a) Channel pattern; (b) channel islands.

group. The exponent $b_1 \approx 0.55$, but b_2 appears to take either the value 0.40 or the value 0.33. The former indicates a proportional increase in forces on the bed as the channels become larger (the condition of 'Froude similarity'), whereas the latter implies that the limit strength of the bed constrains the increase in channel depth. In these conditions bed-forms become important as additional energy dissipators.

These equations would describe the change in channel form downstream in uniform sediments as drainage area—hence discharge—becomes greater. The reference flow is one competent to move sediment, hence to form the channel. Such equations can be developed for all classes of channels discussed in this chapter, but have been investigated systematically only on large chan-

nels. Many small and intermediate channels follow the scale $b_2 = 0.4$, whereas most large channels follow $b_2 = 0.33$.

Flow continuity associates a third equation with these two, a 'kinematic' equation for the flow:

$$\langle v \rangle = (1/a_w a_d) Q_n^{(1-b_1-b_2)}$$

in which $\langle v \rangle$ is the mean velocity for the channel. This equation reveals the necessity for channel size, shape, plan-form pattern, and bed-forms to become mutually adjusted so that the total resistance to flow permits just the velocity $\langle v \rangle$ to be maintained in the stable channel.

The actual relation between sediment calibre and channel pattern depends upon the discharge and the gradient of the channel. Together, they

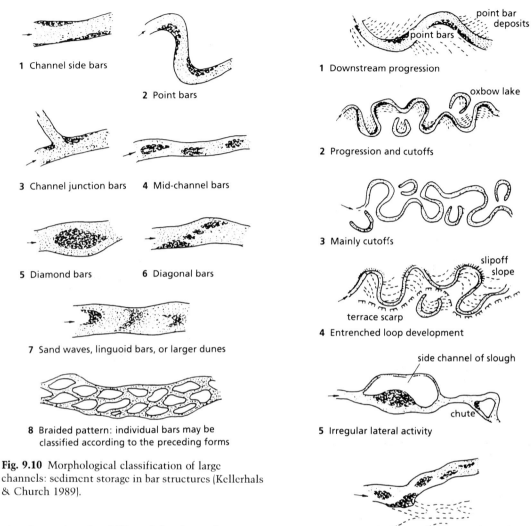

Fig. 9.10 Morphological classification of large channels: sediment storage in bar structures (Kellerhals & Church 1989).

Fig. 9.11 Classification of large channels: lateral activity (Kellerhals & Church 1989).

also determine the ability of the channel to move sediment, the product $\rho Q S$ giving the rate of energy expenditure along the channel. The limit gradient for a river is the valley gradient, S_v. However, the river might take up a lower gradient, S, by doing erosional and sediment transporting work to become sinuous. The sinuosity is the quotient S_v/S = river length/valley length. The sinuosity introduces an additional mechanism for energy dissipation since the water must be accelerated radially through the bends, so that eventually a statistically stable plan-form will be reached for a given formative flow and sediment supply. At this stage, the channel will continue to move sediment and shift laterally, but will not

become more sinuous, on average. If sediment supply becomes relatively large, the river may not transport all of it, even when flowing on the limit gradient, S_v. Then, the river will raise its bed (aggrade) by net deposition of sediment; in the very long term it will thereby increase its gradient. In such a state, the usual pattern of lateral instability is a sudden shift of course, termed an *avulsion*, just as in smaller channels that are episodically blocked by debris. If the imbalance of sediment load is sufficiently great, the river will perform frequent avulsions within a broad channel zone, thereby taking up a braided habit through the material. An empirical summary of conditions for various channel patterns is given in Fig. 9.12. Where major rivers are confined or entrenched, they often are constrained to flow in a single channel with gradient S_v. Such a channel is almost never alluvial.

Where large rivers flow in large valleys, or without constraint on plains or deltas, they characteristically are associated with a 'floodplain'. In hydraulic literature the term floodplain denotes a surface adjacent to a river that is episodically flooded, without implication about its formation. In ecological and geomorphological literature, a floodplain almost always denotes a genetic floodplain—a surface constructed by the present river as the result of sediment deposition during lateral shifting and overbank flooding. Such floodplains sometimes contain lakes or side-channels, seasonally or permanently connected with the main channel, which themselves constitute important aquatic habitats.

Braided and split channels present special habitat conditions. As flows change, the main channel—as in single-thread channels—experiences deeper and faster flows, but additional side-channels tend to become active, so that the low-flow habitat tends to be replicated toward the channel margins. The centre of a main channel often is a relatively hostile environment, where high sediment transport maintains a sterile substrate and high velocities extract large energy tolls from organisms to maintain themselves. The channel margin usually is much more favourable, both hydraulically and for food resources. Side-channels may, then, provide the best habitat. In braided systems, secondary channels change rapidly, but the persistent secondary channels of anastomosed systems may provide the most stable habitat units of all. In circumstances where the coarser part of the sediment load, travelling on or near the bed, is excluded at the side-channel entrance, they may also present conditions that are qualitatively distinct from those in the main channel.

9.5 PHYSICAL BASIS OF AQUATIC HABITAT

The equations of hydraulic geometry given above were developed to describe downstream changes in channel form at some flow of constant return period. However, they can be adapted to provide a description of hydraulic changes in a single cross-section as flow varies. In this context, they are

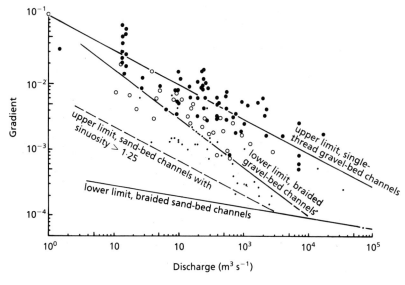

Fig. 9.12 Channel pattern gradient-discharge limits based on data compiled by the writer. Limits for sand-bed channels based on work by Bray (1973) and Neill (1973). • Braided gravel-bed channel; ○ wandering gravel-bed channel; ● braided sand-bed channel.

purely descriptive; coefficients and exponents are not constrained in the manner discussed above, and a single equation may not necessarily describe the entire range of flows. However, they are very helpful in habitat description, since P (or w_s)

approximates available substrate area per unit length of channel, $PR = w_s d. = A$ defines water volume per unit length of channel, and $\langle v \rangle = Q/A$ provides a measure of water velocity.

Figure 9.13(a) shows the at-a-station hydraulic

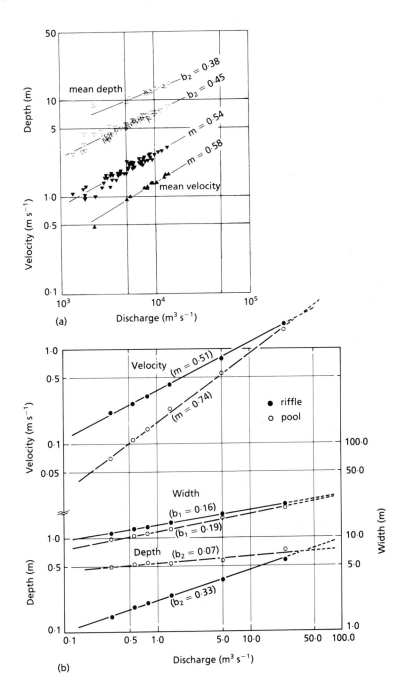

Fig. 9.13 At-a-station hydraulic geometry. (a) Fraser River, British Columbia at Agassiz (gravel bed ($\triangledown$ $\blacktriangledown$)) and at Mission (sand bed ($\triangle$ $\blacktriangle$)); $m = 1 - b_1 - b_2$. (b) Hangover Creek, Queen Charlotte Islands, British Columbia, a boulder–gravel stream of intermediate size, averaged over major morphological units (Hogan & Church 1989).

geometry for two stations on Fraser River, British Columbia. One section is in a steep, gravel-bed reach with wandering channel morphology, whereas the other is in a straight, sand-bed reach with much lower gradient. In both cases, channel width is about the same. The steep reach has much higher velocity and smaller depth at all flows. The at-a-station concept can be extended to consider the average hydraulic geometry based on a number of cross-sections within a reach. Figure 9.13(b) shows the average hydraulic geometry for two major morphological units in an intermediate boulder–gravel channel. Some distinctions between the units are evident; of particular interest, however, is the loss of distinctiveness at high flows.

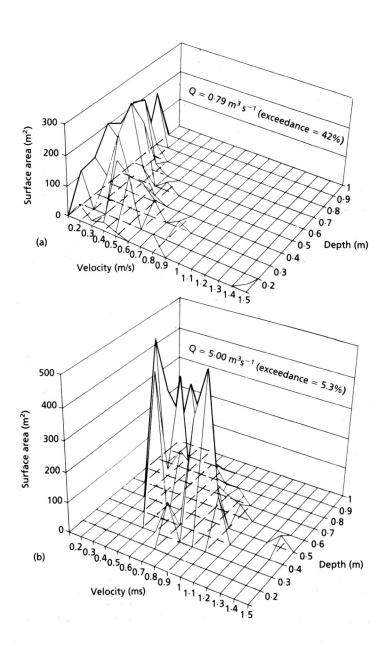

Fig. 9.14 Disaggregated hydraulic geometry; mean depth and mean velocity for Hangover Creek, Queen Charlotte Islands (Hogan & Church 1989).

The summary hydraulic geometry obscures some aspects of channel shape that are significant in habitat appraisal. In a trapezoidal cross-section with steep banks, such as the gauge section at Mission on Fraser River, most of the bed area has a very similar depth at any given stage. In comparison, a triangular cross-section, which the Agassiz section approximates, maintains a varied distribution of depths at all stages, which usually is desirable for habitat maintenance. Natural channels tend often to take up triangular shapes, whereas engineered channels are nearly always trapezoidal.

Complete habitat assessment usually requires a more refined assessment of depth and velocity distributions. Fish prefer particular combinations of velocity and depth in the channel. Accordingly, the best description of physical habitat would be a completely disaggregated bivariate distribution of velocities and depths everywhere in a reach. Figure 9.14 shows the distribution of surface area by mean depth–mean velocity classes along a 260-m reach in a cobble–gravel bed stream. This information is recoverable from a topographic map of the reach provided that water surface elevations are available for each flow and the flow is measured in one section (a gauge section). This is relatively easy to arrange, but it still entails a degree of undesirable averaging within sections. The alternatives require a very onerous measurement programme to determine velocities at many points in the reach at each flow (Collings *et al* 1972; Sullivan 1986), or some means to estimate velocities. In large channels with regular velocity profiles, simulation procedures may allow estimation (Stalnaker *et al* 1989), but in intermediate to small channels with high relative roughness this is not feasible. The question of whether characteristic distributions about the mean of velocity can be assumed in such channels remains a question at the research frontier.

ACKNOWLEDGEMENTS

D.L. Hogan has generously shared many results on rivers dominated by organic debris, a much underappreciated determinant of morphology in smaller channels. He has also provided helpful comments on the text. The regional bias of this text is determined by the writer's experience, but that constrains us all and readers should always be aware of it.

REFERENCES

Ackers P. (1988) Alluvial channel hydraulics. *Journal of Hydrology* **100**: 177–204. [9.4]

Bisson PA, Nielsen JL, Palmason RA, Grove LE. (1981) A system of naming habitat types in small streams with examples of habitat utilization by salmonids during low streamflow. In: Armantrout NB (ed.) *Acquisition and Utilization of Aquatic Habitat Inventory Information*, pp 62–73. American Fisheries Society, Western Division. [9.3]

Bray DI. (1973) Regime relations for Alberta gravel-bed rivers. *Proceedings of the 7th Canadian Hydrology Symposium*, pp 440–52. [9.1, 9.4]

Church M, Jones DP. (1982) Channel bars in gravel-bed rivers In: Hey RD, Bathurst JC, Thorne CR (eds) *Gravel-bed Rivers*, pp 291–336. John Wiley and Sons, Chichester. [9.3]

Collings MR, Smith RW, Higgins GT. (1972) The hydrology of four streams in western Washington as related to several salmon species. *United States Geological Survey Water Supply Paper 1968*. [9.5]

Costa JE. (1984) Physical geomorphology of debris flows. In: Costa JE, Fleisher PJ (eds) *Developments and Applications of Geomorphology*, pp 268–317. Springer-Verlag, Berlin. [9.2]

Dunne T, Leopold LB. (1978) *Water in Environmental Planning*. WH Freeman, San Francisco. [9.1]

Fremling CR, Claflin TO. (1984) Ecological history of the Upper Mississippi River. In: Wiener JG, Anderson RV, McConville DR (eds) *Contaminants in the Upper Mississippi River*, pp 5–24. Butterworth, Stoneham, Massachusetts. [9.1]

Grant GE, Swanson FJ, Wolman MG. (1990) Pattern and origin of stepped-bed morphology in high-gradient streams, Western Cascades, Oregon. *Geological Society of America Bulletin* **102**: 340–52. [9.3]

Hogan DL. (1986) Channel morphology of unlogged, logged and torrented streams in the Queen Charlotte Islands. *British Columbia Ministry of Forests Land Management Report 49*. British Columbia Ministry of Forests, Victoria. [9.3]

Hogan DL, Church M. (1989) Hydraulic geometry in small, coastal streams: progress toward quantification of salmonid habitat. *Canadian Journal of Fisheries and Aquatic Sciences* **46**: 844–52. [9.5]

Keller EA, Swanson FJ. (1979) Effects of large organic material on channel form and fluvial processes. *Earth Surface Processes* **4**: 361–80. [9.3]

Kellerhals R. (1967) Stable channels with gravel-paved

beds. *American Society of Civil Engineers. Proceedings, Journal of the Waterways and Harbours Division* 93 (WW1), pp 63–84. [9.1]

Kellerhals R, Church M. (1989) The morphology of large rivers: characterization and management. In: Dodge DP (ed.) *Proceedings of the International Large River Symposium. Canadian Special Publication of Fisheries and Aquatic Sciences* **106**: 31–48. [9.1, 9.4]

Kellerhals R, Church M. (1990) Hazard management on fans, with examples from British Columbia. In: Rachocki A, Church M (eds) *Alluvial Fans: A Field Approach*, pp 335–54. John Wiley & Sons, Chichester, UK. [9.2]

Kellerhals R, Church M, Bray DI. (1976) Classification and analysis of river processes. *American Society of Civil Engineers Proceedings, Journal of the Hydraulics Division* **102**: 813–29. [9.4]

Leopold LB, Wolman MG. (1957) River channel patterns: braided, meandering and straight. *United States Geological Survey Professional Paper* 282–8.

Lisle TE. (1986) Stabilization of a gravel channel by large streamside obstructions and bedrock bends, Jacoby Creek, northwestern California. *Geological Society of America Bulletin* **97**: 999–1011. [9.3]

Marston RN. (1982) The geomorphic significance of log steps in forest streams. *Association of American Geographers Annals* **72**: 99–108. [9.3]

Mollard JD. (1973) Air photo interpretation of fluvial features. *Proceedings of the 7th Canadian Hydrology Symposium* 341–80. [9.4]

Neill CR. (1973) Hydraulic geometry of sand rivers in Alberta. *Proceedings of the 7th Canadian Hydrology Symposium* 453–61. [9.1, 9.4]

Osterkamp WR, Hedman ER. (1982) Perennial-streamflow characteristics related to channel geometry and sediment in Missouri River basin. *United States Geological Survey Professional Paper 1242.* [9.1]

Schumm SA. (1977) *The Fluvial System.* John Wiley & Sons, New York. [9.1]

Schumm SA. (1985) Patterns of alluvial rivers. *Annual Reviews of Earth and Planetary Science* **13**: 5–27. [9.4]

Shreve RL. (1967) Infinite topologically random channel networks. *Journal of Geology* **75**: 178–86. [9.1]

Stalnaker CB, Milhous RT, Bovee KD. (1989) Hydrology and hydraulics applied to fishery management in large rivers. In: Dodge DP (ed.) *Proceedings of the International Large River Symposium. Canadian Special Publication of Fisheries and Aquatic Sciences* **106**: 13–30. [9.5]

Sullivan K. (1986) *Hydraulics and fish habitat in relation to channel morphology.* PhD dissertation, The Johns Hopkins University, Baltimore. [9.3, 9.5]

Teversham JM, Slaymaker O. (1976) Vegetation composition in relation to flood frequency in Lillooet River valley, British Columbia. *Catena* 3: 191–201. [9.1]

Whittaker JG, Jaeggi MNR. (1982) Origin of step-pool systems in mountain streams. *American Society of Civil Engineers Proceedings, Journal of the Hydraulics Division* **108**: 758–73. [9.2]

10: Floodplain Construction and Erosion

J. LEWIN

There are three major reasons why floodplain development and characteristics need to be understood for river management purposes. At high flows, floodplains become part of the surface flow system, and even at low water, groundwater returns can be a highly significant baseflow component of river discharge. A knowledge both of the surface forms that affect the passage of flooding waters, and of how these forms can change, is thus important as is a knowledge of subsurface floodplain materials that affect groundwater storage, flow and quality. A second reason involves the dynamic nature of rivers and floodplains. In mid-latitudes where there are mobile river channels, a high proportion of river sediments may be derived from the erosion of floodplain materials so that their physical and chemical characteristics are determinants of channel material flows. The same applies to other environments, such as semi-arid systems where vertical cut and fill sequences over a timescale of decades to centuries makes floodplains alternately into sources or sinks for transported sediments. Finally, because floodplains are primarily constructed through past river sedimentation, they can preserve an extended record of the changing suite of formative factors involving climatic change and variable human influence which has affected valley floor environments over some thousands of years.

10.1 FLOODPLAIN CONSTRUCTION

Fluvial deposition results from the major processes of *sedimentation* of finer material which has been suspended in flowing or ponded water, or where material travelling along at or near the river bed (which is usually coarser) comes to rest.

The latter commonly occurs in *accretionary bed forms* or bars. Lag deposits, coarser material left behind or let down on to river beds, may also form significant accumulations in some environments.

Processes of accretion and sedimentation may take place within river channels, or beyond them out on floodplains at periods of high river flows. It is usual to associate suspension deposits with overbank flows and accretionary deposits with within-channel flows, but in practice the depositional environments in which these processes occur can be very varied. Flood gravels can be deposited overbank, and fine sedimentation can take place quite commonly in within-channel slack-water zones. The latter may of course be reactivated again, and there may be seasonal recycling of fines within river channels according to river stage, with low flow build-up and flood flow removal.

River deposition processes of these kinds, which may create more permanent floodplain construction, reflect variability in three groups of factors. The first concerns sediment supply, both in terms of volume and sediment size. Floodplain materials reflect the bedrock and superficial deposits of their catchments, the size and quantity of such materials released into rivers by weathering and erosion (including accelerated soil erosion) on hillslopes, and the upstream erosional activity of river channels themselves. In-stream hydraulic sorting and abrasion processes also reduce the size of materials available downstream.

It is extremely difficult to generalize about the rates at which these processes operate. Figure 10.1 shows some sediment budget studies in which the sources of sediments are identified together

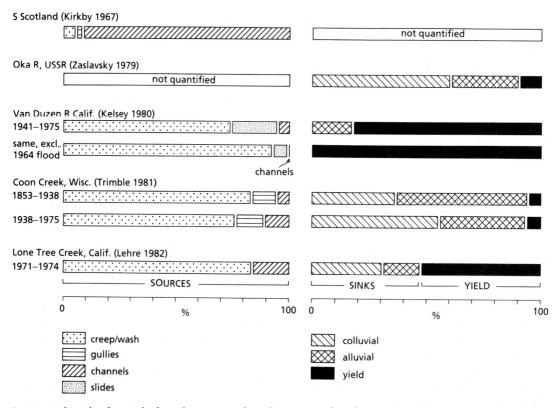

Fig. 10.1 Selected sediment budgets for contrasted catchments. Bar lengths represent the proportions of eroded sediment being derived from the source types indicated, and the percentages which are deposited on slopes (colluvial) or on floodplains (alluvial), or which are transported out of the studied catchments.

with the proportion of such materials which remain in alluvial 'storage'. This may differ according to the time period assessed, with individual flood events having particular and distinctive effects (Van Duzen River, California, USA), and human-induced changes producing effects which differ over time (Coon Creek, Wisconsin, USA). Some catchment materials, such as wind-deposited loess, may give catchments very high sediment throughput and deposition rates. The major example is the Huang He (Yellow River) in China with its tributaries draining thick loess deposits particularly in Shaanxi Province. Loess deposits have also been important sources for floodplain deposition in the southeastern USA and southern Britain. For practical purposes it is therefore necessary to make an assessment of

catchment sediment sources, actual and potential, in order to explain or anticipate floodplain sedimentation response.

A second factor concerns the availability of subaqueous voids which must be available for thick accumulations to occur. On a local scale, this can be provided by repeated overbank flooding such that deposits may accumulate in laminae, layer on layer, as long as overbank flows can be maintained across the thickening deposit. For accretionary deposits to persist, lateral channel mobility is usually required so that bank erosion may create channel enlargement which is balanced by infilling. Both of these processes may be self-limiting, but there can also in the longer run be external controls on sedimentation, such as tectonic subsidence. Surveys of floodplain ge-

ometry, and of channel mobility patterns and rates, are required to set floodplain sedimentation patterns in context.

The third factor concerns hydraulics. These have been discussed in Chapters 8 and 9 and it is important here mainly to appreciate that available river energy may set limits on the sizes of material that can be transported (and hence subsequently deposited). Energy availability may vary on a catchment basis (from high to low in the downstream direction), or locally within sinuous channel patterns, and out on the floodplain itself, with distance away from the channel. Both local and regional hydraulic factors give variety to the size suite of accretionary and suspension deposits that are eventually incorporated into floodplains.

On the assumptions that the sizes of deposited sediments are competence- or transport-limited, and that sediment size or size distributions can be uniquely related to measures of streamflow, it has proved possible to reconstruct former conditions via a range of palaeohydrological equations (see Williams 1983; Church 1987; Maizels 1987). A more generalized approach to floodplain material description has been the characterization of facies (units of uniform sediment type) using a simple coding system developed by A. Miall. A version of this is given in Table 10.1. Sediments are broadly grouped as gravel-sized, sandy or fines, with units being further categorized by sedimentary structure. Using bank sections or borehole data it is also possible to record facies thicknesses, vertical sequences, and possibly lateral extent. Again, without recourse to interpretation, it is also desirable to record measures of floodplain form, for example width, gradient and local relief, possibly using profiles or three-dimensional mapping (see, for example, Lewin & Manton 1975; Chisholm & Collin 1980) on a scale that is much more detailed than usually available from published maps. From these it commonly emerges that floodplains have significant if small-scale relief. For example Lewin and Manton (1975) showed that relief of up to 2.5 m was to be found on the floodplains of Welsh rivers, with the present river banks the highest part of the 'plain'. Figure 10.2 shows a portion of the floodplain of the Afon Tywi at Llandeilo, Dyfed, Wales (SN 6423 and SN 6524). The topographic 'lows' lead to

early entry of floodwaters out on to the floodplain, with depressions being waterlogged for long periods. Such floodplain features are created in a variety of sedimentary environments.

10.2 SEDIMENTARY ENVIRONMENTS

At a scale approximating that of river channel width, it is possible to identify six fluvial sedimentary environments in which sedimentation or accretion may take place (Table 10.2). Over time or as rivers shift their courses, these environments may relocate with the development of enlarging sediment units which come together to make up the floodplain itself. The six environments may each develop characteristic or indeed contrasting facies and geometries, and it is helpful to appreciate what these may be.

On the beds of active channels, *lag deposits* may line the basal erosion surface produced by migrating streams. Because stream depth varies, this surface can be undulating. For example, as meanders develop and tighten their radius of curvature, there may be a corresponding deepening of the channel so that basal erosion surfaces are lowered. Furthermore, confinement of channels against unerodible valley sides or artificial structures also leads to channel deepening, so that coarser lags may be found at increased depths in such locations.

Channel deposits assume a variety of geometries, depending initially on active bar geometry. These may be classified with regard to formation position in the channel (mid-channel, lateral or point bars attached to one bank, or diagonal, passing from one side of the channel to the other); according to relative dimension (longitudinal, where the downstream dimension is several times the cross-stream dimension, or transverse where they are more nearly equivalent); or according to some descriptor of shape (lingoid, crescentic, lobate, rhomboid, scroll etc.). In time, sediment bars may build up as a succession of accreting sheets, or may prograde laterally or downchannel with avalanche fronts producing steeply dipping beds. Accretionary surfaces may thus be at various inclinations, possibly quite steep, as on the inside of curving channels where migration of the channel on one side may

Table 10.1 A lithofacies/structure coding for fluvial sediments (from Miall 1978)

Facies code	Lithofacies	Sedimentary structures	Interpretation
Gms	Massive, matrix-supported gravel	None	Debris flow deposits
Gm	Massive or crudely bedded gravel	Horizontal bedding, imbrication	Longitudinal bars, lag deposits
Gt	Gravel, stratified	Trough crossbeds	Minor channel fills
Gp	Gravel, stratified	Planar crossbeds	Linguoid bars or growths from older bar remnants
St	Sand, medium to very coarse, may be pebbly	Solitary or grouped trough crossbeds	Dunes (lower flow regime)
Sp	Sand, medium to very coarse, may be pebbly	Solitary or grouped planar crossbeds	Linguoid, transverse bars, sandwaves (lower flow regime)
Sr	Sand, very fine to coarse	Ripple marks	Ripples (lower flow regime)
Sh	Sand, very fine to very coarse, may be pebbly	Horizontal lamination, parting or streaming lineation	Planar bed flow (lower and upper flow regime)
Sl	Sand, fine	Low angle (<10°) crossbeds	Scour fills, crevasse splays, anti-dunes
Se	Erosional scours with intraclasts	Crude crossbedding	Scour fills
Ss	Sand, fine to coarse, may be pebbly	Broad, shallow scours including cross-stratification	Scour fills
Sse, She, Spe	Sand	Analogous to Ss, Sh, Sp	Eolian deposits
Fl	Sand, silt, mud	Fine lamination, very small ripples	Overbank or waning flood deposits
Fsc	Silt, mud	Laminated to massive	Backswamp deposits
Fcf	Mud	Massive, with freshwater molluscs	Backswamp pond deposits
Fm	Mud, silt	Massive, desiccation cracks	Overbank or drape deposits
Fr	Silt, mud	Rootlets	Seatearth
C	Peat, carbonaceous mud	Plants, mud films	Swamp deposits
P	Carbonate	Pedogenic features	Soil

be accompanied by lateral accretion on the other, to form so-called 'epsilon' cross-stratification. As bars develop over time, their sediments may be incorporated into floodplains as much more extensive contiguous bodies whose limits are set by switches in channel location and erosion. Accretionary sediments, perhaps crudely stratified sandy gravels or alternatively cross-bedded sands, may thus come to dominate the floodplain materials of some mid-latitude migrating-stream floodplains, where active sediment sources and available stream powers are appropriate.

Many larger and some smaller rivers may possess larger sedimentary bed-forms than the

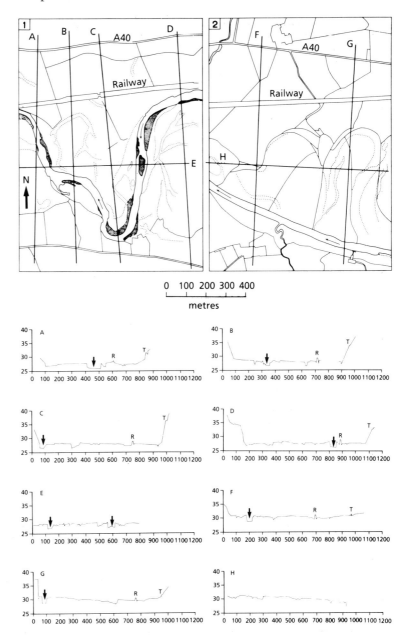

Fig. 10.2 Plan and profiles for parts of the floodplain of the Afon Tywi, near Llandeilo, Dyfed, UK. Arrow marks river; units are in metres. (after Lewin & Manton 1975).

'unit bar' features of about channel-width dimensions so far discussed. For example, Crowley (1983) reported 200–400 m long sandy bed-forms on the Platte river, whilst large sandwaves occur on the lower Brahmaputra (Coleman 1969; Bristow 1987).

Channel fills occupy abandoned channels; their cross-section and plan follows the dimensions of the former channel, but their lengths may vary from short arcuate reaches to very extensive linear traces where much longer channel stretches have been deserted. For example, former courses of the Huang He in China, some hundreds of kilometres in length, have been abandoned and infilled,

Table 10.2 Fluvial sedimentary environments

Environment	Major facies
Active channels, lag deposits	G
Active channel bars	Gm, Gt, Gp, St, Sp, Sr
Abandoned channel fills	Sr, Sh, Fl, Fsc
Channel margin (levee, crevasse, crevasse splay)	Sh, Sr, Sl
Floodplains and backswamps	Fl, Fsc, Fm, C, P
Non-fluvial materials (aeolian, glacial, colluvial)	Sse, She, Spe, Gms

whilst branches of former multiple channel systems may be found in many mid-latitude environments where now only a single channel is to be found (a number of examples are illustrated in Petts 1989). Fill materials can be very varied: rapidly accreting sand plugs adjacent to the new channel, organic fills where the supply of mineral flood sediments is less dominant, or thick laminated flood clays. Such clay plugs may later restrict lateral channel movement across the floodplain, as indeed may fine-grained deposits more generally (Ikeda 1989).

In the nature of things, abandoned channels are ephemeral features which are liable to be infilled over a period of years. Temporarily they may provide important wetland habitats, but under natural conditions such habitats would eventually be infilled, to be replaced by others elsewhere.

Characteristic *channel marginal* environments include levees and crevasse chutes and splays. The former are channel-side ridges, usually with sediments of sand size or coarser, fining away from the river with decreasing elevation, and produced by accretion/sedimentation in the critical zone where overbank flows first pass out on to the floodplain. Hydraulics in this zone have recently been modelled, with travelling voltices in the shearing zone along channel banks being involved in sedimentation processes (see Knight 1989). Levees may be dissected and their deposits (together with direct channel-derived additional material) spread out across the floodplain in crevasse splay features.

In this zone, sedimentation/accretion features may also be crucially related to accretionary bar features, infilling hollows (*swales*) and other topographic depressions. Bar-top deposition thus blurs the surface form of bars with finer infills.

Floodplain/backswamp environments are dominated by sedimentation, generally with materials that are increasingly finer with distance from the channel. Such environments may be temporarily or permanently waterlogged with ponded waters. Four factors crucially affect the nature of sedimentation: overbank suspension loads, the geometry of the floodplain, the sequence of channel developments (see Farrell 1987), and the nature of vegetation growth. It is thus possible to have at one end of the scale relatively planar floodplains rapidly building up with silty materials (Schmudde 1963); at the other, sandy levees or rivers on raised ribbon-like alluvial ridges may be backed by swamps and lakes in which organic or lacustrine sedimentation dominates (Blake & Ollier 1971). Under natural conditions these basins may be very persistent, although in settled landscapes they may be a prime target for land drainage works.

Finally *colluvial* or other non-fluvial sediments may form an appreciable component of floodplain deposits, especially at the margins under conditions of active slope erosion. This is particularly true of steep, narrow headwater valleys (Lattman 1960), and was true of Pleistocene cold-climate environments which, as we shall see, contributed a great deal to the valley fills of many lowland mid-latitude areas. In dry conditions, wind-blown material, either derived from arid-zone deflation, glacial materials or from river deposits, may also be important components of floodplains.

Floodplain sedimentation as a whole reflects very strongly the ways in which these six sediment types are combined within individual floodplain segments; this in turn relates to stream sediment loads and the pattern of channel change and migration. For alluvial rivers (the 'large channels' of chapter 9, section 9.4), sedimentation style is commonly associated with channel pattern prototypes (braided, meandering or anastomosing) which have themselves been further differentiated largely on grounds of sediment size. For example, Miall (1977) suggested six *braided-*

stream facies assemblage types, with dominant sand or gravel components, whilst Jackson (1978) similarly identified five classes for *meandering* streams dominated respectively by muddy, sandy or gravelly sediments and mud/sand or sand/gravel combinations. A couplet of finer sedimentation deposits overlying accretionary deposits is very common for meandering streams. To these may be added an anastomosing prototype (Smith & Smith 1980), a low-gradient, aggrading, multiple-channel and sinuous system dominated by ribbon channel sediments and organic wetland flood basins. Miall (1985) similarly provided 'illustrations' of 12 models of fluvial sedimentation style (which he stressed was not a comprehensive set).

Identification of these numerous prototypes is an advance on the pioneering but simplistic distinction between braided and meandering models of earlier workers, but there may still be other models to consider. For example, Graf (1988) has questioned the applicability of the term 'floodplain' to dryland rivers where the whole valley floor between terraces is ephemerally occupied by a braided channel, with parts of this becoming infilled during periods when a transition to meandering occurs. A meandering style may thus be set within largely braiding sediments. Rust and Nanson (1986) have reported coexisting mudbraids (activated during exceptional floods) and sand-bed anastomosing channels, the latter operating at moderate discharges. Others have also reasonably argued for a continuum of channel patterns (e.g. Ferguson 1987) responding to such factors as bank strength and sediment competence.

From a floodplain point of view, it is also necessary to appreciate that channel deposits are indeed only one of the six sedimentary environments discussed above. Their relative significance is extremely important in determining the nature of floodplains: many are dominated by overbank and backswamp processes beyond a relatively narrow ribbon of lateral channel activity (e.g. Blake & Ollier 1971), although others may largely reflect accretion processes (Wolman & Leopold 1957).

Channel pattern and migration style are of interest in that they determine how floodplains may develop contrasted patterns of lateral and vertical age sequencing, and the 'preservation times' that alluvial units may have before they are re-eroded. For example, meandering streams may migrate in different styles producing facies of contrasted grain sizes as they evolve. Where quality factors are important, as where pollutants may have been incorporated into floodplains during particular time periods, or into particular sediment size ranges, it is important to identify where on floodplains such 'risk' zones are concentrated. Equally, particular floodplain sites may have particular 'lifetimes' which vary according to the rate and pattern of channel migration.

Figure 10.3 shows a reach of the River Teme, UK in 1947 and 1973. The earlier photography, obtained after the major 1947 floods, shows complex evolving loops and cutoffs and the large quantity of overbank flood sediments (white areas). By 1973, further loop development and abandonment occurred, with channel fills and overbank sedimentation being of considerable relative importance. Use of air photographs in this way, and also of historical maps and dating of sediments, may allow age zoning of floodplains to be undertaken for particular sites of interest.

Figure 10.4 shows accretion patterns in the last century for three Welsh river reaches. Although in meandering streams, point bar sedimentation is usually considered to dominate, this is not in practice always the case. For example, cutoff fills occupy 14% of the recent sediments on the Teme, with point bar sediments on both Teme and Dee amounting to only about half the total of recent accretion sediments.

10.3 RATES OF ACTIVITY

The complex store of sediments which make up floodplains may be augmented or eroded at varying rates and with varying patterns. We can identify these rates on three scales. First are the erosion/deposition rates for particular sedimentary environments. Then come input–output budgets for 'architectural assemblages' as they are combined in particular alluvial valley reaches. Finally, there are the responses of alluvial systems within whole drainage basins.

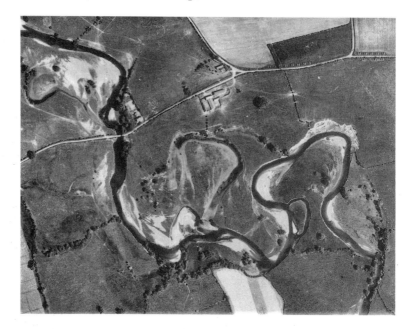

(a)

(b)

Fig. 10.3 The River Teme near Ludlow, Hereford and Worcester, UK, in (a) 1947 and (b) 1973.

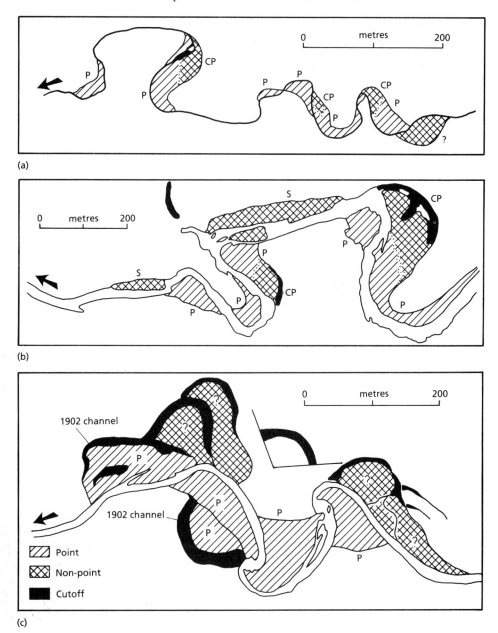

Fig. 10.4 Floodplain accretion in the last century on rivers in Wales. (a) The Afon Twymyn, Dyfed (SN 885998); (b) the Dee, Gwynedd (SH 980369); (c) the Teme, Powys (SO 384730) (from Lewin 1983).

Erosion and deposition rates

Rates of overbank sedimentation that have been recorded were usefully reviewed by Bridge and Leeder (1979); there have been several more recent studies for particular events at specific sites using a range of direct measurement or dating techniques (e.g. Lambert & Walling 1987). A diffusion model developed by Pizzuto (1987) appears to predict quite well both rates of sedimentation and the decrease in grain size which occurs away from channels in the post-settlement alluvial overbank deposits in Pennsylvania, USA. The model predicts a more rapid decrease in sand deposition than is apparent in the deposits, suggesting non-diffusion processes may be additionally important for sand dispersion.

Long-term rates given by Bridge and Leeder (1979) range from 0.01 m year^{-1} to 0.0002 m year^{-1}; Brown (1987) suggested a rate of 0.0014 for the main floodplain of the Lower Severn, UK. Cutoff and within-channel sedimentation fills can be more rapid than this, as can deposition from single extreme events. Stear (1985) recorded 2–3.5 m of laminated fine-grained sand and silt deposited overbank by an extreme flood in a semi-arid environment. The deposits of a spring 1964 flood on the Ohio River, USA ranged from 0.46 m on a levee to less than 3 mm away from the river (Alexander & Prior 1971). Rates of sedimentation are very clearly both site and event specific.

Lateral accretion rates can in some circumstances be determined by meander migration rate assessment (Wolman & Leopold 1957), either through direct measurement using ground survey, or from historical maps (see also Lewin *et al* 1977; Hooke 1980). Migration rates (and thus rates of floodplain accretion) have been linked to stream power, outer bank height and a resistance coefficient in a regression model developed by Hickin and Nanson (1984). Such rates may be variously expressed, as m year^{-1} or as a percentage of channel width per year, for example, with annual rates in the order of metres per year being common in higher lateral accretion rate environments.

Cutoff and avulsion rates are significant for the termination of meandering sequences. Cutoffs may be of several types: in particular neck cutoffs produced by the breaching of narrow gaps between adjacent channel bends, and chute cutoffs which involve more extensive scouring of new channels across floodplains and point bars. High sinuosity and low floodplain slope appear to favour the former, steep slopes and erodible non-cohesive and poorly vegetated surfaces the latter. In a study of some 1000 km of valley floor in Wales and the Borderland, Lewis and Lewin (1983) found some 145 recent cutoffs of which 55% were simple

chute, and 16% simple neck types. Others involved cutoffs related to bar development and multiloop forms.

Avulsions are important in the development of larger floodplains, for example in the Holocene development of the lower Mississippi Valley, USA, in the evolution of anastomosing systems (Smith *et al* 1989), and also in other types of multichannel system (Ferguson & Werritty 1983). However, such avulsions are extremely difficult to predict. Avulsion may be accompanied by a transformation in channel patterns (Brizga & Finlayson 1990).

Input–output budgets

It is possible to model valley floors and floodplains as complexes of sediment storage units of different ages. For example, a study of recent floodplain history of a reach of the Little Missouri involved the age zoning of the valley floor by tree age (Everitt 1968), with a negative exponential relationship between area and age of deposition. Reservoir theory may be applied to such data (see Dietrich *et al* 1982), and computation of mean residence time, age distribution of sediment leaving storage, and sediment transport rates may be undertaken. These may vary for different valley floor components (Kelsey *et al* 1986; Nakamura *et al* 1987). Breaks in cumulative age–area distribution curves may be interpreted as resulting from significant events reordering parts of the valley floor. 'Inset' valley floors of different age, some on different topographic levels, may also be distinguished, as shown in Fig. 10.5.

Drainage basin alluvial responses

At the catchment scale, downstream sediment transfers and upstream recession of erosion phases come into effect. Upstream incision may lead to downvalley aggradation, whilst there may be downvalley movement of 'slugs' of sediment. These kinds of 'complex response' activities (Schumm 1977) may operate at a variety of timescales. Rains and Welch (1988) have discussed the existence of non-synchronous Holocene alluviation which could be explained in this way. On a shorter timescale, alluviation and incision in

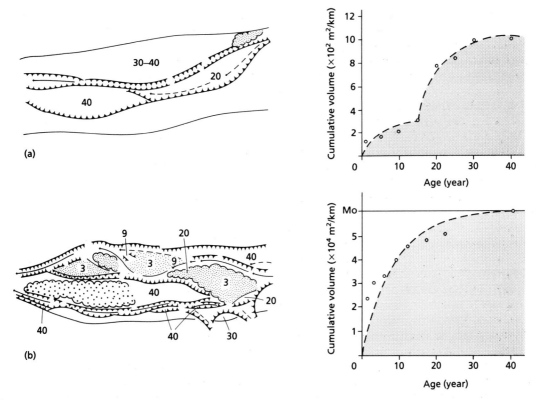

Fig. 10.5 Valley floor area/age relationships for two valley types on the River Furano, Hokkaido, Japan (after Nakamura *et al* 1987).

different parts of catchment systems have been documented with respect to recent historical alluviation (e.g. Trimble 1981; Knox 1987).

10.4 LATE QUATERNARY HISTORIES

Whilst floodplain sediments accumulate, environments may change. Over the past several thousand years, there have been six important external factors that have modified floodplain processes to the extent that floodplain sediments themselves record a considerable variety of process system changes. These are:

1 Major climatic oscillations between glacial and interglacial conditions. The most recent maximum advance of ice sheets in mid-latitudes some 18 000 years ago directly over-ran many valleys and provided large volumes of sediment in the form of glacigenic materials. Peripheral to ice

sheets, periglacial conditions also led to considerable slope sediment production, infilling smaller valleys and aggrading larger ones. Low sea levels steepened and incised rivers close to the sea.

2 Subsequent climatic recovery re-established forest vegetation with a developing soil cover. Rising sea levels drowned the lower parts of valleys. These filled, in part with fluvial and marine sediments at a rate determined by local postglacial sediment inputs. Some drowned valleys remain as estuaries; others have infilled and are now floodplains. Removal of the ice-sheet load also led to isostatic (tectonic) movements, and the complex interaction of sea-level rise, tectonics and sediment supply has produced complex alluvial histories. On the whole, postglacial conditions have led to decreasing sediment supply rates under natural conditions (see Church & Rider 1972).

3 Within the Holocene (the last 10 000 years) there have been second-order climatic fluctuations. For example, alternating flood- and drought-dominated river regimes have been identified in coastal rivers of New South Wales, Australia (Erskine & Warner 1988), whilst dry and humid periods can also be identified in Europe (Probst 1989). The so-called Litte Ice Age, *c.* 1600–1850, led to recorded increases in floods, slope erosion and glacier advance, and these achieved transformations in channel pattern (Bravard 1989). El Nino events have produced phases of aggradation in South America (Wells 1990). Finally, humid–arid transitions in North America have produced marked changes in alluviation patterns (see Knox 1983; Schumm & Brackenridge 1987).

4 In tectonically unstable orogenic environments, rivers have been affected by uplift and subsidence. This may provide both deepening alluvial basins and steepening or incising rivers where they are accommodating to uplift (see Schumm 1986).

5 Forest clearance, agriculture and urbanization, over the past several thousand years in some places but only decades or less in others, have severely modified sediment systems. 'Post-settlement' alluvium is a common feature of settled landscapes, resulting particularly from accelerated soil erosion. This has been widely recorded in Europe in the longer term (so-called Haugh loams) and North America in the shorter (see review by Butzer 1981).

6 To such inadvertent environmental effects of human activities, must be added deliberate efforts to improve land drainage and to control river runoff and erosion. This includes field underdrainage, ditching for upland afforestation, wetland drainage, cutoff infilling and other landfill exercises, river channelization, and river impounding for water supply, power generation, flood control, navigation and irrigation. In few settled environments, therefore, are rivers now free to continue the processes of sedimentation which generated the floodplains through which they run.

The nature and timing of these environmental changes and constraints has varied very considerably, and even at well-studied sites it may be an open question as to whether a set of deciphered changes in sedimentation style is a response to 'natural' or 'human' causes in the catchment. In some studies climatic causes have been pinpointed (e.g. Brackenridge 1980). By contrast, for the Upper Thames, UK, Robinson and Lambrick (1984) were convinced that human activities were an 'entirely adequate' explanation for Holocene changes. They suggested a lower unit of Late Devensian (last glaciation) gravels that had not subsequently been reworked, with a relatively dry valley floor existing until the late Neolithic and Bronze Age. This was followed by a period with increased evidence of flooding (by around 2000 BP), and then the deposition of alluvial clays sealing in prehistoric remains. Later alluviation phases followed in Saxon and medieval times to produce the floodplain we see today.

In the Lower Severn also, Brown (1987) has suggested stable earlier Holocene channels, possibly anastomosing in pattern. Simplification of this pattern by abandonment with some avulsion and cutoff activity then followed, together with deposition of a thick layer of overbank sediments of anthropogenic origin. In what are now the Norfolk Broads, UK, Holocene organic sedimentation dominated in backswamp environments, and the peat deposits were cut and then the depression left drowned to create the lakes we see today (Lambert *et al* 1960).

In upland Britain, with higher power streams and steeper gradients, Holocene rivers have been incised into glacigenic and periglacial materials, and they have continued to rework them. This has created vertical flights of valley-floor terraces at some locations (e.g. Macklin & Lewin 1986), the most recent of which may be associated with 19th century aggradation. Activity in such gravel-bed rivers may be concentrated also in limited 'sedimentation zones' separated by more stable reaches, with a spatially rather intricate response to environmental change (Macklin & Lewin 1989). Deforestation can also produce sediment waves moving downvalley (Roberts & Church 1986). Extreme events can also have a greater effect in steepland headwater catchments (see for example Carling 1986; Harvey 1986). Some environments may indeed be dominated by such events, as in parts of south-east Australia where

Nanson (1986) has documented a catastrophic stripping of large parts of floodplains during extreme floods, followed by rebuilding and sedimentation over hundreds or even thousands of years. Useful instances and reviews of flood effects are provided by Beven and Carling (1989).

In the light of all these environmental changes, and the interpretations briefly reviewed above, three conclusions are possible. The first is that present channel activity may give only a poor guide as to the ways in which adjacent floodplains have developed. This is especially true of long-settled and managed environments, ones where small environmental perturbations can trigger large alluvial effects, and where the record of alluviation is an extended one. Second, rivers and their floodplains are not, contrary to some assumptions, environmentally equilibrated. Channels may be incising, aggrading, sedimenting and accreting in response to longer-term patterns of environmental change to which the record of floodplain sediments provides the key. Third,

analysis of individual valley-floor reaches is required because the history that any location has undergone is as yet not 'retrodictable', nor is its future predictable, on the basis of present knowledge. Figure 10.6 illustrates, for example, the vertical sequence that a Welsh river is believed to have undergone since Late Devensian times. This sequence is site specific; it draws attention to the potential for vertical instability as well as the lateral channel mobility that floodplains have been created by (see for example Fig. 10.3). It illustrates both the alluvial 'recovery' from partial infilling of the valley with glacial outwash, but also the significant effects of human-induced effects. Figure 10.7 shows part of the San River valley near Jaroslaw in Poland. Here considerable changes in sedimentation style can again be seen: large-scale meandering in the Late Glacial (*c.* 10 000 BP), smaller meanders in the earlier Holocene, and then braiding most recently. The present river channel is incised into these deposits.

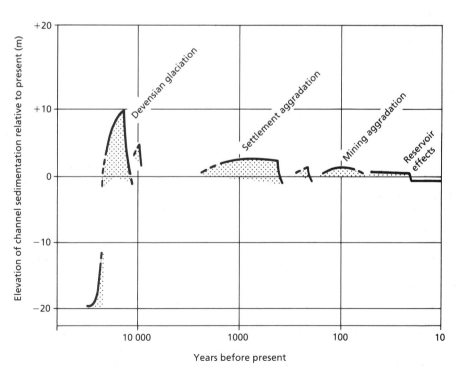

Fig. 10.6 Vertical tendencies in alluviation of the Afon Rheidol, Dyfed, UK, based on data from Macklin and Lewin (1986). Note the logarithmic timescale.

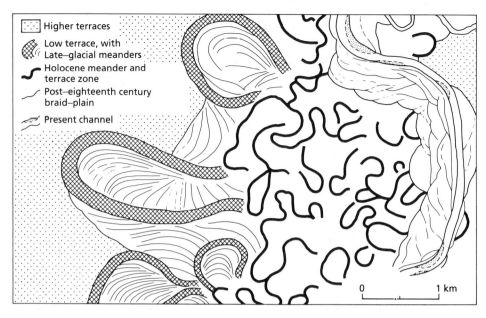

Fig. 10.7 A part of the San River valley near Jaroslaw, Poland (after Szumanski 1983).

10.5 FLOODPLAIN QUALITY

Floodplains can be important accumulation zones for pollutants that are dispersed by fluvial activity from point or areal sources. This can lead to soil pollution *in situ*, or to further river contamination as floodplain sediments are mechanically re-eroded and incorporated into active river systems. Metal mining pollution has been a particular focus of concern and study, but other substances derived from agricultural treatments, urban effluent and industrial and mining activities may also be incorporated into floodplain materials. Under 'natural' conditions floodplains benefit from nutrient enrichment, but pollution effects may become significant in certain environments downvalley from source areas.

Metal dispersal patterns in alluvial sediments have been widely studied in the UK (Lewin & Macklin 1987), in Belgium and the Netherlands (Leenaers *et al* 1988), in Poland (Klimek & Zawilinska 1985), in Australia (East *et al* 1988), and North America (Miller & Wells 1986; Knox 1987; Marron 1989; Graf 1990). A prime aim has been to establish downvalley distance/concentration relationships together with the sediment size

ranges or sedimentary environments which may be particularly enriched. Knowledge of the historical development of floodplains in relation to the timing of pollution inputs is also helpful. In addition to the passive dispersal of metals, the volume of wastes may also transform floodplains, causing aggradation and then possibly incision once mining activity and waste input ceases (Lewin *et al* 1983). Particularly dramatic were the effects of hydraulic mining during the 19th century California gold rush (James 1989).

Figure 10.8 shows a schematic cross-section of the South Tyne, UK, with levels of sedimentation and alluvial metal content illustrated. This valley aggraded in the 19th century when metal mining was active in the catchment; the river has subsequently become incised, with a decrease in the metal content of deposited sediment (Macklin & Lewin 1989). Downstream from urban, industrial or mining activities, floodplains are liable to suffer quality problems of this kind. Fine suspension deposits may be particularly affected, to the extent that remobilization by river erosion or indeed the use of particular soils for activities such as market gardening may be undesirable.

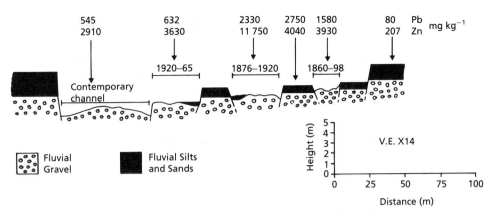

Fig. 10.8 A sketch section of the South Tyne valley floor at Broomhouse, Northumberland, UK (NY 6962), showing alluvial sediments and metal concentrations (after Macklin & Lewin 1989).

10.6 CONCLUSIONS

Floodplains are in large measure produced by physical processes of river deposition in a range of sedimentary subenvironments whose varying dominance produces a considerable global and regional variety to floodplain forms. Studies in recent years have also stressed that floodplains have a history, recording environmental changes of recent centuries and millenia, and not least the effects of human actions on geomorphological activity and sediment quality. There are no simple global models for floodplain formation: instead it is important that individual site investigations should be undertaken and the variety of forms, sediments and processes be recorded. Present and potential future interactions between rivers and floodplains may then be properly understood.

REFERENCES

Alexander CS, Prior JC. (1971) Holocene sedimentation rates in overbank deposits in the Black Bottom of the Lower Ohio River, Southern Illinois. *American Journal of Science* **270**: 361–72. [10.3]

Beven K, Carling P. (1989) *Floods: Hydrological, Sedimentological and Geomorphological implications.* John Wiley & Sons, Chichester. [10.4]

Blake DH, Ollier CD. (1971) Alluvial plains of the Fly River, Papua. *Zeitschrift für Geomorphologie* **12** (Suppl.): 1–17. [10.2]

Brackenridge GR. (1980) Widespread episodes of stream erosion during the Holocene and their climatic cause. *Nature* **283**: 655–6. [10.4]

Bravard JP. (1989) La metamorphose des rivières des Alpes Françaises à la fin du moyenage et à l'époque moderne. *Bulletin de la Société Géographique du Liège* **25**: 145–57. [10.4]

Bridge JS, Leeder MR. (1979) A simulation model of alluvial stratigraphy. *Sedimentology* **26**: 617–44. [10.3]

Bristow CS. (1987) Brahmaputra River: channel migration and deposition. In: Ethridge FG, Flores RM, Harvey MD (eds) *Recent Developments in Fluvial Sedimentology*, pp 63–74. Society of Economic Paleontologists and Mineralogists, Tulsa, Oklahoma. [10.2]

Brizga SO, Finlayson BL. (1990) Channel avulsion and river metamorphosis: the case of the Thomson River, Victoria, Australia. *Earth Surface Processes and Landforms* **15**: 391–404. [10.3]

Brown AG. (1987) Holocene floodplain sedimentation and channel response of the lower River Severn, United Kingdom. *Zeitschrift für Geomorphologie* **31**: 293–310. [10.3, 10.4]

Butzer KW (1981) Holocene alluvial sequences: problems of dating and correlation. In: Cullingford RA, Davidson DA, Lewin J (eds) *Time-scales in Geomorphology*, pp 131–42. John Wiley & Sons, Chichester. [10.4]

Carling PA (1986) The Noon Hill flash floods: July 1983. Hydrological and geomorphological aspects of a major formative event in an upland catchment. *Transactions of the Institute of British Geographers* **NS 11**: 105–18. [10.4]

Chisholm NWT, Collin RL. (1980) Topographic definition for washland mapping. *Photogrammetric Record* **10**: 215–27. [10.1]

Church M. (1987) Palaeohydrological reconstructions from a Holocene valley fill. In: Miall AD (ed.) *Fluvial Sedimentology*, pp 743–72. Canadian Society of Petroleum Geologists, Calgary. [10.1]

Church M, Rider JM. (1972) Paraglacial sedimentation; a consideration of fluvial processes conditioned by glaciation. *Geological Society of America Bulletin* **83**: 3059–72. [10.4]

Coleman JM. (1969) Brahmaputra River: channel processes and sedimentation. *Sedimentology and Geology* **3**: 129–239. [10.2]

Crowley KD. (1983) Large-scale bed configurations (macroforms), Platte River Basin, Colorado and Nebraska: primary structures and formative processes. *Geological Society of America Bulletin* **94**: 117–33. Washington, DC. [10.2]

Dietrich WE, Dunne T, Humphrey NF, Reid LM. (1982) Construction of sediment budgets for drainage basins. In: Swanson FJ, Jands RJ, Dunne T, Swanson DN (eds) *Sediment Budgets and Routing in Forested Drainage Basins*, pp 5–23. US Department of Agriculture, Forest Service General Technical Report PNW-141. [10.3]

East TJ, Cull RF, Murray AS, Duggan K. (1988) Fluvial dispersion of radioactive mill tailings in the seasonally wet tropics, northern Australia. In: Warner RF (ed.) *Fluvial Geomorphology of Australia*, pp 303–22. Academic Press, Sydney. [10.5]

Erskine WD, Warner RF. (1988) Geomorphic effects of alternating flood- and drought-dominated regimes on NSW coastal rivers. In: Warner RF (ed.) *Fluvial Geomorphology of Australia*, pp 223–44. Academic Press, Sydney. [10.4]

Everitt BL. (1968) Use of the cottonwood in an investigation of the recent history of a floodplain. *American Journal of Science* **266**: 417–39. [10.3]

Farrell KM. (1987) Sedimentology and facies architecture of overbank deposits of the Mississippi River, False River Region Louisiana. In: Ethridge FG, Flores RM, Harvey MD (eds) *Recent Developments in Fluvial Sedimentology*, pp 111–20. Society of Economic Paleontologists and Mineralogists, Tulsa, Oklahoma. [10.2]

Ferguson RI. (1987) Hydraulic and sedimentary controls of channel pattern. In: Richards KS (ed.) *River Channels: Environment and Process*, pp 129–58. Blackwell Scientific Publications, Oxford. [10.2]

Ferguson RI, Werritty A. (1983) Bar development and channel changes in the gravelly River Feshie, Scotland. In: Collinson JD, Lewin J (eds) *Modern and Ancient Fluvial Systems*, pp 181–93. Blackwell Scientific Publications, Oxford. [10.3]

Graf WL. (1988) *Fluvial Processes in Dryland Rivers*. Springer-Verlag, New York. [10.2]

Graf WL. (1990) Fluvial dynamics of thorium-230 in the Church Rock Event, Puerco River, New Mexico. *Annals of the Association of American Geographers* **80**: 327–42. [10.5]

Harvey A. (1986) Geomorphic effects of a 100-year storm in the Howgill Fells, northwest England. *Zeitschrift für Geomorphologie* **30**: 71–91. [10.4]

Hickin EJ, Nanson GC. (1984) Lateral migration rates of river bends. *Journal of Hydraulic Engineering* **110**: 1557–67. [10.3]

Hooke JM. (1980) Magnitude and distribution of rates of river bank erosion. *Earth Surface Processes* **5**: 143–57. [10.3]

Ikeda H. (1989) Sedimentary controls on channel migration and origin of point bars in sand-bedded meandering rivers. In: Ikeda S, Parker G (eds) *River Meandering*, pp 51–68. American Geophysical Union, Water Resources Monograph No. 12. Washington, DC. [10.2]

Jackson RG. (1978) Preliminary evaluation of lithofacies models for meandering alluvial streams. In: Miall AD (ed.) *Fluvial Sedimentology*, pp 543–76. Canadian Society of Petroleum Geologists, Calgary. [10.2]

James LA. (1989) Sustained storage and transport of hydraulic gold mining sediment in the Bear River, California. *Annals of the Association of American Geographers* **79**: 570–92. [10.5]

Kelsey HM. (1980) A sediment budget and an analysis of geomorphic processes in the Van Duzen River basin, north coastal California, 1941–1975. *Geological Society of America Bulletin Part II*, **91**: 1119–216. [10.1]

Kelsey HM, Lamberson R, Madej MA. (1986) Modeling the transport of stored sediment in a gravel bed river, northwestern California. In: *Basin Sediment Delivery*, pp 367–91. IASH Publication No. 159. Wallingford, UK. [10.3]

Kirkby MJ. (1967) Measurement and theory of soil creep. *Journal of Geology* **75**: 359–78.

Klimek K, Zawilinska I. (1985) Trace elements in alluvium of the Upper Vistula as indicators of palaeohydrology. *Earth Surface Processes and Landforms* **10**: 273–80. [10.5]

Knight DM. (1989) Hydraulics of flood channels. In: Beven K, Carling P (eds) *Floods: Hydrological, Sedimentological and Geomorphological Implications*, pp 83–105. John Wiley & Sons, Chichester. [10.2]

Knox JC. (1983) Responses of river systems to Holocene climates. In: Wright HE Jr (ed.) *Late Quaternary Environments of the United States*, pp 26–41. University of Minnesota Press, Minneapolis. [10.4]

Knox JC. (1987) Historical valley floor sedimentation in the Upper Mississippi valley. *Annals of the Association of American Geographers* **77**: 224–44. [10.3, 10.5]

Lambert CP, Walling DE. (1987) Floodplain sedimentation: a preliminary investigation of contemporary deposition rates within the lower reaches of the River Culm, Devon, UK. *Geografiska Annaler* **69A**: 393–404. [10.3]

Lambert JM, Jennings JN, Smith CT, Green C, Hutchinson JN. (1960) *The Making of the Broads*. R.G.S. Research Series 3. [10.4]

Lattman LH. (1960) Cross section of a floodplain in a moist region of modest relief. *Journal of Sedimentary Petrology* 30: 275–82. [10.2]

Leenaers H, Schouten CJ, Rang MC. (1988) Variability of the metal content of flood deposits. *Environmental Geology and Water Science* 11: 95–106. [10.5]

Lehre AK. (1982) Sediment budget of a small Coast Range drainage basin in north-central California. In: Swanson FJ, Janda RJ, Dunne T, Swanston DN (eds) *Sediment Budgets and Routing in Forested Drainage Basins*, pp 67–77. US Department of Agriculture, Forest Service General Technical Report PNW-141. Washington, DC. [10.1]

Lewin J. (1983) Changes of channel patterns and floodplains. In: Gregory KJ (ed.) *Background to Palaeohydrology*, pp 303–19. John Wiley & Sons, Chichester. [10.2]

Lewin J, Macklin MG. (1987) Metal mining and floodplain sedimentation in Britain. In: Gardiner V (ed.) *International Geomorphology 1986*, Part I, pp 1009–27. John Wiley & Sons, Chichester. [10.5]

Lewin J, Manton MMM. (1975) Welsh floodplain studies: the nature of floodplain geometry. *Journal of Hydrology* 25: 37–50. [10.1]

Lewin J, Hughes D, Blacknell C. (1977) Incidence of river erosion. *Area* 9: 177–80. [10.3]

Lewin J, Bradley SB, Macklin MG. (1983) Historical valley alluviation in mid-Wales. *Geological Journal* 18: 331–50. [10.5]

Lewis GW, Lewin J. (1983) Alluvial cutoffs in Wales and the Borderland. In: Collinson JD, Lewin J (eds) *Modern and Ancient Fluvial Systems*, pp 145–54. Blackwell Scientific Publications, Oxford. [10.3]

Macklin MG, Lewin J. (1986) Terraced fills of Pleistocene and Holocene age in the Rheidol Valley, Wales. *Journal of Quaternary Science* 1: 21–34. [10.4]

Macklin MG, Lewin J. (1989) Sediment transfer and transformation of an alluvial valley floor: the River South Tyne, Northumbria, UK. *Earth Surface Processes and Landforms* 14: 233–46. [10.4, 10.5]

Maizels J. (1987) Large-scale flood deposits associated with the formation of coarse-grained braided terrace sequences. In: Ethridge FG, Flores RM, Harvey MD (eds) *Recent Developments in Fluvial Sedimentology*, pp 135–48. Society of Economic Paleontologists and Mineralogists, Tulsa, Oklahoma. [10.1]

Marron DC. (1989) Physical and chemical characteristics of a metal contaminated overbank deposit, west-central South Dakota, USA. *Earth Surface Processes and Landforms* 14: 419–32. [10.5]

Miall AD. (1977) A review of the braided river depositional environment. *Earth Science Reviews* 13: 1–62. [10.2]

Miall AD. (1978) Lithofacies types and vertical profile models in braided river deposits: a summary. In: Miall AD (ed.) *Fluvial Sedimentology*, pp 597–604. Canadian Society of Petroleum Geologists, Calgary. [10.1]

Miall AD. (1985) Architectural-element analysis: a new method of facies analysis applied to fluvial deposits. *Earth Science Reviews* 22: 261–308. [10.2]

Miller JR, Wells SG. (1986) Types and processes of short term sediment and uranium tailings storage in arroyos: an example from the Rio Puerco of the West, New Mexico. In: *Basin Sediment Delivery*, pp 335–53. IASH Publication No. 159. Wallingford, UK. [10.5]

Nakamura F, Araya T, Higashi S. (1987) Influence of river channel morphology and sediment production on residence time and transport distance. In: *Erosion and Sedimentation in the Pacific Rim*, pp 355–64. IASH Publication No. 165. Wallingford, UK. [10.3]

Nanson GC. (1986) Episodes of vertical accretion and catastrophic stripping: a model of disequilibrium flood-plain development. *Geological Society of America Bulletin* 97: 1467–75. [10.4]

Petts GE. (ed.) (1989) *Historical Change of Large Alluvial Rivers: Western Europe*. John Wiley & Sons, Chichester. [10.2]

Pizzuto JE. (1987) Sediment diffusion during overbank flows. *Sedimentology* 34: 301–17. [10.3]

Probst J-L. (1989) Hydroclimatic fluctuations of some European rivers since 1800. In: Petts GE (ed.) *Historical Changes of Large Alluvial Rivers — Western Europe*, pp 41–55. John Wiley & Sons, Chichester. [10.4]

Rains B, Welch J. (1988) Out-of-phase Holocene terraces in part of the North Saskatchewan River basin, Alberta. *Canadian Journal of Earth Sciences* 25: 454–64. [10.3]

Roberts RG, Church M. (1986) The sediment budget in severely disturbed watersheds, Queen Charlotte Ranges, British Columbia. *Canadian Journal of Forest Research* 16: 1092–106. [10.4]

Robinson MA, Lambrick GH. (1984) Holocene alluviation and hydrology in the upper Thames basin. *Nature* 308: 809–14. [10.4]

Rust BR, Nanson GC. (1986) Contemporary and paleochannel patterns and the Late Quaternary stratigraphy of Cooper Creek, Southwest Queensland, Australia. *Earth Surface Processes and Landforms* 11: 581–90. [10.2]

Schmudde JH. (1963) Some aspects of the landforms of the lower Missouri river floodplain. *Annals of the Association of American Geographers* 53: 60–73. [10.2]

Schumm SA. (1977) *The Fluvial System*. John Wiley & Sons, New York. [10.3]

Schumm SA. (1986) Alluvial river response to active tectonics. In: Wallace R (ed.) *Active Tectonics*, pp 80–94. Geophysical Research Forum, National Academic Press, Washington. [10.4]

Schumm SA, Brackenridge GR. (1987) River responses.

Table 11.1 Human influences on channel change

Indirect change
Land use change:
   Deforestation
   Afforestation
   Agricultural (e.g. conversion of grazing to arable)
   Urbanization
   Mining

Land drainage:
   Agricultural drainage
   Surface water sewers

Direct change
Regulation:
   Impoundment of water
   Water diversions (e.g. for irrigation)

Channel management:
   Gravel extraction
   Straightening
   Flood control
   Bank erosion protection
   Dredging

produce bank erosion and sediment transport. These factors tend to determine the cross-sectional shape, pool-riffle formation and meander shape of alluvial river channels. A natural channel is neither straight nor uniform. At the catchment scale, its size increases downstream as tributaries join and increase the flow; as width and depth increase downstream the slope decreases; the size of the sediment load on the bed decreases, typically with cobbles and gravels in the middle reaches and sand or silt farther downstream. Even within a single reach a natural channel is non-uniform: the bed undulates as a series of pools (topographically low areas) and riffles (high areas); the flow varies considerably over short distances and through time over a range from high to low discharge; the bed material of a pool is typically fine-grained sand, whilst a riffle has a concentration of larger rock sizes (gravels).

Pools and riffles result from the interaction of flowing water and mobile sediments. At very high discharges the flow tends to converge at pools, causing scour, whilst the flow is divergent at riffles, corresponding to deposition. In contrast, at low flows a pool is characterized by relatively deep, slow-moving water, and a riffle by fast-flow and a steep water-surface gradient. The cross-section is often symmetrical at a riffle. A meander bend typically has an area of deposition on the inside and an area of erosion at the outer bank. Deposition of material produces a point bar adjacent to a pool, forming an asymmetrical channel shape. The shapes of natural river channels may be complicated further by a variety of bar forms, either attached to the bank or formed mid-channel. A classification of more complex morphological types by Kellerhalls *et al* (1976), includes bars, channel patterns (or planforms) and islands which may form in braided river channels.

Although a natural river channel is formed and maintained by the flow it carries, it is never of sufficient size to contain all discharges. Larger floods invariably overflow the normal channel banks onto the adjacent floodplain, on average about once per year, and this process is important in building a floodplain through deposition of sediment. It is also natural for alluvial channels with sufficiently powerful streams to erode laterally at various rates across their floodplains.

This hydraulic and morphological variability determines different ecological habitats in various parts of a river channel and adjacent floodplain. At low discharges there are a variety of conditions for feeding, breeding and cover, ranging from slow, deep flow in pools to fast, shallow water on riffles. Sufficient water depth is maintained during dry periods to support aquatic life. Habitat diversity and fish diversity are positively associated (Wesche 1985).

Pools, riffles and bars are composed of different bedload materials which provide environments for a variety of benthic organisms. Substrate size, heterogeneity, frequency of turnover of sediment, and the rates of erosional or depositional processes are all thought to determine invertebrate diversity and abundance (Petts 1984). In particular the size of substrate has been related by numerous investigators to the standing crops of benthic invertebrates. According to Pennak and Van Gerpen (1947) benthic invertebrates decrease in number in the series rubble, bedrock, gravel and sand. A similar decrease in the series rubble, coarse gravel, sand, fine gravel and silt was reported by Kimble and Wesche (1975). Larger

Table 11.1 Human influences on channel change

Indirect change
Land use change:
  Deforestation
  Afforestation
  Agricultural (e.g. conversion of grazing to arable)
  Urbanization
  Mining

Land drainage:
  Agricultural drainage
  Surface water sewers

Direct change
Regulation:
  Impoundment of water
  Water diversions (e.g. for irrigation)

Channel management:
  Gravel extraction
  Straightening
  Flood control
  Bank erosion protection
  Dredging

produce bank erosion and sediment transport. These factors tend to determine the cross-sectional shape, pool-riffle formation and meander shape of alluvial river channels. A natural channel is neither straight nor uniform. At the catchment scale, its size increases downstream as tributaries join and increase the flow; as width and depth increase downstream the slope decreases; the size of the sediment load on the bed decreases, typically with cobbles and gravels in the middle reaches and sand or silt farther downstream. Even within a single reach a natural channel is non-uniform: the bed undulates as a series of pools (topographically low areas) and riffles (high areas); the flow varies considerably over short distances and through time over a range from high to low discharge; the bed material of a pool is typically fine-grained sand, whilst a riffle has a concentration of larger rock sizes (gravels).

Pools and riffles result from the interaction of flowing water and mobile sediments. At very high discharges the flow tends to converge at pools, causing scour, whilst the flow is divergent at riffles, corresponding to deposition. In contrast, at low flows a pool is characterized by relatively deep, slow-moving water, and a riffle by fast-flow and a steep water-surface gradient. The cross-section is often symmetrical at a riffle. A meander bend typically has an area of deposition on the inside and an area of erosion at the outer bank. Deposition of material produces a point bar adjacent to a pool, forming an asymmetrical channel shape. The shapes of natural river channels may be complicated further by a variety of bar forms, either attached to the bank or formed mid-channel. A classification of more complex morphological types by Kellerhalls *et al* (1976), includes bars, channel patterns (or planforms) and islands which may form in braided river channels.

Although a natural river channel is formed and maintained by the flow it carries, it is never of sufficient size to contain all discharges. Larger floods invariably overflow the normal channel banks onto the adjacent floodplain, on average about once per year, and this process is important in building a floodplain through deposition of sediment. It is also natural for alluvial channels with sufficiently powerful streams to erode laterally at various rates across their floodplains.

This hydraulic and morphological variability determines different ecological habitats in various parts of a river channel and adjacent floodplain. At low discharges there are a variety of conditions for feeding, breeding and cover, ranging from slow, deep flow in pools to fast, shallow water on riffles. Sufficient water depth is maintained during dry periods to support aquatic life. Habitat diversity and fish diversity are positively associated (Wesche 1985).

Pools, riffles and bars are composed of different bedload materials which provide environments for a variety of benthic organisms. Substrate size, heterogeneity, frequency of turnover of sediment, and the rates of erosional or depositional processes are all thought to determine invertebrate diversity and abundance (Petts 1984). In particular the size of substrate has been related by numerous investigators to the standing crops of benthic invertebrates. According to Pennak and Van Gerpen (1947) benthic invertebrates decrease in number in the series rubble, bedrock, gravel and sand. A similar decrease in the series rubble, coarse gravel, sand, fine gravel and silt was reported by Kimble and Wesche (1975). Larger

11: River Channel Change

A. BROOKES

11.1 INTRODUCTION

The morphology of a river channel is formed by the movement of water and sediment in relation to the material locally available in the bed and banks (see Chapter 9). Hydraulic and morphological variability through space and time determine the different habitats found both within a given river channel and also in the adjacent riparian and floodplain zones. Channels change in a variety of ways through the processes of erosion and deposition, processes that are affected by catchment characteristics. Surface runoff and sediment load may be expected to change in response to land use alterations such as deforestation, agriculture, grazing, urbanization and other influences. However, a change in the discharge and sediment load rarely produces an immediate response but instead initiates a change, or sequence of changes, which may extend over a long period of time. Geomorphological studies have concentrated mainly on understanding the effects of climatic change over the period 10 000–15 000 years (Gregory *et al* 1987) or human impacts over the last 50 years (e.g. Gregory 1977a). Relatively little attention has been given to changes over the historic timescale of 100–500 years, a period of intensifying human impact (Petts 1989a).

Channel changes resulting from the historic development of land and water resources have undoubtedly been extensive (see Petts 1994). Over the last 2000 years, and particularly during the last 300 years, human impacts have had an increasing influence on drainage basins and their watercourses. Table 11.1 lists some of the major human impacts on river channels, which continue today. At present, dams over 15 m high are being completed throughout the world at a rate of about 500 per year. By the year 2000 it is estimated that 60% of the total streamflow in the world will be regulated (Petts 1989a). Understanding the ways in which river channels have changed through historical time may be important for ecologically-sound management. It may be desirable for conservation purposes to restore a river channel to a condition existing prior to development. It may also be necessary for the purposes of environmental impact assessment to determine the magnitude and rate of change arising from a variety of impacts. Also, analyses of historical channel changes should be an integral part of the concept and design of alternative management strategies (see Calow & Petts 1994).

Alterations of channel size and shape, and substrate characteristics, have been shown to have major impacts on benthic invertebrates (e.g. Keefer & Maughan 1985; Petts & Greenwood 1985), fish (e.g. Milhous 1982) and aquatic macrophytes (e.g. Brookes 1987). Changes of river channel mobility and pattern significantly influence instream and floodplain habitats. This chapter outlines some of the main causes and types of channel change and the biological impacts which may arise from such perturbations. The final section identifies the problems that need to be addressed further if ecologically-sound management techniques are to be explored further and improved.

11.2 MORPHOLOGICAL VARIABILITY

It is important to appreciate that hydraulic factors such as depth, slope and velocity of flow directly

In: Ruddiman WF, Wright HE (eds) *North America and Adjacent Oceans during the Last Deglaciation*, pp 221–40. Geological Society of America, Boulder, Colorado. [10.4]

Smith DG, Smith ND. (1980) Sedimentation in anastomosed river systems: examples from alluvial valleys near Banff, Alberta. *Journal of Sedimentary Petrology* **50**: 157–64. [10.2]

Smith ND, Cross TA, Dufficy JP, Clough SR. (1989) Anatomy of an avulsion. *Sedimentology* **36**: 1–23. [10.3]

Stear WM. (1985) Comparison of the bedform distribution and dynamics of modern and ancient sandy ephemeral flood deposits in the south-western Karoo region, South Africa. *Sedimentary Geology* **45**: 209–30. [10.3]

Szumanski A. (1983) Paleochannels of large meanders in the river valleys of the Polish lowlands. *Quaternary Studies in Poland* **4**: 207–16. [10.4]

Trimble SW. (1981) Changes in sediment storage in the Coon Creek basin, Driftless area, Wisconsin 1953–1975. *Science* **214**: 181–3. [10.3]

Wells LE. (1990) Holocene history of the El Niño phenomenon as recorded in flood sediments of northern coastal Peru. *Geology* **18**: 1134–7. [10.4]

Williams GP. (1983) Paleohydrological methods and some examples from Swedish fluvial environments, 1 Cobble and boulder deposits. *Geografiska Annaler* **65A**: 227–43. [10.1]

Wolman MG, Leopold LB. (1957) River floodplains: some observations on their formation. *US Geological Survey Professional Paper 282-C, 83–109*. [10.2, 10.3]

Lattman LH. (1960) Cross section of a floodplain in a moist region of modest relief. *Journal of Sedimentary Petrology* **30**: 275–82. [10.2]

Leenaers H, Schouten CJ, Rang MC. (1988) Variability of the metal content of flood deposits. *Environmental Geology and Water Science* **11**: 95–106. [10.5]

Lehre AK. (1982) Sediment budget of a small Coast Range drainage basin in north-central California. In: Swanson FJ, Janda RJ, Dunne T, Swanston DN (eds) *Sediment Budgets and Routing in Forested Drainage Basins*, pp 67–77. US Department of Agriculture, Forest Service General Technical Report PNW-141. Washington, DC. [10.1]

Lewin J. (1983) Changes of channel patterns and floodplains. In: Gregory KJ (ed.) *Background to Palaeohydrology*, pp 303–19. John Wiley & Sons, Chichester. [10.2]

Lewin J, Macklin MG. (1987) Metal mining and floodplain sedimentation in Britain. In: Gardiner V (ed.) *International Geomorphology 1986*, Part I, pp 1009–27. John Wiley & Sons, Chichester. [10.5]

Lewin J, Manton MMM. (1975) Welsh floodplain studies: the nature of floodplain geometry. *Journal of Hydrology* **25**: 37–50. [10.1]

Lewin J, Hughes D, Blacknell C. (1977) Incidence of river erosion. *Area* **9**: 177–80. [10.3]

Lewin J, Bradley SB, Macklin MG. (1983) Historical valley alluviation in mid-Wales. *Geological Journal* **18**: 331–50. [10.5]

Lewis GW, Lewin J. (1983) Alluvial cutoffs in Wales and the Borderland. In: Collinson JD, Lewin J (eds) *Modern and Ancient Fluvial Systems*, pp 145–54. Blackwell Scientific Publications, Oxford. [10.3]

Macklin MG, Lewin J. (1986) Terraced fills of Pleistocene and Holocene age in the Rheidol Valley, Wales. *Journal of Quaternary Science* **1**: 21–34. [10.4]

Macklin MG, Lewin J. (1989) Sediment transfer and transformation of an alluvial valley floor: the River South Tyne, Northumbria, UK. *Earth Surface Processes and Landforms* **14**: 233–46. [10.4, 10.5]

Maizels J. (1987) Large-scale flood deposits associated with the formation of coarse-grained braided terrace sequences. In: Ethridge FG, Flores RM, Harvey MD (eds) *Recent Developments in Fluvial Sedimentology*, pp 135–48. Society of Economic Paleontologists and Mineralogists, Tulsa, Oklahoma. [10.1]

Marron DC. (1989) Physical and chemical characteristics of a metal contaminated overbank deposit, west-central South Dakota, USA. *Earth Surface Processes and Landforms* **14**: 419–32. [10.5]

Miall AD. (1977) A review of the braided river depositional environment. *Earth Science Reviews* **13**: 1–62. [10.2]

Miall AD. (1978) Lithofacies types and vertical profile models in braided river deposits: a summary. In: Miall AD (ed.) *Fluvial Sedimentology*, pp 597–604.

Canadian Society of Petroleum Geologists, Calgary. [10.1]

Miall AD. (1985) Architectural-element analysis: a new method of facies analysis applied to fluvial deposits. *Earth Science Reviews* **22**: 261–308. [10.2]

Miller JR, Wells SG. (1986) Types and processes of short term sediment and uranium tailings storage in arroyos: an example from the Rio Puerco of the West, New Mexico. In: *Basin Sediment Delivery*, pp 335–53. IASH Publication No. 159. Wallingford, UK. [10.5]

Nakamura F, Araya T, Higashi S. (1987) Influence of river channel morphology and sediment production on residence time and transport distance. In: *Erosion and Sedimentation in the Pacific Rim*, pp 355–64. IASH Publication No. 165. Wallingford, UK. [10.3]

Nanson GC. (1986) Episodes of vertical accretion and catastrophic stripping: a model of disequilibrium flood-plain development. *Geological Society of America Bulletin* **97**: 1467–75. [10.4]

Petts GE. (ed.) (1989) *Historical Change of Large Alluvial Rivers: Western Europe.* John Wiley & Sons, Chichester. [10.2]

Pizzuto JE. (1987) Sediment diffusion during overbank flows. *Sedimentology* **34**: 301–17. [10.3]

Probst J-L. (1989) Hydroclimatic fluctuations of some European rivers since 1800. In: Petts GE (ed.) *Historical Changes of Large Alluvial Rivers — Western Europe*, pp 41–55. John Wiley & Sons, Chichester. [10.4]

Rains B, Welch J. (1988) Out-of-phase Holocene terraces in part of the North Saskatchewan River basin, Alberta. *Canadian Journal of Earth Sciences* **25**: 454–64. [10.3]

Roberts RG, Church M. (1986) The sediment budget in severely disturbed watersheds, Queen Charlotte Ranges, British Columbia. *Canadian Journal of Forest Research* **16**: 1092–106. [10.4]

Robinson MA, Lambrick GH. (1984) Holocene alluviation and hydrology in the upper Thames basin. *Nature* **308**: 809–14. [10.4]

Rust BR, Nanson GC. (1986) Contemporary and paleochannel patterns and the Late Quaternary stratigraphy of Cooper Creek, Southwest Queensland, Australia. *Earth Surface Processes and Landforms* **11**: 581–90. [10.2]

Schmudde JH. (1963) Some aspects of the landforms of the lower Missouri river floodplain. *Annals of the Association of American Geographers* **53**: 60–73. [10.2]

Schumm SA. (1977) *The Fluvial System.* John Wiley & Sons, New York. [10.3]

Schumm SA. (1986) Alluvial river response to active tectonics. In: Wallace R (ed.) *Active Tectonics*, pp 80–94. Geophysical Research Forum, National Academic Press, Washington. [10.4]

Schumm SA, Brackenridge GR. (1987) River responses.

Table 11.2 Key geomorphological factors influencing river ecology over different time-scales

Factor	Typical time-scale of change	Method used to determine change of river morphology	Knowledge of biological response over time	Knowledge of biological response using space as a substitute for time
Cross-sectional shape (e.g. bank collapse)	Hours, days, months, years (e.g. during extreme flood events)	Direct observation and measurement	Limited	Fair
Cross-sectional size	As above	As above	Limited	Fair
Pool and riffle sequence	As above	As above	Limited	Fair
Point bars	As above	As above	Limited	Some
Islands	As above	As above	Some	Fair
Substrate (including sedimentation)	As above	As above	Good	Good
Pattern (e.g. lateral migration)	100–50 years	Historical sources such as maps and documents	Limited	Fair
Pattern (e.g. braided to meandering)	250–15 000 years +	Sedimentary evidence; dating techniques (floral and faunal evidence; artifactual remains)	Not applicable	Some

substrate provides insects with a firm surface to hold on to and also provides some protection from the force of the current.

At higher discharges shelter areas are available as protection from higher water velocities. Overhanging banks cut on the outside of a bend adjacent to a pool may provide natural cover for fish. Overbank flooding and sedimentation are important processes affecting floodplain wetlands and wetland vegetation. Wetlands may be important for waterfowl, amphibians and reptiles and some mammals. Cutoff channels may support a number of breeding, feeding and nesting habitats.

Table 11.2 summarizes some key geomorphological factors of natural alluvial river channels which are important for biological and aesthetic diversity. Typical time-scales over which these factors have been observed to change are also shown. Whilst there are many studies which have examined the biological response of changed substrate, for example as a result of sedimentation, there is little knowledge of the impacts in changes of other factors. By contrast, a greater amount of knowledge of biological response is evident from the method of substituting space for time, which is sometimes referred to as the ergodic hypothesis (Table 11.2). Local changes within a single river are often inferred by comparison of biological populations between rivers or between 'natural' and 'changed' sections along individual rivers. The differences between the actual measured values and the expected ones form the basis for the estimation of the nature and amount of change arising from human activities such as land-use change, channelization or dam construction (but see Underwood 1994).

11.3 PRINCIPLES OF CHANNEL EQUILIBRIUM AND RESPONSE

A river channel transports both water and sediment and the flow is typically non-uniform, unsteady and turbulent. Under conditions of equilibrium, channels tend to be morphologically stable, transporting the water and sediment load imposed from the catchment upstream without

enlarging or aggrading. *It is necessary to understand that a channel which is described as stable may naturally erode its bed and banks.* This erosion may be a relatively slow process which may only be pronounced after a major flood discharge. As the channel migrates across the floodplain it maintains a stable average form. There are a whole range of channel types with different degrees of activity (see Chapters 9 & 10).

Channels that maintain a stable average form are described by geomorphologists as graded or in equilibrium, whereas engineers usually refer to them as 'in regime' or stable. It is only if different bank materials are encountered during migration, or a new discharge regime is imposed, that the channel becomes unstable. The equilibrium concept implies that stable width, depth, slope (and probably planform) can be expressed as functions of the controlling variables: discharge, sediment supply and channel bed and bank materials (Fig. 11.1). Fluctuations of channel form about an equilibrium condition are termed autogenic changes. Where the sequence of water flow and sediment transported by rivers is changed, then alterations in processes operating within the river channel occur as a direct result. These adjustments are referred to as allogenic changes.

Although a set of determinate equations fully describing stream behaviour have not yet been developed, research has succeeded in defining some of the functional relationships between stream variables. Used with caution, such relationships can suggest the types of channel adjustment to be expected from changes in the upstream catchment or from change in one or more stream characteristics. Some of the most useful relationships which can be used in planning and design are listed in Table 11.3, together with examples of the types of impact which cause river channels to change. The following discussion concentrates on those changes induced by human impact, which has undoubtedly been the strongest influence over the past 5000 years, rather than the effects of climatic change.

The types of channel change which may occur range from the sedimentary bedforms (e.g. pools and riffles), through the cross-section of the river channel, the river channel planform and the drainage network (Gregory 1976). For each of these

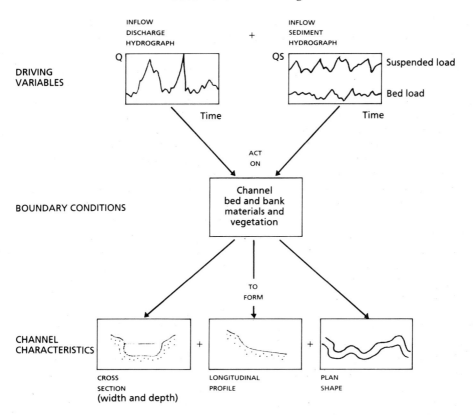

Fig. 11.1 Equilibrium concept for erodible channels.

scales there can be adjustments of size, shape and composition. Examples are given in Table 11.4.

11.4 LAND USE CHANGES AND SEDIMENT YIELD

Changes in the drainage basin may alter rainfall–runoff and sediment yield relationships and may have an indirect effect on channel characteristics. The most extensive changes affecting streams are land use changes attributable to agriculture, forestry, mining, grazing and urbanization. Deforestation for the purpose of grazing or cultivation began in Europe more than 3000 years ago. A major impact of deforestation is accelerated erosion on hillslopes, together with gully erosion and increased supply of sediment to the streams. Runoff characteristics are also affected by the conversion of forest to farming. Indeed, any change that eliminates or reduces vegetative

cover is likely to increase sediment discharge proportionately more than the water discharge. Mining (Graf 1979), forest fires (White & Wells 1979) and construction activities (Wolman 1967) are all examples of this type of disturbance. The conversion of forest to farmland in Wisconsin in the USA began in about 1820, leading to much higher sediment loads, generally causing channel-bed aggradation and overbank deposition on floodplains to depths in excess of 1 m (Meade 1976). Channels have tended to become wider, shallower and less sinuous and much of the eroded sediment is still stored in the channel and floodplains. Conversely, afforestation, improved land management or conservation practices reduce soil erosion and have the opposite effect.

In the Southern Piedmont of the USA rapid expansion of intensive crop agriculture from about 1800 until 1920 increased runoff and sediment yield (Trimble 1974). Equation 8 (Table 11.3)

Table 11.3 Functional relationships (after Schumm 1969)

Equations		Examples of change
$Qs^+, Qw^{++} \approx S^-, d_{50}{}^+, D^+, W^+$	(1)	Long-term effect of urbanization. Increased frequency and magnitude of discharge. Channel erosion (increasing width and depth)
$Qs^0, Qw^+ \approx S^-, d_{50}{}^+, D^+, W^+$	(2)	
$Qs^-, Qw^+ \approx S^-, d_{50}{}^+, D^+, W^-$	(3) }	Intensification of vegetation cover through afforestation and improved
$Qs^{--}, Qw^- \approx S^-, d_{50}{}^+, D^+, W^\pm$	(4) }	land management reduces sediment loads
$Qs^-, Qw^+ \approx S^-, d_{50}{}^+, D^+, W^\pm$	(5)	Diversion of water to a river
$Qs^+, Qw^+ \approx S^\pm, d_{50}{}^\pm, D^\pm, W^+$	(6) }	Commensurate changes of water and sediment discharge with
$Qs^-, Qw^- \approx S^\pm, d_{50}{}^\pm, D^\pm, W^-$	(7) }	unpredictable changes of slope, flow depth and bed material
$Qs^{++}, Qw^+ \approx S^+, d_{50}{}^-, D^-, W^+$	(8)	Land use change from forest to crop production. Sediment discharge increasing more rapidly than water discharge. Bed changes from gravel to sand, wider shallower channels
$Qs^+, Qw^0 \approx S^+, d_{50}{}^-, D^-, W^+$	(9)	
$Qs^0, Qw^- \approx S^+, d_{50}{}^-, D^-, W^-$	(10) }	
$Qs^-, Qw^{--} \approx S^+, d_{50}{}^-, D^-, W^-$	(11) }	Extraction of water from river resulting in narrower stream
$Qs^+, Qw^- \approx S^+, d_{50}{}^-, D^-, W^\pm$	(12)	Increased water and sediment discharge

D = flow depth; d_{50} = median particle size of bedmaterial; Qs = bedload (expressed as a percentage of total load); Qw = water discharge (e.g. mean annual discharge); S = channel slope; W = flow width.
0 indicates no change; $+/-$ indicates increase or decrease respectively; $++/--$ indicates a change of considerable magnitude.

Table 11.4 Potential river channel adjustments (after Gregory 1976)

Potential adjustments of	Fluvial landform		
	River channel cross-section	River channel pattern	Drainage network
Size	*Increase or decrease of river channel capacity* Erosion of bed and banks can produce a larger channel which maintains the same shape. Sedimentation can produce a smaller channel which maintains the same shape	*Increase or decrease of size of pattern* Increase or decrease of meander wavelength whilst preserving the same planform shape	*Increase or decrease of network extent and density* Extension of channels or shrinkage of perennial, intermittent and ephemeral streams
Shape	*Adjustment of shape* Width/depth ratio may be increased or decreased	*Alteration of shape of pattern* A change from regular to irregular meanders	*Drainage pattern changed in shape* Inclusion of new stream channels after deforestation
Composition	*Change in channel sediments* Alteration of grain size of sediments in bed and banks possibly accompanied by development of berms or bars	*Planform metamorphosis* Change from single to multi-thread channel or converse	*Network composition changed* The replacement of channels with no definite stream channel by a clearly defined channel

describes the situation, with sediment discharge (Qs) increasing more than water discharge (Qw). This predicts increased slope, decreased median sediment size, decreased depth of flow and increased flow width. The resulting channels are wider and shallower. These predictions conform to the findings of Trimble (1974) who noted that the composition of channel beds changed from gravel to sand. The increased frequency of flooding arising from the channel aggradation created extensive swamp and marshy areas.

After the 1920s improved conservation practices and reforestation caused drastic reductions in Qs and a moderate increase in Qw (equation 3). Trimble (1974) found that the streams began to erode their accumulated sediments, and channels became entrenched and narrower. Erosion of sandy sediments uncovered coarser bed gravels.

Changes such as these in turn may trigger adjustments in stream systems and associated biological and physical systems. One of the major adverse effects of inorganic sediments on aquatic ecosystems relates to sediment covering the substrate (Hynes 1973). The impacts on fish and aquatic invertebrates of sediments arising from land use changes such as deforestation have been reviewed by Cordone and Kelley (1961). The effects of finer sediments can be direct, by causing abrasive injuries to external organs such as gills, protective mucous coverings and fins, or by smothering nests and eggs. Indirect effects can range from elimination of a preferred food source to elimination of preferred reproductive habitats for fish and aquatic invertebrates. The problem of excess fine sediment within a river channel may be temporary and with the completion of a particular land disturbance activity, there may be

a tendency for reduced sediment loads and the gradual removal of deposited sediment.

11.5 URBANIZATION

Developing an area for housing or other urban purposes increases the area of low or zero infiltration capacity and increases the speed of water transmission in channels or surface water sewers (Leopold 1968; Hammer 1972; Dunne & Leopold 1978). Increased discharges arising from paved urban areas often cause enlargement of natural channels downstream in both temperate (e.g. Graf 1975; Richards & Wood 1977) and tropical environments (Whitlow & Gregory 1989). There may also be changes of substrate composition (Thoms 1987).

The impact of urbanization is complicated by the fact that the channel degradation itself, together with erosion of the catchment surface during construction, can cause large amounts of sediment to be released initially to the channel, which must subsequently be carried out of the system (Fig. 11.2). Wolman and Schick (1967) note that sediment which had accumulated within a channel was still present 7 years after construction activities had ended. It is only after the phase of temporary aggradation that the channel itself begins to enlarge in response to increased runoff from paved areas. The rate of enlargement also depends on the type, size and location of the urban area within the catchment.

As soon as a catchment is completely urbanized or development is stable, it is possible that the stream may become adjusted to the changed hydrological state (Fig. 11.2). The immediate results of increased storm runoff may be

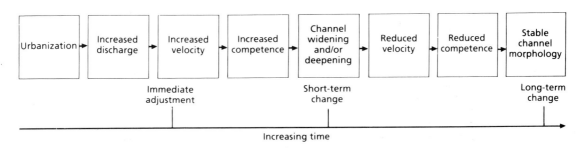

Fig. 11.2 Impact of urbanization (after Morisawa 1985).

increased velocities, leading to erosion of the channel bed and/or banks (see equation 1, Table 11.3). However, negative feedback may eventually occur, leading to a new equilibrium between the changed hydrological state, sediment load and morphology of the channel (Fig. 11.2). Associated with changes in the size of the channel there may be changes in channel shape, and below urban areas it has been suggested that width may change more easily than depth because of the armouring of the channel bed.

For the Avondale stream basin in Harare, Zimbabwe, Whitlow and Gregory (1989) showed how downstream reaches had widened by an average of 1.7 times over a period of 12 years, involving average rates of bank erosion of 0.33 m per year. This was in response to increased peak discharges, estimated to be two times former

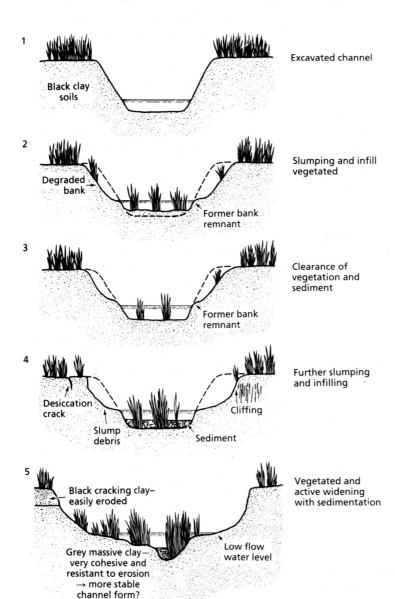

Fig. 11.3 Sequence of channel change in the lower part of the urbanized Avondale stream, Zimbabwe (from Whitlow & Gregory 1989, by permission of John Wiley & Sons).

values. Figure 11.3 depicts the sequence of change in the lower reaches. Essentially the original excavated channel has been modified progressively by human activity and natural processes to produce a broad, irregular channel which is still in a phase of active widening.

Studies of urban river sediments are limited, particularly those concerned with the textural characteristics of the channel bed sediments (Douglas 1985). A study of the River Tame, within and below the Birmingham conurbation, UK, provided information on the structural properties of the river bed (Thoms 1987). The study suggested that whilst the accumulation of fine matrix sediments arising from construction has an immediate impact, in the longer term, as a consequence of the changed hydrological regime, the river not only reduces this impact but also establishes a substrate containing less fine sediment. This impact can influence relatively long reaches. The ecological effects of such changes are little known as yet (Thoms 1987; Gilbert 1989). In general, and in unpolluted water, density of invertebrates tends to fall off as erosion increases (Percival & Whitehead 1929). The unusual patterns of erosion and siltation in urban channels coupled with pollution combine together in a variety of ways to reduce invertebrate numbers (Gilbert 1989).

11.6 ADJUSTMENTS OF RIVER CHANNEL CROSS-SECTIONS BELOW DAMS AND RESERVOIRS

Construction of the Thomson Dam, Australia, caused siltation of the channel substrate for a distance of 25 km downstream (Davey *et al* 1987). However, following construction the magnitude and frequency of peak discharges is often decreased below dams because of the effect of water storage. The sediment load becomes trapped in the reservoir. The channel below a dam typically scours and this has to be allowed for in the design of a dam to prevent subsequent undermining and failure (Komura & Simons 1967). Because reservoirs trap up to 95% of the bed load and suspended sediment carried by the river, this encourages scour immediately below the dam. Thus below Glen Canyon dam on the Colorado River, which

was constructed between 1956 and 1963, approximately 10 million m³ of sediment was scoured from below the dam (Pemberton 1976). However, the bed subsequently became armoured and stabilized between 1965 and 1975.

Williams and Wolman (1984) document changes that have occurred within alluvial river channels downstream of 21 dams in the USA over a period of up to 70 years. The reduction in flood peaks for the affected rivers ranged from 3% to 91%, with an average of 39% of the pre-dam values. Erosion of the channel bed had been experienced at all sites, with a maximum value of 7 m occurring immediately below the dam and decreasing further downstream. The response of channel width to dam closure was found to be more variable. Widths decreased, increased or remained constant depending on the site characteristics. These findings for dams in the USA are supported by information obtained from other geographical regions (see Petts 1984).

The rate of channel change below dams varies but studies suggest that the majority of bed degradation occurs within the first 10–15% of the period of adjustment. Bed degradation can continue and migrate downstream for several decades after dam closure at rates of up to 50 km per year. However, variations in the rate of migration of the zone of bed degradation are dependent on numerous factors, including the nature of the bed and bank materials and the discharge regime. Degradation below a dam may provide increasing sediment load in a downstream direction (Meade & Trimble 1974).

Deposition within regulated rivers may also occur because the regulation of high-magnitude floods artificially slows sediment transport, whilst the sediment supplied by tributaries is unaffected or possibly increased (Petts 1984). Coarse sediments introduced by tributary streams with relatively high gradients are therefore deposited in the main channel, which invariably has a lower slope and wide channel. An extensively documented study is that of the Colorado River below Glen Canyon Dam, where deposition has occurred at every major tributary confluence for distances of up to 24 km and beyond (Howard & Dolan 1981).

Benthic invertebrates are influenced by a

number of factors which are likely to change immediately upon dam closure. However, channel morphology and the composition and stability of the substrate are known to change slowly and it can be anticipated that changes in the flora and fauna are likely to occur over a period of many years. Sedimentation in the Thomson River, southeastern Australia, during construction of a dam reduced the total numbers of macroinvertebrate species and the density of species and individuals (Doeg *et al* 1987). Immediately below the Navajo Dam, Upper Colorado River, USA, densities of invertebrates were observed to increase during a 5-year period following closure of the dam, from 828 m^{-2} to 6727 m^{-2} (Mullan *et al* 1976). This change may be due to improved

bed stability following degradation and channel armouring. Increased benthos observed immediately below many dams may be a consequence of improved stability and removal of finer material (Pearson *et al* 1968; Petts 1984). However Voelz and Ward (1990) observed that species which feed on coarse sedimentary detritus (shredders) were absent from below a dam on the Blue River, Colorado.

Further downstream the invertebrate communities may relate to either erosional or sedimentational processes. Thus, 13 km below the Navajo Dam the density of invertebrates was 6727 m^{-2} and the dominant species were of the Ephemeroptera, Diptera, Hydropsychidae, Gastropoda and Amphipoda, all typical of er-

Table 11.5 Channel change and invertebrate distributions within the Afon Rheidol, below Nant-y-Moch Reservoir, Wales, UK (after Petts & Greenwood 1985)

Reach	Control channel	1	2	3	4
Number of samples	10	8	17	18	13
Mean substrate particle size (mm)	54	75	40	13	Gravel and boulders
Proportion less than 2 mm (%)	18.1	6.6	11.3	30.1	19.6
Channel width (m)	14	8	16	22	14
Substrate stability	Unstable	Stable	Unstable	Unstable	Stable
Proportion of sites with moss (%)	0	10	18	0	63
Average number of individuals (per m^2)	336	402	177	313	437
Number of species	10	13	12	21	23
Species diversity	0.917	1.239	1.833	2.250	2.298
Diptera (no. m^{-2})	275	309	91	76	153
Chironomidae	258	116	45	50	114
Simuliidae	16	189	42	2	19
Dicranota sp.	0	4	4	20	7
Ephemeroptera (no. m^{-2})	10	5	21	11	14
Plecoptera (no. m^{-2})	6.5	41.5	29	48	73
Diura bicaudata	5	1.5	5	0	3
Protonemura meyeri	0	7	13	3	41
Leuctra hippopus	1.5	33	11	45	29
Trichoptera (no. m^{-2})	39	41	42	116	163
Polycentropus sp.	1	8	+	+	8
Rhyacophilidae	18	10	5	7	28
Cased caddis	20	23	36	89	116
Oligochaeta (no. m^{-2})	4	2	12	31	8

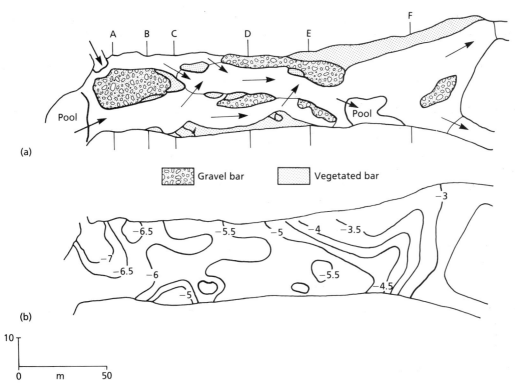

Fig. 11.4 Sedimentation within the River Rheidol below Nant-y-Moch Dam, UK: morphology (a) and mean particle size variations (phi values) of the surface layer (from Petts 1984, by permission of John Wiley & Sons).

osional situations (Mullan *et al* 1976). At a distance of 30 km downstream sedimentation processes were predominant, the increased turbidity reducing the benthic species to between 32 and 192 m⁻². These processes also eliminated a trout fishery. Petts (1988) found localized siltation of the substrate below two dams in the UK, particularly below tributary confluences.

The effect of channel changes on benthic macroinvertebrates has been investigated below the Nant-y-Moch reservoir, on the River Rheidol, Wales, UK (Petts & Greenwood 1985). Channel changes had been caused by sedimentation arising from point-source inputs of sediment to the mainstream from tributary streams. Four distinct successive morphological and sedimentological units were identified downstream from the tributary confluence. The invertebrate distributions associated with the channel changes are shown in

Table 11.5. The first reach immediately below the confluence junction produced reduced channel dimensions and a coarse, stable substrate (Fig. 11.4). Below this reach (Reach 2) the channel had experienced less change and was braided with an unstable substrate. The third reach was found to be characterized by progressive sedimentation, whilst the fourth reach had an existing substrate stabilized by flow regulation, and mosses and algae covering much of the channel bed. This reach had the largest number of benthic species and greatest species diversity. It is anticipated that reach 1 will extend downstream as the processes of sedimentation change the channel morphology. Reaches 2 and 3 will shift downstream, whilst reach 4 will be progressively eliminated. Changes in the macroinvertebrate populations may be anticipated in accordance with this change, which may take many years.

11.7 DIRECT INFLUENCES

Direct modification of river channels includes dam and reservoir construction, channelization and irrigation diversions. There may be changes which occur as a consequence of a change at a specific location or along a length of river.

Channelization

Since channelization involves manipulation of one or more of the dependent hydraulic variables of slope, depth, width and roughness, then feedback effects may be initiated which proceed to promote a new state of equilibrium (Brookes 1988). Channelization may induce instability not only in the improved channel reach but also upstream and/or downstream of the managed reach. Perhaps the most dramatic adjustments occur in response to slope changes associated with channel straightening or regrading, or to extended bottom widths. For the East and West Prairie rivers in Alberta, Canada, Parker and Andres (1976) found that channelization of a meandering stream increased the slope by providing a shorter channel path. This increase of slope enabled the transport of more sediment than was supplied at the upstream end of the channelized reach and the difference was obtained from the bed, causing degradation which progressed upstream as a nickpoint. An excess of load was then supplied to the downstream part of the channelized reach and because the flatter natural reach downstream could not transport this sedi-

Table 11.6 Examples of the morphological consequences of river channelization

| Area applied | Dates | Extent | Morphological effects | | | Source |
			Downstream	Within	Upstream	
Willow R, Iowa	1906–20			440% capacity increase between 1920 and 1958	Series of nickpoints	Daniels (1960); Ruhe (1970)
Blackwater R, Missouri	1910			1000% increase of cross-sectional area		Emerson (1971)
East and West Prairie rivers, Canada	1953–71		Deposition of sediment	500% increase of cross-sectional area	4 m of degradation (1964–74)	Parker & Andres (1976)
Tisza R, Hungary	1850s	32% reduction of length	No effect – it enters the Danube	Bed lowered by a maximum of 2.3 m		Szilagy (1932)
Mississippi R, Baton Rouge	1929–47	35% reduction of length	No effect due to lack of time since completion			Lane (1947)
Colorado R, Laguna Dam		9.6 km length reduction	Insufficient sediment to raise bed level	Bed lowered	Lowering limited by dam located upstream	Lane (1947)
Peabody R, New Hampshire				Reach lowered by a maximum of 3.6 m; width increase of 400%		Yearke (1971)
103 streams in North America	1960–70	Lengths of 0.07–4.2 km		Degradation at only 17 sites; bank stability poor at 7 sites		Brice (1981)

ment it was deposited on the bed. Degradation within the channelized reach caused bank collapse. A similar process operated on the Willow River in Iowa where a 440% increase of capacity was achieved in 38 years following channelization (Daniels 1960) and in the Blackwater River, Missouri, where the cross-sectional area increased by a maximum of 1000% over a 60-year period (Emerson 1971). Table 11.6 provides some further examples of the morphological consequences of artificial straightening.

Adjustments of channel morphology have been noted within resectioned reaches and the problems of overwide channels have been exemplified for part of a flood alleviation scheme constructed on the River Tame in England (Nixon 1966). The channel reverted to its original capacity in less than 30 years in the absence of maintenance; this is due to the enlarged channel being in equilibrium with the design flood flow but out of equilibrium with the normal range of flows which predominate for most of the time. The low flows deposit sediment because of reduced velocities in the overwide reach and these deposits may become stabilized to form more permanent morphological features.

Enlarged cross-sections in the lower reaches of projects sometime function as sediment traps. Flow depths and channel sizes may be reduced as the channels fill with sediment. In the lower reaches of streams the problem may be compounded by low slopes and backwater effects; for instance, within 10 years of the completion of a flood control channel on the San Lorenzo River in California, 350 000 m^3 of sediment had been deposited (Griggs & Paris 1982).

Channelization may have both direct and indirect effects on aquatic habitat and associated organisms (Brookes 1988). Traditional channelization practices may dramatically change the channel morphology to the detriment of the biology of watercourses. In North America the number of published works on the effects of channelization on the flora and fauna of rivers far exceeds those concerned solely with the physical effects. Figure 11.5 compares the channel morphology and hydrology of a natural stream with the morphology of a channel changed by channelization. Channelization can change the

original dimensions and shape of a channel, the slope and the planform, changing a heterogeneous system into a homogeneous one (McClellan 1974; Neuhold 1981). The most commonly stated reason for a change in fish populations in channelized rivers is the loss of a natural pool-riffle sequence which provides a variety of flow conditions suitable as cover for fish and for the organisms on which fish feed. Shelter areas are required at high flows to protect fish from abnormally high velocities; these conditions are absent in a channelized river, where a meandering stream, with an abundance of long pools separated by short riffles, may be converted into a straight stream composed mainly of run habitat. Alteration of the width and depth variables in a channel may create shallow and unnatural hydraulic conditions which create an unsuitable habitat for fish and may present topographical difficulties for fish migration.

Disruption of the substrate can have significant impacts on benthic invertebrates, and in an extreme case natural bed materials may be replaced by bedrock (Huggins & Moss 1974). An unstable substrate following channelization may lead to excessive sediment loads, as may other forms of morphological adjustment, and this is detrimental to the fish populations (Graesser 1979). The effect of a changed substrate may be complex; Schmal and Sanders (1978) found a marked seasonal trend with low invertebrate populations coincident with high spring flows and an unstable substrate, and high populations related to a stable substrate at other flows.

The processes of biological recovery take time, and may be naturally or artificially induced. Tarplee *et al* (1971), in a study of North Carolina's coastal plains' streams, suggested that species diversity of fish may recover in 15 years provided no further alterations of the stream bed occur. On a resectioned part of the Chariton River in Missouri it was found that after 30 years the channel had tended to revert from the uniform width and depth of the constructed channel back to natural conditions with meanders and a corresponding improvement in fish populations (Congdon 1971). Further examples of the biological impacts of channelization are given in Table 11.7.

Channelization of the Twentymile Creek in

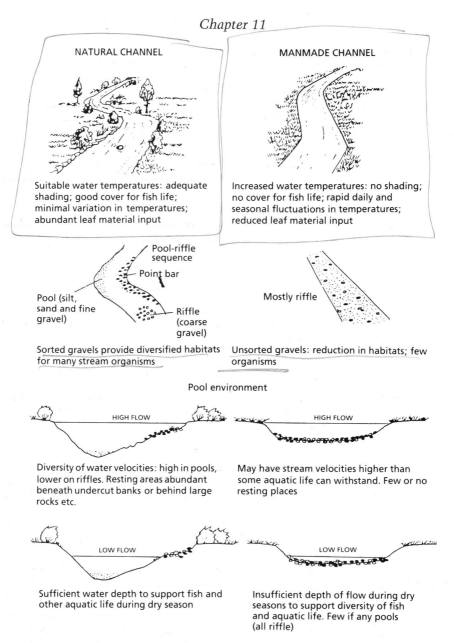

NATURAL CHANNEL

Suitable water temperatures: adequate shading; good cover for fish life; minimal variation in temperatures; abundant leaf material input

MANMADE CHANNEL

Increased water temperatures: no shading; no cover for fish life; rapid daily and seasonal fluctuations in temperatures; reduced leaf material input

Pool-riffle sequence
Point bar
Pool (silt, sand and fine gravel)
Riffle (coarse gravel)

Mostly riffle

Sorted gravels provide diversified habitats for many stream organisms

Unsorted gravels: reduction in habitats; few organisms

Pool environment

HIGH FLOW

HIGH FLOW

Diversity of water velocities: high in pools, lower on riffles. Resting areas abundant beneath undercut banks or behind large rocks etc.

May have stream velocities higher than some aquatic life can withstand. Few or no resting places

LOW FLOW

LOW FLOW

Sufficient water depth to support fish and other aquatic life during dry season

Insufficient depth of flow during dry seasons to support diversity of fish and aquatic life. Few if any pools (all riffle)

Fig. 11.5 Comparison of the channel morphology and hydrology of a natural stream with a channelized watercourse (based on Corning 1975, by permission of Virginia Department of Game and Inland Fisheries).

northern Mississippi took place at various dates between 1910 and 1966 for drainage and flood control purposes (Shields & Hoover 1991). Straightening and channel enlargement caused a nickpoint to progress upstream, for about 14 km,

doubling the channel cross-section over a period of 12 years, causing degradation and bank failure. However, between 1982 and 1986 grade control structures were installed to check the slope and prevent further degradation. Rock riprap and

Table 11.7 Examples of the ecological impacts of river channelization

Location	Type	Date	Variables affected	Reduced parameters	Recovery	Source
Missouri R, Nebraska		1950s	Brush and piles and pools eliminated	Benthic area reduced by 67%; standing crop of drift reduced 88%	Study represents 15 years after channelization	Morris *et al* (1968)
Little Sioux R, Iowa			Lack of suitable substrate	Fewer macroinvertebrates but more drift organisms		Hansen & Muncy (1971)
Buena Vista Marsh, Portage Co, Wisconsin	Dredging	Various	Seasonal effect; substrate only unstable at high flow; at other times vegetation/silt	High invertebrate populations when substrate stable; vegetation and silt favours snails/ midges. Elimination of stonefiles		Schmal & Sanders (1978)
Mud Creek, Douglas Creek, Kanas	Enlarged	1971	Destruction of pool-riffle sequence	Number of fish and species reduced; biomass reduced by 82% and diversity by 49%	2 years after	Huggins & Moss (1974)
Rush Creek, Modoc County, California	Realigned	1969	Destruction of pool-riffle sequence	Trout biomass reduced by 86%; total number by 36% and total biomass by 69%	5 years after construction	Moyle (1976)
Missouri River	Various channel works		Lack of niches (especially pools)	Annual catch reduced by 47%; harvest rate by 48% and standing crop by 63%		Groen & Schmulbach (1978)

vegetation were used to protect stream banks. Engineering stabilization has been shown to have a positive effect on a degraded channel. Bank stability enabled riparian vegetation to increase and fish assemblages near to the grade control structures were shown to be more diverse than at other sites, partly due to an improved substrate (Shields & Hoover 1991).

11.8 PLANFORM AND NETWORK CHANGES

The discussion of changes arising from specific impacts (Sections 11.3 to 11.7) has concentrated mainly on changes of channel cross-section and composition of the substrate. Two areas of study which have received relatively little attention are changes of the planform and network characteristics of river systems. Two types of change of channel planform have been distinguished (Lewin 1977), namely autogenic, which are inherent in the river regime, and allogenic, which occur in response to changes in the river system, including the influence of human activities. A detailed study in England and Wales by Hooke and Redmond (1989) revealed that almost 35% of channels which drain upland areas had shown some change of planform between 1870 and 1950 and that at least some of these changes were either due to the direct (channelization) or indirect (urbanization, deforestation) influences of humans. With many studies it has been difficult to separate the influence of human interference from other variables (e.g. Hooke, 1977). A study along the Dunajee River in southern Poland revealed that the channel had become more sinuous and branching over the period 1787 to 1955 (Klimek & Trafas 1972). This change of shape was attributed to deforestation and the increase of potato cultivation in the 19th century because this allowed for high rates of runoff.

Ecological changes resulting from such planform changes have received even less attention. A recent study of the French Upper Rhône River has elucidated ecological changes which have occurred since 1750 (Roux *et al* 1989). This was achieved by using historical maps, floristic and

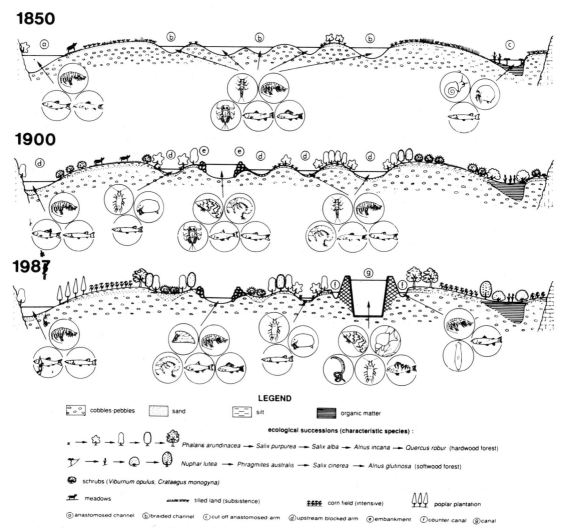

Fig. 11.6 Transect of part of the Upper Rhône River showing channel changes between the years 1850, 1900 and 1987. Certain animals are plotted to illustrate similarities and differences between aquatic communities (from Roux *et al* 1989, by permission of John Wiley & Sons).

faunistic evidence from relic channels and documentary sources. Figure 11.6 is a typical transect which shows the numerous braided and anastomosed channels in 1850 before civil engineering works; in 1900 after the construction of embankments; and in 1987 following canalization and regulation of the river for hydro-electric power generation in the 1980s. The corresponding ecological changes are complex. The short-term impacts of embanking in the last century increased the structural and functional diversity of the ecosystems in the braided sections. The embankments provided new biotopes and the new hydrological and sedimentological processes allowed the development of several ecological sequences which were previously inhibited. The diversity further increased with the canalization and regulation undertaken in the 1980s. However, it is anticipated that the diversity of habitats and communities will eventually decrease due to the lowering of the water table and the reduction of the influence of hydrological and sedimentological processes on the floodplain.

Changes of the drainage network are also widespread as a consequence of both direct and indirect human intervention. New drainage ditches have been dug (e.g. Marshall *et al* 1978; Ovenden & Gregory 1980) and land use changes have probably led to erosion of new channels or rills. For example, Eyles (1977) showed that in the Southern Tablelands of New South Wales, Australia, tree removal and grazing pressure caused dramatic changes between 1840 and 1950. A sequence was observed from a chain of scour ponds, to discontinuous gully, to incised channel and eventually to a permanent stream.

11.9 TIME DIMENSION

A particular problem is the complexity of ecological adjustments to channel changes. Many of the ecological studies may represent transient states and not final equilibria (Petts 1987). For example, changes caused by regulation may be conceptualized as a hierarchy of responses (Petts 1980). First-order impacts occur at the time of dam closure, or shortly afterwards, and affect the transfer of energy and material into and within the downstream river. Second-order impacts are

the channel changes and changes of the floodplain dynamics and primary production arising from the local effects of the first-order impacts (Fig. 11.7). Third-order impacts on benthic invertebrates, fish and floodplain fauna result from a combination of all the first- and second-order impacts as well as from biotic interactions between populations. The complete adjustment of biological populations must be preceded by the adjustment of abiotic factors, of which channel change can take a considerable length of time, even hundreds of years.

For channelized rivers the period between the commencement and the attainment of a new state of equilibrium, the relaxation time, has been found to vary. Values have been given ranging from 9 to 15 years for three Mississippi Rivers (Schumm *et al* 1984) to as much as 1000 years elsewhere (Brookes 1988). Recovery of fish populations is dependent on an improvement of habitat, which may be offset by maintenance of channelized reaches. Studies in North America have shown that substantial recovery of fish populations may occur in periods of between 30 and 86 years (Brookes 1988).

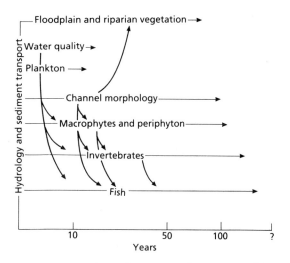

Fig. 11.7 Interrelationships among the major variables and appropriate time-scales for the consideration of adjustment to river regulation (after Petts 1989b, by permission of CRC Press).

11.10 THRESHOLDS AND COMPLEX RESPONSE

In addition to the time sequence for change it is also important to note that a particular change, as evidenced by erosion or deposition, may only occur after a critical threshold level has been exceeded (Schumm 1973). For example, large floods may be responsible for triggering a sequence of change. Perhaps best known are the threshold velocities required to entrain sediment of a certain size. Thresholds can either be extrinsic or intrinsic. Extrinsic thresholds exist within a system but change will not occur without the influence of an external variable. Intrinsic thresholds may be exceeded where the external variables remain relatively constant, yet progressive change within the system renders it unstable and change occurs.

Responses to change may be instantaneous, progressive, delayed or complex. The problems of complex response were outlined by Schumm (1973) who described how two river basins can react in different ways, or at differing rates, to the same stimulus for change. The response is likely to be complex, with spatial and temporal lags. For example, a period of forest clearance may be replaced by conservation measures, and yet it may take a considerable amount of time for the wave of sediment arising from the forest clearance to work through the system.

It is further apparent that the interpretation of river channel change must consider the inter-relationships between the channel cross-section, channel pattern and drainage network (Gregory 1976). Dumaresq Creek in New South Wales has undergone an increase in the number of stream channels through extension of the drainage network, reduced channel sizes following reservoir construction and enlargement due to urban runoff (Gregory 1977b).

11.11 PROSPECT

This chapter has illustrated a variety of channel adjustments arising from changes in the discharge and sediment loads of rivers. Channel changes may extend over periods of up to several hundred years. Because many of the responses of a river cannot be predicted by theory alone, an important means for evaluating impact is through documented case studies. The size, shape and composition of channels may change (Table 11.4). Most biological studies have concentrated on the impacts arising from a change of composition of the channel substrate; few have examined the impact of changes of channel size or shape (cf. Fig. 11.3) on the ecology of a watercourse. The precise biological impacts of different types of channel change have yet to be fully quantified. The loss of lotic habitat as a result of straightening, or floodplain modification, is another area which has received relatively little attention. Such studies are important so that knowledge can be extended and improved management strategies developed and implemented.

Discussion of biological impacts is difficult without understanding the intricate relationships between the biotic and abiotic elements of the stream ecosystem. Both abiotic and biotic factors vary seasonally and geographically. If changes caused by regulation are conceptualized as a hierarchy of responses (Petts 1980), then the third-order impacts on benthic invertebrates, fish and floodplain fauna result from a combination of all the first- and second-order impacts as well as from biotic interactions between populations (see section 11.9). However, this is often complicated because channel changes along a specific length of river may be the result of more than one impact. For example, urbanization often includes channelization to accommodate anticipated increases of peak discharge. In other circumstances channelization may have been undertaken on channels affected by increased sedimentation arising from land use changes in the upstream catchment. Examples of the complex nature of channel changes arising from a variety of impacts have been documented by Brookes and Gregory (1988). Channel changes are further complicated by the presence of thresholds, both intrinsic and extrinsic.

Substantial efforts by geomorphologists have elucidated changes occurring in the 10–50-year time period. However, relatively few studies have so far identified the significance of changes occurring over longer time spans of, say, up to 500 years, the period of intensifying human impact in temperate zones. The lack of study is partly a result of the absence of documentary sources, including maps. However, archaeology is likely to be invaluable as a means of extending our knowledge of this period, and hence the likely ecological repercussions. Through reconstruction of the past it will be possible to determine the rate and magnitude of channel change in response to single and composite impacts. It should also enable improved prediction of the effects of more recent impacts and the development of ecologically-sound management tools (Brookes 1991).

Understanding how river channels have changed is vital if restoration strategies which are both morphologically stable and ecologically desirable are to be developed further.

ACKNOWLEDGEMENTS

The views expressed in this paper are those of the author and not necessarily those of his employer, the National Rivers Authority (Thames Region). Permission to publish this paper is gratefully appreciated.

REFERENCES

Brice JC. (1981) *Stability of Relocated Stream Channels.* Technical Report No. FHWA/RD-80/158, Federal Highways Administration, US Department of Transportation, Washington, DC. [11.7]

Brookes A. (1987) Recovery and adjustment of aquatic vegetation within channelization works in England and Wales. *Journal of Environmental Management* **24**: 365–82. [11.1]

Brookes A. (1988) *Channelized Rivers: Perspectives for Environmental Management.* John Wiley & Sons, Chichester. [11.7, 11.9]

Brookes A. (1991) Geomorphology. In: Gardiner JL (ed) *River Projects and Conservation: A Manual for Holistic Appraisal.* pp. 57–66. John Wiley & Sons, Chichester. [11.11]

Brookes A, Gregory KJ. (1988) Channelization, river engineering and geomorphology. In: Hooke JM (ed) *Geomorphology in Environmental Planning*, pp. 145–67. John Wiley & Sons, Chichester. [11.11]

Calow P, Petts GE. (eds) (1994) *The Rivers Handbook*, vol. 2. Blackwell Scientific Publications, Oxford. [11.1]

Congdon JC. (1971) Fish populations of channelized and unchannelized sections of the Chariton River, Missouri. In: Schneberger E, Funk JL (eds) *Stream Channelization: A Symposium*, pp. 52–83. Special Publication No. 2, North Central Division, American Fish Society, Bethesda, Maryland. [11.7]

Cordone AJ, Kelley DW. (1961) The influence of inorganic sediment on aquatic life of streams. *California Fish and Game* **47**: 189–228. [11.4]

Corning RV. (1975) Channelization: shortcut to nowhere. *Virginia Wildlife* **6**: 8. [11.7]

Daniels RB. (1960) Entrenchment of the Willow Creek Drainage Ditch, Harrison County, Iowa, *American Journal of Science.* **258**: 161–76. [11.7]

Davey GW, Doeg TJ, Blyth JD. (1987) Changes in benthic sediment in the Thomson River, Victoria, during construction of the Thomson Dam. *Regulated Rivers* **1**: 71–85. [11.6]

Doeg TJ, Davey GW, Blyth JD. (1987) Response of the aquatic macroinvertebrate communities to dam construction on the Thomson River, Southeastern Australia. *Regulated Rivers* **1**: 195–209. [11.6]

Douglas I. (1985) Urban sedimentology. *Progress in Physical Geography* **9**: 255–80. [11.5]

Dunne T, Leopold LB. (1978) *Water in Environmental Planning.* W.H. Freeman, San Francisco. [11.5]

Emerson JW. (1971) Channelization: a case study. *Science* **173**: 325–6. [11.7]

Eyles RJ. (1977) Birchams Creek: the transition from a chain of ponds to a gully. *Australian Geographical Studies* **15**: 146–57. [11.8]

Gilbert OL. (1989) *The Ecology of Urban Habitats.* Chapman and Hall, London. [11.5]

Graesser NWC. (1979) How land improvement can damage Scottish salmon fisheries. *Salmon and Trout Magazine* **215**: 39–43. [11.7]

Graf WL. (1975) The impact of suburbanization on fluvial geomorphology. *Water Resources Research* **11**: 690–2. [11.5]

Graf WL. (1979) Mining and channel response. *Annals of the Association of American Geographers* **69**: 262–75. [11.4]

Gregory KJ. (1976) Changing river basins. *Geographical Journal* **142**: 237–47. [11.3, 11.10]

Gregory KJ. (1977a) The context of river channel changes. In: Gregory KJ (ed) *River Channel Changes*, pp. 1–12. John Wiley & Sons, Chichester. [11.1]

Gregory KJ. (1977b) Channel and network metamorphosis in northern New South Wales. In: Gregory KJ (ed) *River Channel Changes*, pp. 389–410. John Wiley & Sons, Chichester. [11.10]

Gregory KJ, Lewin J, Thornes JB. (1987) *Palaeohydrology in Practice: a River Basin Analysis.* John Wiley & Sons, Chichester. [11.1]

Griggs GB, Paris L. (1982) Flood control failure: San Lorenzo River, California. *Environmental Management* **6**: 407–19. [11.7]

Groen CL, Schmulbach JC. (1978) The sport fishery of the unchannelized Middle Missouri River. *Trans-*

actions of the American Fisheries Society **107**: 412–18. [11.7]

Hammer TR. (1972) Stream channel enlargement due to urbanization. *Water Resources Research* **8**: 1530–40. [11.5]

Hansen DR, Muncy RJ. (1971) *Effects of Stream Channelization on Fish and Bottom Fauna in the Little Sioux River. Iowa.* Iowa State Water Resources Institute, Ames, Iowa. [11.7]

Hooke JM. (1977) The distribution and nature of changes in river channel pattern. In: Gregory KJ (ed) *River Channel Changes*, pp. 265–80. John Wiley & Sons, Chichester. [11.8]

Hooke JM, Redmond CE. (1989) River channel changes in England and Wales. *Journal of the Institution of Water and Environmental Management* **3**: 328–35. [11.8]

Howard AD, Dolan R. (1981) Geomorphology of the Colorado River in the Grand Canyon. *Journal of Geology* **89**: 307–21. [11.6]

Huggins DG, Moss RE. (1974) Fish population structure in altered and unaltered areas of a small Kansas stream. *Transactions of the Kansas Academy of Science* **77**: 18–30. [11.7]

Hynes HB. (1973) The effects of sedimentation on the biota in running water. In *Proceedings of the 9th Canadian Hydrology Symposium: Fluvial Processes and Sedimentation*, pp. 653–62. Inland Waters Directorate, Canadian Department of the Environment, Canada.

Keefer L, Maughan OE. (1985) Effects of headwater impoundment and channelization on invertebrate drift. *Hydrobiologia* **127**: 161. [11.1]

Kellerhalls R, Church M, Bray DI. (1976) Classification and analysis of river processes. *Journal of the Hydraulics Division. American Society of Civil Engineers* **102**: 813–29. [11.2]

Kimble LA, Wesche TA. (1975) Relationships between selected physical parameters and benthic community structure in a small mountain stream. Water Resources Series No. 55. University of Wyoming, Laramie, Wyoming. [11.2]

Klimek K, Trafas K. (1972) Young Holocene changes in the course of the Durajec River in the Beskid Sadecki Mountains (Western Carpathians). *Studia Geomorphologica Carpatho-Balcanica*, **VI**: 85–92. [11.8]

Komura S, Simons DB. (1967) River bed degradation below dams. *Journal of the Hydraulics, American Society of Civil Engineers* **93**: 1–14. [11.6]

Lane EW. (1947) The effects of cutting off bends in rivers. In: *Proceedings of the Third Hydraulics Conference*. Bulletin No. 31, University of Iowa, Iowa. [11.7]

Leopold LB. (1968) Hydrology for urban land planning: a guidebook on the hydrologic effects of urban land use.

US Geological Survey Circular No. 554. [11.5]

Lewin J. (1977) Channel pattern changes. In: Gregory KJ (eds) *River Channel Changes*, pp. 167–84. John Wiley & Sons, Chichester. [11.8]

McClellan TJ. (1974) *Ecological Recovery of Realigned Stream Channels: Portland, Oregon.* Technical Report, Federal Highways Administration, US Department of Transportation, Portland, Oregon. [11.7]

Marshall EJP, Wade PM, Clare P. (1978) Land drainage channels in England and Wales. *Geographical Journal* **144**: 254–63. [11.8]

Meade RH. (1976) Sediment problems in the Savannah River Basin. In: Dillman BL, Steep JM (eds) *The Future of the Savannah River*, pp. 105–29. Water Resources Research Institute, Clemson, South Carolina. [11.4]

Meade RH, Trimble SW. (1974) Changes in sediment loads in rivers of the Atlantic drainage of the United States since 1900. In: *Effects of Man on the Interface of the Hydrological Cycle with the Physical Environment*, pp. 99–104. IASH Publication, No. 113. [11.6]

Milhous RT. (1982) Effect of sediment transport and flow regulation on the ecology of gravel-bed rivers. In: Hey RD, Bathurst JC, Thorne CR (eds) *Gravel-Bed Rivers*, pp. 819–92. John Wiley & Sons, Chichester. [11.1]

Morisawa M. (1985) *Rivers: Form and Process.* Longman, London. [11.5]

Morris LA, Langemeirs RM, Russell TR, Witt A. (1968) Effects of main stem impoundments and channelization upon the limnology of the Missouri River, Nebraska. *Transactions of the American Fisheries Society* **97**: 380–8. [11.7]

Moyle PB. (1976) Some effects of channelization on the fisheries and invertebrates of Rush Creek, Modoc County, California. *California Fish and Game* **63**: 179–86. [11.7]

Mullan JW, Starostka VJ, Stone JL, Wiley RW, Wiltzius WJ. (1976) Factors affecting Upper Colorado River Reservoir tailwater trout fisheries. In: Orsborn JF, Allman CE (eds) *Instream Flow Needs*, Vol. II, pp. 405–23. American Fisheries Society, Bethesda, Maryland. [11.6]

Neuhold JH. (1981) Strategy of stream ecosystem recovery. In: Barrett GW, Rosenburg R (eds) *Stress Effects of Natural Ecosystems*. John Wiley & Sons, New York. [11.7]

Nixon M. (1966) Flood regulation and river training. In: Thorn RB (ed) *River Engineering and Water Conservation Works*, pp. 293–7. Butterworths, London. [11.7]

Ovenden JC, Gregory KJ. (1980) The performance of stream networks in Britain. *Earth Surface Processes* **5**: 47–60. [11.8]

Parker G, Andres D. (1976) Detrimental effects of river channelization. In: *Proceedings of Conference: Rivers 1976, American Society of Civil Engineers*, pp. 1248–66. [11.7]

Pearson WD, Kramer RH, Franklin DR. (1968) Macroinvertebrates in the Green River below Flaming Gorge Dam, 1964–65 & 1967. *Proceedings of the Utah Academy of Science, Arts and Letters*, 45: 148–67. [11.6]

Pemberton EL. (1976) Channel changes in the Colorado River below Glen Canyon Dam. In: *Proceedings of the Third Interagency Sedimentation Conference*, pp. 5–61 to 5–73. Washington, DC. [11.6]

Pennak RW, Van Gerpen ED. (1947) Bottom fauna production and physical nature of the substrate in a northern Colorado trout stream, *Ecology* 28: 42–8. [11.2]

Percival E, Whitehead H. (1929) A quantitative study of the fauna of some types of stream bed. *Journal of Ecology* 17: 282–314. [11.5]

Petts GE. (1980) Long-term consequences of upstream impoundment. *Environmental Conservation* 7: 325. [11.9, 11.11]

Petts GE. (1984) *Impounded Rivers: Perspectives for Ecological Management*. John Wiley & Sons, Chichester. [11.2, 11.6]

Petts GE. (1987) Time-scales for ecological change in regulated rivers. In: Craig JF, Kemper JB (eds) *Regulated Streams: Advances in Ecology*, pp. 257–66. Plenum Press, New York. [11.9]

Petts GE. (1988) Accumulation of fine sediment within substrate gravels along two regulated rivers, UK. *Regulated Rivers* 2: 141–53. [11.6]

Petts GE. (1989a) Historical analysis of fluvial hydrosystems. In: Petts GE, Moller H, Roux AL (eds) *Historical Change of Large Alluvial Rivers: Western Europe*, pp. 1–18. John Wiley & Sons, Chichester. [11.1]

Petts GE. (1989b) Perspectives for ecological management of regulated rivers. In: Gore JA, Petts GE (eds) *Alternatives in Regulated River Management*, pp. 3–24. CRC Press, Boca Raton, Florida. [11.9]

Petts GE. (1994) Rivers: dynamic components of catchment ecosystems. In: Calow P, Petts GE (eds) *The Rivers Handbook*, vol. 2, pp 3–22. Blackwell Scientific Publications, Oxford. [11.1]

Petts GE, Greenwood MG. (1985) Channel changes and invertebrate faunas below Nant-y-Moch Dam, River Rheidol, Wales, UK. *Hydrobiologia* 122: 65 [11.1, 11.6]

Richards KS, Wood R. (1977) Urbanization, water redistribution, and their effect on channel processes. In: Gregory KJ (eds) *River Channel Changes*, pp. 369–88. John Wiley & Sons, Chichester. [11.5]

Roux AL, Bravard J-P, Amoros C, Pautou G. (1989) Ecological changes of the French Upper Rhône River since 1750. In: Petts GE, Moller H, Roux AL (eds) *Historical Change of Large Alluvial Rivers: Western Europe*, pp. 323–50. John Wiley & Sons, Chichester. [11.8]

Ruhe RV. (1970) Stream regimen and man's manipulation. In: Coates DR (ed) *Environmental Geomorphology*, pp. 9–23. State University of New York, New York. [11.7]

Schmal RN, Sanders DF. (1978) Effects of stream channelization on aquatic macroinvertebrates. Buena Vista Marsh, Portage County, Wisconsin. Report No. FWS/OBS-78/92. Office of Biological Services, Fish and Wildlife Service, US Department of the Interior, Washington, DC. [11.7]

Schumm SA. (1969) River metamorphosis. *Journal of the Hydraulics Division, American Society of Civil Engineers* 95: 255–73. [11.3]

Schumm SA. (1973) Geomorphic thresholds and complex response of drainage systems. In: Morisawa ME (ed) *Fluvial Geomorphology*, pp. 299–310. State University of New York, Binghamton, New York. [11.10]

Schumm SA, Harvey MD, Watson CC. (1984) *Incised Channels: Morphology, Dynamics and Control*. Water Resources Publications, Littleton, Colorado. [11.9]

Shields FD Jr, Hoover JJ. (1991) Effects of channel restabilization on habitat density, Twentymile Creek, Mississippi. *Regulated Rivers* 6: 163–82. [11.7]

Szilagy J. (1932) Flood control on the Tisza River. *Military Engineer* 24: 632. [11.7]

Tarplee WH, Louder DE, Weber AJ. (1971) *Evaluation of the Effects of Channelization on the Fish Populations in North Carolina's Coastal Streams*. North Carolina Wildlife Resources Commission, North Carolina. [11.7]

Thoms MC. (1987) Channel sedimentation within the urbanized River Thame, UK. *Regulated Rivers* 1: 229–46. [11.5]

Trimble SW. (1974) *Man-induced Soil Erosion on the Southern Piedmont*. Soil Conservation Society of America, Ankeny, Iowa. [11.4]

Underwood AJ. (1994) Spatial and temporal problems with monitoring. In: Calow P, Petts GE (eds) *The Rivers Handbook*, vol. 2, pp 101–23. Blackwell Scientific Publications, Oxford. [11.2]

Voelz NJ, Ward JV. (1990) Macroinvertebrate responses along a complex regulated stream environment gradient. *Regulated Rivers* 5: 365–74. [11.6]

Wesche TA. (1985) Stream channel modifications and reclamation structures to enhance fish habitat. In: Gore JA (ed) *The Restoration of Streams and Rivers*, pp. 103–63. Butterworths, Boston. [11.2]

White WD, Wells SG. (1979) Forest fire devegetation and drainage basin adjustments in mountain terrain. In: Rhodes DD, Williams GP (eds) *Adjustments to* the *Fluvial System*, pp. 199–223. Kendall-Hunt, Dubuque, Iowa. [11.4]

Whitlow JR, Gregory KJ. (1989) Changes in urban stream channels in Zimbabwe. *Regulated Rivers* **4**: 27–42. [11.5]

Williams GP, Wolman MG. (1984) Downstream Effects of Dams in Alluvial Rivers. United States Geological Survey Professional Paper 1286. [11.6]

Wolman MG. (1967) A cycle of sedimentation and erosion in urban river channels. *Geografiska Annaler* **49A**: 385–95. [11.4]

Wolman MG, Schick AP. (1967) Effect of construction on fluvial sediment, urban and suburban areas of Maryland. *Water Resources Research* **3**: 451–64. [11.5]

Yearke LW. (1971) River erosion due to channel relocation. *Civil Engineering* **41**: 39–40. [11.7]

12: Hydrology and Climate Change

N. W. ARNELL

12.1 INTRODUCTION

The catchment manager has always had to deal with a changing physical and social environment. The pressures on rivers and catchments have changed over time, the required standards for environmental protection have (generally) been rising and catchment management has increasingly been seen as a multi-disciplinary activity involving hydrology, geomorphology, ecology and engineering. Since the late 1980s there has also been a need to deal with possible changes in those climatic inputs previously assumed to be constant.

The threat of global warming due to an increase in the concentration of so-called greenhouse gases has attracted widespread scientific, public and political interest. One of the most significant impacts of climate change is likely to be on the hydrological system. The objectives of this chapter are therefore (i) to summarize how climate changes might affect hydrological processes and river flow characteristics, and (ii) to review how the impacts on hydrological regimes, and therefore ecological, geomorphological and water resources systems, can be estimated in a particular catchment.

Table 12.1 summarizes some of the implications of climate change for a catchment. First, however, it is useful to review briefly the processes behind the enhanced greenhouse effect and the climate changes that might result.

The greenhouse effect

The basic principles behind the greenhouse effect are well established, and are reviewed in reports published by the United Nations Intergovernmental Panel on Climate Change (IPCC 1990, 1992). A few gases within the earth's atmosphere are transparent to incoming shortwave solar radiation, but they block a proportion of the outgoing longwave radiation radiated from the earth's surface so heating up the lower atmosphere. The most important of these radiatively-active gases are water vapour, carbon dioxide (CO_2) and methane, and without them the temperature at the earth surface would be around 33°C cooler than at present. The greenhouse effect is so called because the panes of glass in a greenhouse have a similar effect to the radiatively-active gases in the atmosphere (although in practice much of the effect of a real greenhouse arises from the prevention of the movement of air).

The concentrations of carbon dioxide and methane, however, have increased since the mid-19th century as industrial and agricultural development has expanded. New greenhouse gases. such as nitrous oxide and the man-made chlorofluorocarbons (CFCs), have been added and, molecule for molecule, these are more effective at blocking outgoing longwave radiation than CO_2 (although the 1992 IPCC report suggested that the radiative effects of CFCs were offset to a large extent by the effects of the depletion of stratospheric ozone caused by CFC emissions).

There are several major difficulties in estimating the impact that increasing concentrations of greenhouse gases might have on global and regional climates.

1 There are large uncertainties in forecast rates of emission of greenhouse gases, because they depend on rates of economic development and technological change. The IPCC considered a

Table 12.1 Some implications of changes in hydrological regimes induced by climate change

Water resources
    Change in reliability of supply
    Change in quality of supply
    Change in risk of flooding
    Change in potential for power generation

Geomorphology
    Change in sediment yields
    Change in erosion and sedimentation

Ecology
    Change in instream habitats
    Change in seasonal inundation of riverine areas
    Change in groundwater levels

range of emissions scenarios in 1990 and 1992, reflecting different assumptions about economic and technological change as well as a number of policies for limiting future increases.

2 The emission of some greenhouse gases depends to an unknown degree on the actual changes in climate that might take place. Methane emissions from decomposable organic matter in high-latitude tundra, for example, might rise as increases in temperatures cause permafrost to thaw: such a positive feedback would exaggerate global warming.

3 The rates at which greenhouse gases are taken up by plants and oceans, and interact in the atmosphere, are currently not well understood.

4 The global and, particularly, regional climatic consequences of increasing concentrations of greenhouse gases are very uncertain. How, for example, will changes in the radiative properties of the atmosphere affect the global temperature? How would changes in cloud cover exacerbate or ameliorate increases in surface temperature? These, and other, questions are tackled with the aid of computer models of the climate system (IPCC 1990).

A consensus has evolved amongst atmospheric scientists that the effect of increasing greenhouse gas concentrations would be to increase global temperatures and change regional climates (IPCC 1990, 1992). The IPCC calculated that, under 'business-as-usual' emissions scenarios, increasing greenhouse gas concentrations would lead to a rise in global mean temperature of approximately 0.3°C per decade, with a range of 0.2 to 0.5°C/decade (IPCC 1990, 1992). This would mean that the global average temperature by 2050 would be around 1.8°C higher than at present, which would be considerably higher than at any time in human history. The rate of change too is unprecedentedly high, and may be higher than the rate at which ecosystems (and management systems) can respond.

Table 12.2 Uncertainty in estimating the implications of climate change (from Fiering & Matalas 1990)

Proposition	Degree of belief			
	Certain	Less certain	Speculative	Uncertain
1 CO_2 is increasing	XXXX	X		
2 Global mean temperature increases	XXX	XX	X	
3 Global magnitude of change in temperature	X	XXX	XX	
4 Regional magnitude of change in temperature		XXXX	XX	
5 Mean global change in precipitation	X	XX	XXX	
6 Magnitude and sign of change in global precipitation		XX	XXX	X
7 Magnitude and sign of change in regional precipitation			XXXX	XX
8 Hydrological change given climate scenario		X	XXXX	X
9 Societal adaption		X	XXX	XX
10 Adequacy of existing and projected resource systems		X		XXXX

The predicted consequences of climate change become even more uncertain if the focus switches from global mean temperature to regional impacts and measures of climate such as precipitation. Table 12.2 (Fiering & Matalas 1990) summarizes how uncertainty increases as questions about the future become increasingly demanding.

Detecting the greenhouse signal

Analysis of observational temperature records has shown an increase in global average temperature since 1900 of around 0.5°C (IPCC 1990). This does not, however, prove that the enhanced greenhouse effect is leading to global warming, for several reasons:

1 The evidence is circumstantial: there is no proven link between increasing greenhouse gas concentrations and rising temperatures, although a physical explanation has been proposed.

2 The variations in the observed temperature record could reflect 'simple' decade-to-decade variability about an unstable mean, or might be due to long-time-scale variability associated with changes in solar radiation receipts (although the changes in solar radiation over the last few hundred years are small compared with the radiative effects of increased concentrations of greenhouse gases; IPCC 1990).

3 The observed temperature record may contain some systematic bias. Records tend to be taken in or near urban areas, but Jones *et al* (1989) estimate a maximum bias of 0.1°C per 100 years due to urban warming.

The detection of greenhouse gas-induced global warming is made very difficult by the fact that any signal will be masked by very large year-to-year variability. The IPCC (1990) predicted that the unequivocal detection of global warming will not be possible for at least a decade.

It will be even harder to find evidence for global warming in hydrological records. Not only do river flow series show the effects of year-to-year climatic variability, but they may also be affected by changes in catchment vegetation and water resource developments. Such changes may in some cases overwhelm any signal from global warming.

12.2 CLIMATE CHANGE AND HYDROLOGICAL PROCESSES

The hydrological and climate systems are interdependent. Hydrological regimes are determined by the combination of climate inputs with catchment physiography, geology and vegetation: catchment vegetation is itself dynamically dependent on both climatic and hydrological conditions. Hydrological processes control the return of energy and moisture to the atmosphere.

The dependence of hydrological systems on climatic inputs means that a change in climate (and associated change in vegetation) will affect hydrological behaviour, and the influence of catchment characteristics means that impacts will vary between catchments. This chapter concentrates on the impact of climatic changes on hydrological processes: it must be remembered that changes in hydrological processes and regimes feed back into effects on climate.

Figure 12.1 attempts to summarize the impact of a change in greenhouse gas concentrations on the hydrological system, emphasizing the interrelationships between all the elements of the system (the Figure excludes management intervention). The increase in greenhouse gas concentrations results in an increase in net radiation at the earth surface. This leads to an increase in temperature, which results in changes in atmospheric moisture contents and circulation patterns. These in turn produce (regionally-variable) changes in rainfall and evaporation regimes, and hence soil moisture regimes. Temperature, rainfall, evaporation and soil moisture all affect plant growth and therefore the ecosystem, as do changes in radiation and the atmospheric CO_2 concentration. Changes in rainfall and evaporation (compounded by changes in vegetation) result in changes in groundwater recharge and river flow which, together with changes in temperature, impact upon stream water chemistry and biology. Finally, a rise in sea level will affect hydrological characteristics in lowlying areas. The rest of this chapter examines some of these changes in more detail.

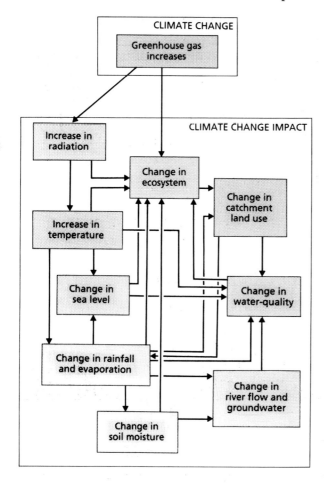

Fig. 12.1 The impact of climate change on the hydrological system.

Precipitation

Precipitation drives the hydrological system. Changes in its amount, intensity, duration and timing during the year will impact directly upon flow regimes and groundwater. The extent of impact depends, of course, on the degree of change, and different components of the hydrological regime are sensitive to different aspects of precipitation. Changes in low flows in a catchment, for example, might be most sensitive to changes in the intervals between rainfall events. In very responsive catchments (i.e. those with little storage), changes in short-duration rainfall intensities might have the greatest impact on flood characteristics, whilst in other less responsive catchments flood frequencies are more af-

fected by changes in the rate of occurrence of prolonged wet periods. In many parts of the world summer flows are dependent on the melt of snow stored on the catchment over the winter season; flows are profoundly affected both by changes in the volume of snow stored and the rate at which water is released by melt (see Gleick 1987, for example).

The impact of a given change in rainfall depends also on the characteristics of the catchment. This is illustrated quantitatively in Fig. 12.2, which shows the effect of an increase in winter rainfall and a reduction in summer rainfall on average monthly flows in three UK catchments (Arnell *et al* 1990). The changes were simulated using a monthly water balance model, and although the assumed percentage change in

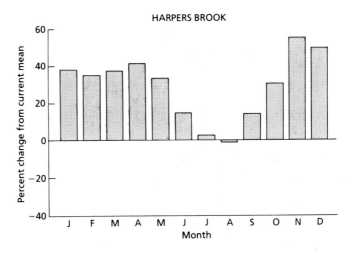

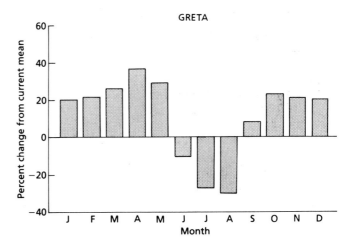

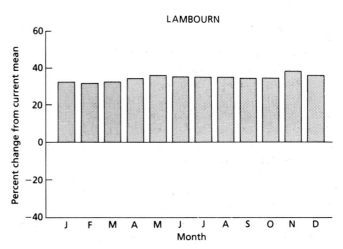

Fig. 12.2 The effect on mean monthly flow regimes in three UK catchments of increasing winter rainfall (by 20%) and reducing summer rainfall (by up to 15%). Potential evaporation is assumed unchanged (from Arnell *et al* 1990).

rainfall is the same in each catchment, differing catchment characteristics mean the effects of the change vary. The Greta is a small upland catchment (area 86 km^2) in Northumbria with a fine balance in summer between rainfall and evapotranspiration. A reduction in rainfall therefore leads to a large percentage reduction in the amount of water becoming streamflow. The Harpers Brook is an eastern lowland catchment (area 74 km^2), where rainfall in summer is insufficient to satisfy the demands of evaporation: little of the summer rainfall produces streamflow, so a reduction in rainfall in that season has little effect. Flows in the Lambourn (area 233 km^2) are dominated by a chalk aquifer, and during the drier summer are maintained by the slow drainage out of the aquifer of the extra winter groundwater recharge.

At the annual scale, the effect of a given percentage change in rainfall on runoff depends on the runoff coefficient, or the ratio of long-term average runoff to long-term average rainfall (Wigley & Jones 1985; Arnell 1992). The lower the proportion of average annual rainfall going to runoff, the more sensitive the runoff to a given change in rainfall. Studies in Britain have shown that, other things being equal, a 10% increase in average annual rainfall would result in an increase in average annual runoff of between 12

and 30% (with the highest increase in the driest regions: Arnell *et al* 1990). The greater the concentration of the additional rainfall in winter, the greater the effect on runoff.

Evapotranspiration

Potential evapotranspiration is defined as the evaporation and transpiration that would occur from an extensive short grass crop with an unlimited supply of water. If there is a shortage of water the *actual* evapotranspiration falls below the potential rate. Changes in actual evapotranspiration will therefore depend on both changes in potential evapotranspiration and changes in the availability of water through the year, which depend in turn on changes in precipitation inputs. It is possible that an increase in potential evapotranspiration at a site may be outweighed by such a large reduction in water availability that actual evapotranspiration reduces.

Potential evapotranspiration is controlled by inputs of net radiation, the ability of the air to hold moisture, the rate of renewal of air above an evaporating surface, and plant physiological properties. The plant properties of interest include aerodynamic resistance, which affects the rate of transfer of air across a leaf surface, and stomatal conductance, which controls the rate at

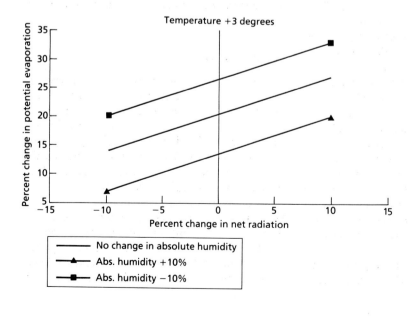

Fig. 12.3 The effect of changes in temperature, net radiation and absolute humidity on potential evapotranspiration in a wheat field in Nebraska over 10 summer days (data from Martin *et al* 1989).

which water can be transpired from the plant leaf. The leaf surface area determines the amount of rainfall that is intercepted and made available for evaporation. Climate change will affect catchment evaporation directly, through changes in hydrometeorological characteristics, and indirectly through changes in catchment vegetation.

The primary direct effect of climate change is likely to arise from the increase in the ability of air to hold moisture as air temperature rises, and Budyko (1980) estimated that potential evapotranspiration would increase by approximately 4% for every degree Celsius rise in temperature. Changes in net radiation, windspeed and plant properties, however, may serve to mitigate or exaggerate the effects of the temperature increase. Figure 12.3 shows the simulated effect of changes in net radiation and absolute humidity on the sensitivity of potential evapotranspiration from a Nebraska wheat field in summer to a 3°C increase in temperature (Martin *et al* 1989). The effect of a particular change in one of the controls on potential evapotranspiration tends to be dependent on the initial values of that and the other controls, so making generalizations difficult or impossible to express.

The characteristics and mix of catchment vegetation will change as rainfall, temperature and soil moisture regimes alter, and will also be affected by the concentration of atmospheric CO_2. Some plants grow much more vigorously in a high CO_2 atmosphere, but the effect depends on how the plant responds to the intake of CO_2. The C_3 plants, which include almost all UK plant species, are most sensitive to increased CO_2 concentrations, whilst C_4 plants (which include maize) are least sensitive (CCIRG 1991; Drake 1992). The different sensitivities among different plant types means that the mix of plants in an ecosystem and catchment may change. Increased temperatures and changed precipitation patterns can also be expected to alter the duration and timing of the plant growth cycle, and an earlier onset to the growing season could lead to significant increases in spring evapotranspiration (Allen *et al* 1991).

Experimental evidence shows that plant stomatal conductance reduces as CO_2 levels increase,

and plant transpiration – and hence water use – therefore declines (Drake 1992). Kimball and Idso (1983), for example, concluded that evapotranspiration would be reduced by up to a third if the atmospheric CO_2 content was to double. However, the extra leaf area due to improved plant growth might compensate for the reduced stomatal conductance. Gifford (1988) conducted laboratory experiments that showed that plant water-use remained virtually constant as CO_2 concentrations increased, and inferred that plants tried to maintain some optimum water use. In the field, the physiological consequences of increased CO_2 concentrations might be limited by nutrient availability and changes in soil moisture regimes, and at the catchment scale may be less important than changes in the catchment vegetation mix.

Catchment land cover might also evolve in indirect response to climatic change, following the responses of farmers and policy-makers to real or perceived climate changes. Changes in the European Community's Common Agricultural Policy resulting from increased desertification in southern Europe, for example, might have implications for UK farmers and hence catchment land use. Land use change can, of course, also occur independently of climate change. Indeed in many places such change is the rule rather than the exception and its effect may be more rapid than any due to climate change; an example is catchment afforestation.

Runoff generation

Rainfall intensities in humid temperate environments rarely exceed the rate at which water can infiltrate into the soil (except in disturbed areas), and river runoff is generated by water flowing through the soil and by rainfall falling directly on saturated ground. Soil properties influence the rate at which rainfall infiltrates, the rate at which water moves through the soil, and the time over which soil becomes saturated.

Most pedological processes operate over a very long time-scale, but changes in organic matter content and soil structure may become apparent over a time-scale of less than 10 years (CCIRG 1991). Higher temperatures and increased rainfall would lead to a loss of soil organic matter and hence a decrease in the ability of the soil to hold moisture; higher temperatures would also encourage clayey soils to shrink and crack, thus assisting the passage of water into and through the soil profile (CCIRG 1991). Increased waterlogging could encourage the development of gleyed profiles, although such a change would occur over a longer time period. Changes in soil structure will probably only have a noticeable effect at the small catchment scale (unless the changes are particularly large following, for example, desertification).

Groundwater regimes

Groundwater regimes would of course be affected by changes in aquifer recharge due to alterations in precipitation characteristics, infiltration and evapotranspiration. Aquifers in most humid temperate environments are recharged seasonally. Recharge typically commences in autumn after soil moisture deficits have been removed, and ceases in spring when evapotranspiration rates reach such a level that soil moisture deficits begin to develop. Both the amount of winter rainfall and the length of the recharge season therefore influence the amount of recharge. A shorter recharge season (due to increased evapotranspiration) might offset the effect of any increase in winter rainfall.

Water in aquifers is held in fissures and pores of varying sizes. The smaller the pore, the slower it is to fill and the slower it is to drain. The recharge season might be so truncated that the smallest pores are not filled and groundwater levels therefore decline more rapidly in spring.

Finally, an increase in the proportion of winter rainfall that falls in high-intensity storm events could mean that more of the winter rainfall leaves the catchment as rapid streamflow, and therefore that less goes to recharge. The effects of climate change on groundwater recharge have so far been studied far less intensively than effects on streamflow.

It is difficult to conceive of climatically-induced changes in aquifer properties, except where saline intrusion due to sea-level rise alters local hydraulic gradients.

Water quality

Changes in climate can be expected to lead to changes in stream-water chemical and biological characteristics. Increased stream-water temperatures will affect the rate of operation of chemical and biological processes. A rise in temperature, for example, would increase the rate of biological activity (and de-oxidation) and the rate of re-aeration, but because the effect on de-oxidation is greater the dissolved oxygen content in the watercourse would reduce (Jacoby 1990). Similarly, the rate of denitrification is more sensitive to temperature than the rate of nitrification so, other things being equal, higher temperatures would lead to lower nitrate concentrations (Jenkins 1991). However, in practice other things are not equal, and the effects of a change in temperature may be overwhelmed by changes in dilution, concentration and residence time associated with changes in river discharge. Lower discharges would tend to increase concentrations and lengthen residence times.

Changes in flow pathways due to alterations in

Table 12.3 A matrix of methods for estimating the hydrological impacts of climate change: most studies fall into the bottom left corner

	Change in climate inputs	Change in hydrological characteristics
Arbitrary change		
Temporal analogue		
Spatial analogue		
Use of climate model simulations·		

soil structure would change the amount of time water remains within the soil profile. Minerals could be released into soil water through sulphur and nitrate mineralization as soil structure changes (Jenkins 1991).

A change in climate might also be associated with changes in geochemical inputs to the catchment. Farmers may use more – or fewer – agricultural chemicals in response to climate change, or may change land use completely. Atmospheric deposition rates might also change as the frequency of occurrence of different air masses varies (Wilby 1989).

Sediment yields may also be susceptible to climate change, and particularly to changes in land cover and rainfall intensities and frequencies. Sediment supply in upland catchments is controlled by thresholds in slope stability and rainfall intensity, and the rate at which these thresholds are exceeded could change as climate evolves (Harvey 1991).

12.3 ESTIMATING THE IMPACTS OF CLIMATE CHANGE IN A PARTICULAR CATCHMENT

There are two main obstacles to determining quantitatively the implications of climate change for a specific catchment. Firstly, it is currently very difficult to estimate what the future climate will be in a specific catchment. Secondly, practically every aspect of the hydrological system is expected to be affected by climate change (Fig. 12.1), and the effects of change will be very dependent upon both the current characteristics of the catchment and the interactions between different parts of the system. It may therefore be very risky to transfer quantitative conclusions from another catchment.

Practical assessments of climate change impacts make use of climate change 'scenarios', which are defined to be feasible, internally consistent estimates of future climate. A climate change scenario is not to be confused with a forecast. A range of scenarios must be considered in a study, because it is valuable to investigate how different scenarios translate to differences in hydrological implications.

The four basic groups of methods of creating climate change scenarios are shown in Table 12.3.

Each method can be used to estimate directly the implications of climate change for hydrological systems, or can be used to define climate inputs (such as precipitation, temperature and potential evapotranspiration) which are subsequently fed into a hydrological model. The vast majority of climate change impact studies fall into the bottom left corner in Table 12.3, although examples appear in the literature for each area.

This section first reviews briefly the four different methods of creating scenarios, and concludes by considering the types of hydrological model which are needed to convert climate inputs into hydrological outputs. For catchment management purposes, these hydrological outputs may need to be fed into ecological, geomorphological or water resource systems models: these models are not considered explicitly in this section.

Arbitrary changes

The simplest way of determining the impact of possible climatic change is to define a number of arbitrary changes in rainfall and temperature (or potential evaporation), and estimate the implications for river flow regimes (e.g. Nemec & Schaake 1982). It is also possible to define arbitrary changes in flow regimes and determine the impacts on, for example, reservoir reliability. It is better to view such investigations as sensitivity studies rather than assessments of the implications of possible future climate change.

Temporal analogues

Climate change scenarios can be based on the transfer of data from some period in the past. Palaeoclimatic analogues use information from early warm epochs in the earth's history. Budyko (1989), for example, has proposed that the temperature and rainfall patterns pertaining during the Holocene climatic optimum around 5000 to 6000 BP provide an analogue for conditions in the year 2000. However, such palaeoclimatic anomalies were probably caused by variations in planetary orbit, and it is considered unlikely that increasing greenhouse gas concentrations will have the same climatic effect (IPCC 1990). It is also difficult to obtain hydrological information from a palaeoclimatic analogue, although geomorphological information may give insights into channel characteristics and sediment regimes.

Historical analogues based on past instrumental data (or at the least documentation) are more useful, but there are still several important problems. The period in the past with the highest temperatures locally may not coincide with the warmest period globally, and the difference between the warmest and coolest historical periods may be considerably less than the amount of warming expected due to an increase in greenhouse gas concentrations. The method is perhaps most valuable when information from individual extreme events, such as the 1975/76 drought in the UK, is used to provide insights into how hydrological, ecological, geomorphological and water resources systems might react to future climate anomalies.

Spatial analogues

Climate change scenarios can also be based on 'spatial analogues', which often appear in the popular climate change literature: it might be claimed, for example, that the climate of southern England will in the future be like that currently experienced in the Mediterranean region. However, the transfer of climate data from one region to another is only appropriate when local influences on climate are negligible. The climate of southern England will never be more than superficially similar to that of southern France because of differences in the influence of local and regional topography, the relationship between land, sea and ocean currents, and, also, the length of day. Spatial analogues might be more useful in the centre of a continent where large-scale controls on climate are dominant. Transfer of hydrological data from one area to another should not be attempted because of the importance of local catchment physical characteristics in controlling flow regime (Arnell *et al* 1990).

Use of climate model output

The final method of estimating the effects of climate change on hydrological systems uses the output from climate simulation models. Climate model output can be used in several ways. Firstly,

it is possible to use directly the climate model estimates of change in river runoff and soil moisture contents (Miller & Russell 1992). Secondly, it is possible to use the climate model estimates of changes in rainfall, temperature and potential evaporation as inputs to an independent catchment hydrological model (e.g. Bultot *et al* 1988; Cohen 1991). Finally, it is possible to use climate model output to guide the development of 'generalized' estimates of possible changes in rainfall, temperature and potential evaporation (e.g. Arnell *et al* 1990; McCabe & Wolock 1992).

However, although global climate models are capable of simulating relatively accurately both global climate and general circulation patterns, they are less good at simulating regional climates (IPCC 1990). This is to a large extent due to their coarse spatial resolution: the highest resolution global climate model currently works on a grid approximately 250×350 km. Such a scale is far too coarse for regional hydrological investigations.

It is, therefore, unwise to take the regional output – whether of climatic characteristics such as rainfall and temperature or, especially, river runoff – from any one climate model too literally at present. Scenarios should be based on the results from several different models. As climate models improve, the potential for using climate model output directly will increase, although it is naive to expect that global climate model output will in the foreseeable future satisfy

Table 12.4 Climate change scenarios for 2050 (from CCIRG 1990)

Rainfall	Winter		
Summer	0	+8%	+16%
−16%	Driest		
0		'Best'	
+16%			Wettest

Temperature

Summer
  +2.1°C
Winter
  +2.3−2.5°C

the needs of a detailed, localized water resources impact study.

Table 12.4 reproduces the set of climate change scenarios for the year 2050 used by the UK Climate Change Impacts Review Group (CCIRG 1991). The scenarios, based largely on the interpretation of climate model output, indicate that 'winter' is likely to be wetter in the UK than at present, whilst there is greater uncertainty over change in summer rainfall. The CCIRG did not estimate changes in potential evapotranspiration, but Budyko's (1980) figure of 4% per extra degree Celsius implies that annual potential evapotranspiration in the UK by the middle of the next century could be between 8 and 10% higher than at present.

Hydrological models for impact studies

Whilst a few climate change impact studies have used hydrological data from past climate anomalies to estimate directly the impacts of climate change (e.g. Palutikof 1987), the vast majority have applied catchment hydrological models with climate change scenarios derived largely from climate model output (e.g. Gleick 1987; Bultot *et al* 1988; Cohen 1991).

At one extreme, it would be possible to use a physically-based distributed hydrological model which represents all the processes involved and their interactions, and which has measurable parameters. Such a model requires very detailed input data, and its use in a catchment-scale impact assessment cannot really be justified at present given the highly generalized nature of current climate change scenarios. A physically-based model, however, can be used to determine the sensitivity of soil moisture regimes and river flows to particular changes in process or catchment structure.

At the other end of the scale are empirical models, ranging from simple relationships which estimate average annual runoff from annual climatic data to monthly flow simulation models whose parameters are estimated by multiple regression. The major disadvantage of empirical models is that model parameters are wholly dependent on current conditions, and that slight changes in either the raw data used to define the model or of model form can result in very different implied sensitivities to climate change (see Arnell *et al* 1990, for example).

Between the 'white box' physically-based models and the 'black-box' empirical models lie 'grey-box' conceptual models, which incorporate some representation of the hydrological processes operating in a catchment with a small number of parameters which must be calibrated for a specific catchment. Models used in climate change impacts studies range from two- or three-parameter monthly water balance models (Gleick 1987; Arnell *et al* 1990), to more complicated multiple-parameter daily models (Bultot *et al* 1988; Lettenmaier & Gan 1990). The key assumption is that the model calibration remains appropriate under a changed climate, but this assumption can be partially tested using data from past 'extreme' years.

The choice of a model for impact assessment should be based on the hydrological or water resources problem to be studied. If interest lies primarily in river flows, groundwater recharge or general patterns of stream-water quality, then a relatively simple conceptual model will be adequate. If, however, interest lies in a much more highly detailed, process-based problem, such as changes in soil moisture dynamics in an ecologically-important wetland, for example, then it will be necessary to adopt a physically-based model and make assumptions about possible changes in parameters and short-duration inputs.

Most climate change impact studies so far have run a hydrological model with an observed climatological time series representing current conditions, and then run the model again after perturbing the time series according to some climate change scenario (Gleick 1987; Bultot *et al* 1988; Arnell *et al* 1990). An alternative method would be to generate stochastic synthetic time series representing current and future conditions: the future time series are determined by perturbing the parameters of the stochastic model.

A stochastic approach has two main advantages. Firstly, it is possible to generate a record of any length, and the simulated time series will not necessarily be dominated by one particularly extreme event. The drought of 1975/76, for example, dominates rainfall records over the last 50 years across a large part of the UK. Secondly, it is possible to generate repeated samples and thus get some idea of the sampling variability surrounding a particular hydrological or water resources statistic as estimated under current and future conditions. It is essential to ensure, of course, that the stochastic generation model reproduces realistically the characteristics of current climatic time series, including the sum of departures from average over periods relevant to water resources assessment.

12.4 CONCLUSIONS: COPING WITH CLIMATE CHANGE

Climate drives the hydrological system, and changes in climate can be expected to have consequences for hydrology, geomorphology, ecology and water resources. The significance of these consequences, however, depends on the type and magnitude of climate change, the characteristics of the catchment and hydrological system, and the sensitivity to change of the water resources and ecological systems. The actual impact of change, expressed for example in monetary terms or square metres of lost habitat, may not be very closely related to the percentage change in climate. The more resilient the ecological or water resources system, the greater the climate change it will be able to absorb for a given degree of impact; a fragile system – where demand for water is very close to supply, for example – might be very profoundly affected by only a very small change in climatic inputs. Meanwhile, of course, catchments, habitats and water resource systems are subject to other evolving influences which might be indirectly related to or totally independent of climate change.

So what should the catchment manager do about the possible implications of an evolving climate change? The uncertainty currently surrounding climate change means that it is not appropriate to initiate at present any changes to operational routines. It is also inappropriate to throw away past data. Scheme design should continue to be based on the analysis of all the data available, using currently accepted procedures. However, it is prudent to consider the implications of climate change when designing new schemes, and it is essential to consider cli-

mate change when engaged in long-term planning. Such an assessment would ideally contain the following stages:

1 Define a range of feasible climate change scenarios from inspection of climate model output (or, better, adopt scenarios developed by other 'responsible' organizations, such as the UK Climate Change Impacts Review Group: CCIRG 1991).

2 Run the climate change scenarios through an 'appropriate' hydrological model (bearing in mind the limitations of regression-type models and the extra data requirements of physically-based models).

3 Consider the implications of the scenarios by, for example, running the hydrological output from the previous stage through water resources or ecological models. How large are the differences between the scenarios? How do the changes compare with historical experience? Would one scenario have particularly catastrophic consequences? The scheme or strategy may also be sensitive to factors other than climate change, including for example economic assumptions.

4 Consider whether the scheme or strategy can be designed to minimize its sensitivity to climate change. What would be the implications of designing for the 'worst-case'? Is it possible to design the scheme so that it is not seriously adversely affected by possible changes in climate? Robustness to an uncertain future can be a criterion for selecting amongst alternatives.

5 Investigate the impact on the hydrological, ecological or water resource system of past extreme events.

There are, in conclusion, three main implications of climate change for the catchment manager. Firstly, it is no longer feasible to assume that climate will be constant over a planning period: this adds another dimension to the already complicated long-term planning process. Secondly, simulating the potential impacts of climate change requires realistic models of the processes operating within a catchment. These models must represent not just hydrological and water resources systems, but also the geomorphological and ecological systems. The threat of climate change is therefore yet another stimulus to multi-disciplinary model development. Thirdly, the uncertain nature of future climate change favours the adoption of flexible schemes and plans that are either robust to changing conditions or are easily adapted. In this sense too, the threat of climate change supports current developments in flexible, adaptable catchment management.

ACKNOWLEDGEMENTS

The author gratefully acknowledges the helpful comments of Max Beran, Frank Law, Alan Jenkins and Nick Reynard, all from the Institute of Hydrology.

REFERENCES

Allen RG, Gichuki FN, Rosenzweig C. (1991) CO_2-induced climatic changes in irrigation-water requirements. *Journal of Water Resources Planning and Management* **117**: 157–78. [12.2]

Arnell NW. (1992) Factors controlling catchment response to climate change in a humid temperate environment. *Journal of Hydrology* **132**: 321–42. [12.2]

Arnell NW, Brown RPC, Reynard NS. (1990) *Impacts of Climate Variability and Change on River Flow Regimes in the UK*. Institute of Hydrology Report 107. Wallingford, Oxon. [12.2, 12.3]

Budyko M. (1980) *Climate in the Past and the Future*. Gidrometeoizdat Publishing House, Leningrad. [12.2, 12.3]

Budyko M. (1989) Climatic conditions of the future. In: *Conference on Climate and Water, Helsinki, September 1989*, vol. 1, pp. 9–30. World Meteorological Society, Publications of the Academy of Finland. [12.3]

Bultot F, Coppens A, Dupriez GL, Gellens D, Meulenberghs F. (1988) Repercussions of a CO_2-doubling on the water cycle and on the water balance – a case study for Belgium. *Journal of Hydrology* **99**: 319–47. [12.3]

CCIRG (Climate Change Impacts Review Group). (1991) *The Potential Effects of Climate Change in the United Kingdom*. CCIRG for the Department of the Environment. HMSO, London. [12.2, 12.3, 12.4]

Cohen SJ. (1991) Possible impacts of climatic warming scenarios on water resources in the Saskatchewan River Sub-Basin, Canada. *Climatic Change* **19**: 291–318. [12.3]

Drake BG. (1992) The impact of rising CO_2 on ecosystem production. *Water, Air and Soil Pollution* **64**: 25–44. [12.2]

Fiering MB, Matalas NC. (1990) Decision-making under uncertainty. In: Waggoner PE (ed) *Climate Change and U.S. Water Resources*, pp. 75–84. John Wiley, New York. [12.1]

Gifford RM. (1988) Direct effect of higher carbon dioxide concentrations on vegetation. In: Pearman GL (ed) *Greenhouse: Planning for Climatic Change*, pp. 506–19. CSIRO, Melbourne. [12.2]

Gleick PH. (1987) Regional hydrologic consequences of increases in atmospheric CO_2 and other trace gases. *Climatic Change* **10**: 137–61. [12.2, 12.3]

Harvey AM. (1991) The influence of sediment supply on the channel morphology of upland streams: Howgill Fells, North West England. *Earth Surface Processes and Landforms* **16**: 675–84. [12.2]

IPCC (Intergovernmental Panel on Climate Change). (1990) *Climate Change. The IPCC Scientific Assessment*: Houghton JT, Jenkins GJ, Ephraums JJ (eds). Cambridge University Press, Cambridge. [12.1, 12.3]

IPCC (Intergovernmental Panel on Climate Change). (1992) *Climate Change 1992. The Supplementary Report to the IPCC Scientific Assessment*: Houghton JT, Callander BA, Varney SK (eds). Cambridge University Press, Cambridge. [12.1]

Jacoby HD. (1990) Water quality. In: Waggoner PE (ed) *Climate Change and U.S. Water Resources*, pp. 307–28. John Wiley, New York. [12.2]

Jenkins A. (1991) Climate change and water resources. In: Battarbee RW, Patrick ST (eds) *The Greenhouse Effect: Consequences for Britain*. Environmental Change Research Centre, University College London. [12.2]

Jones PD, Kelly PM, Goodess GB, Karl TR. (1989) The effect of urban warming on the Northern Hemisphere temperature average. *Journal of Climatology* **2**: 285–90. [12.1]

Kimball BA, Idso SB. (1983) Increasing atmospheric CO_2: effects on crop yield, water use and climate. *Agricultural Water Management* **7**: 55–72. [12.2]

Lettenmaier DP, Gan TY. (1990) Hydrologic sensitivities of the Sacramento–San Joaquin River basin, California, to global warming. *Water Resources Research* **26**: 69–86. [12.3]

McCabe GJ, Wolock DM. (1992) Effects of climatic change and climatic variability on the Thornthwaite moisture index in the Delaware River basin. *Climatic Change* **20**: 143–53. [12.3]

Martin P, Rosenberg NJ, McKenney MS. (1989) Climatic change and evapotranspiration: simulation studies of a wheat field, a forest and a tall grass prairie. *Climatic Change* **14**: 117–51. [12.2]

Miller JR, Russell GL. (1992) The impact of global warming on river runoff. *Journal of Geophysical Research* **97**: 2757–64. [12.3]

Nemec J, Schaake JC. (1982) Sensitivity of water resource systems to climate variation. *Hydrological Sciences Journal* **27**: 327–43. [12.3]

Palutikof JP. (1987) Some possible impacts of greenhouse gas-induced climatic change on water resources in England and Wales. In: *The Influence of Climate Change and Climatic Variability on the Hydrological Regime and Water Resources*. International Association of Hydrological Sciences, Publication 168, pp. 585–96. [12.3]

Wigley TML, Jones PD. (1985) Influences of precipitation changes and direct CO_2 effects on streamflow. *Nature* **314**: 149–52. [12.2]

Wilby R. (1989) Changing synoptic weather patterns, rainfall regimes and acid inputs in the East Midlands, UK. In: *Conference on Climate and Water, Helsinki, September 1989*, vol. 1, pp. 209–18. World Meteorological Society, Publications of the Academy of Finland. [12.2]

Index

257